1001 Questions Answered About the Weather

by
Frank H. Forrester

Drawings by James Macdonald

DOVER PUBLICATIONS, INC.
NEW YORK

To

Michael

who is too young

to ask questions

A sense sublime
Of something far more deeply interfused
Whose dwelling is the light of setting suns,
And the round ocean and the living air,
And the blue sky, and in the mind of man;
A motion and a spirit, that impels
All thinking things, all objects of all thought
And rolls through all things.

WILLIAM WORDSWORTH

Published in Canada by General Publishing Company, Ltd., 30 Lesmill Road, Don Mills, Toronto, Ontario.
Published in the United Kingdom by Constable and Company, Ltd.

This Dover edition, first published in 1981, is an unabridged republication of the work originally published in 1957 by Dodd, Mead & Company, N.Y. A new Preface has been written by the author especially for this edition, which is published by special arrangement with Dodd, Mead & Company.

International Standard Book Number: 0-486-24218-8
Library of Congress Catalog Card Number: 81-67041

Manufactured in the United States of America
Dover Publications, Inc.
180 Varick Street
New York, N.Y. 10014

PREFACE TO THE DOVER EDITION

Routinely, each day, weather-satellite photos or radar views are used in the presentation of weather conditions and forecasts by television reporters. This reflects, at just one level, the veritable explosion in technology and research that has occurred since this book was written more than a generation ago.

Earth-orbiting satellites, only in the planning stages in the 1950's, are now operationally at work, providing views of cloud cover over the entire globe. The unwinking eye of the satellite can detect and track tropical storms and other disturbances from their embryonic stages over source regions through their growth into mature stages, and to their ultimate dissipation. By radiometric techniques, it is now possible to estimate temperatures and other properties of the atmosphere, of the ocean, of ice cover over the planet, and other characteristics of important geophysical interest.

Satellites also play an important role in the periodic transmission of observations from hundreds of weather stations and other observing platforms to weather centers in many other countries, providing visual confirmation of the state of the atmosphere at a given time and place.

Electronic computers, evolved from scientific and engineering genius of the last few decades, have virtually revolutionized the atmospheric sciences. With computers and related equipment, scientists can, in relatively swift time, obtain data from stations all over the Earth, analyze them and depict worldwide patterns of wind, pressure and other phenomena. Computers have permitted scientists to cope, numerically, with equations of motion of the atmosphere, and have, perhaps more than any other tool, helped elevate forecasting from an art to a science.

Research into the nature, monitoring and prediction of severe weather systems, such as thunderstorms, tornadoes and hurricanes, has increased markedly within the past decade. Recent research carried out by federal and university scientists, for example, has focused on hailstorms and on weather systems that produce severe storms over the Great Plains states.

In other areas of research, several programs and investigations have concentrated on cloud physics and weather modification. Coaxing needed rain from clouds by artificial techniques, or, conversely, attempting to reduce the severity of local storms, is one of the most exciting—and challenging—areas of meteorological investigations. In recent years, there has been a mixture of significant progress and disappointments on the road to the development of a reliable capability for modifying the weather.

Efforts also have been intensified to learn more about the upper air and its properties and processes. The recent International Magnetospheric Study was launched to gain new knowledge of the highest reaches of the atmosphere in which energetic, charged particles and the magnetic fields of the Earth dominate.

Ozone in the upper air has become a matter of particular concern. Possible effects on the ozone layer from aircraft-engine emissions and even from spray-can propellants are targets for research because there is concern that the amount of ozone about 15 to 30 miles above the Earth's surface may be diminished by manmade substances.

In the 1960's, there developed a greater awareness of and concern for the hazards of environmental pollution. This has led to improved techniques for monitoring contamination in the air. Networks of stations now make systematic observations of the chemical make-up of air and rainwater.

Another area of modern research is aimed at learning more about the reasons for long-term, worldwide climatic and flood and drought patterns and trends. For example, a gradual increase of atmospheric carbon dioxide has been observed, caused largely by the combustion of fossil fuels. The increase of carbon dioxide may be affecting global climate and thus could change climatic patterns significantly, with impact upon man and his activities.

The 1960's saw the inception of the World Weather Programs, including the World Weather Watch, a program carried out by the weather services of many nations of the world under the support of the World Meteorological Organization. An important research part of the program is the Global Atmosphere Research Program (GARP), an effort aimed at extending the period and accuracy of weather forecasts and developing a better understanding of global climate. In the 1970's, parts of the GARP program exended into a large area of the tropical ocean between South America and Africa where data were gathered to learn more about the interaction of the atmosphere and the oceans and the role of the tropics in maintaining air-circulation patterns around the world.

These and many other research projects and investigations will serve us well. There will be tangible benefits—better forecasts and other weather-information services. And there will be other benefits accruing from the increasing body of atmospheric data in the areas of agricultural planning, water-supply management, oceanic and atmospheric transport, extraction of resources from the sea and maintenance of a healthy environment.

Indeed, much has been done in the last generation, but much more needs to be done. The road of research is difficult; the ocean of air is slow to surrender its mysteries.

Meanwhile, the blue of the sky, the formation of a raindrop or a snowflake, the magic of wind—these, and a host of other exciting and wondrous phenomena and occurrences within the atmosphere, continue to fire the imagination and curiosity of young and old alike. The present book was an effort to respond to this curiosity over a generation ago, and continues to answer some of the unchanging questions.

FOREWORD

People naturally ask questions about the weather. Some of these stem from pure curiosity while others originate from a downright practical interest, since weather is all important in personal lives, in business and even in wars. It becomes a tremendous task for an individual to find answers to questions about the weather when he has to consult the large variety of books on the subject. Mr. Frank H. Forrester has, in this book, put together an encyclopedia about the weather in question and answer form which will facilitate greatly the search for the "whys" and "wherefores" of weather and climate. Over a thousand questions have been asked and the answers given in a realistic and concise manner. An absolute and definite answer cannot be given to all questions which may be raised about why certain things happen in the production of weather, and in these cases, Mr. Forrester has made frank statements based on the latest available information. The range of questions and answers is such that it covers areas from the basic part played by the sun to man's attempt at weather modification by artificial means, from individual physical reactions caused by the weather to the reasons why business firms need to pay attention to it, from the climate of numerous countries in this world of ours to the measurement of weather by radar and other instruments. This book will be a valuable reference for the amateur weatherman and the professional alike and should become very popular with the inquiring public in libraries, in schools and at places of business.

ERNEST J. CHRISTIE
Meteorologist in Charge
U.S. Weather Bureau Office
New York, N.Y.

PREFACE TO THE FIRST EDITION

This book is the result of questions about the weather put to me by my family and friends, by students, pilots, boatmen, barbers, bakers, photographers, artists, engineers, housewives, farmers, advertisers, news editors, writers, commentators, motorists, doctors and lawyers.

It is concerned with one of man's greatest environmental factors, the atmosphere. I hope this book will provide some measure of the profound enjoyment which I have experienced in trying to understand the ways of the air ocean under whose ever-restless depths we lead our lives.

I have been extremely fortunate in having received invaluable assistance from many persons and institutions without whose guidance this book would not have been possible. Particular thanks are due to the following: Dr. Herbert Greenberg, New York City Board of Education; Ernest J. Christie, James Osmun, James McCloy and Norman Hagen, U.S. Weather Bureau; Thomas D. Nicholson and Henry M. Neely, American Museum–Hayden Planetarium; Dr. Solomon Schussheim and Dr. Harry S. Levine, Kings County Medical Society; Dr. Frederick Sargent II, Department of Physiology, University of Illinois; Dr. Jacques M. May, American Geographical Society; Commander Sarkis S. Sarkisian, U.S. Naval Hospital, St. Albans, N.Y.; Frances Ashley and Malcolm Rigby, American Meteorological Society; Edmund Dews and Ralph Huschke, Geophysics Research Directorate, Air Force Cambridge Research Center; and Raymond T. Bond, Dodd, Mead and Company. My gratitude to Phoebe L. Pierce and Mary Ann Rucidlo for their typing of the manuscript. For any factual errors that may have crept into the script despite the efforts of these authorities, I must assume sole responsibility. A final, but not minimal, acknowledgment of thanks must be made to my wife, Mary, who has provided the right kind of atmosphere.

CONTENTS

PHOTOGRAPHS

(Following page 144)

I. THE SUN—
WHERE WEATHER BEGINS

Introduction. Like so many other natural forces, the Sun's vital influence is rarely evaluated properly or understood. It is taken for granted like the air we breathe.

Earliest man did not take it for granted. To him the Sun was a super-god. He realized that the glowing ball in the sky was a great force of fertility that meant food and heat—the very core of life's necessity. He observed the Sun carefully. He watched it apprehensively when it sank low in the sky in the late fall and rejoiced when it climbed higher in the spring. Many of our significant festivals are linked to these early feelings of fear and awe concerning the Sun.

Today, of course, most of the supernatural aura of the Sun has been dissolved by scientific discovery. But science has only confirmed what the ancients instinctively surmised—that the Sun is a vast center of dynamic power. Our winds, our weather, our very lives are tied to this giant sphere of incandescent gases. It is a fantastic laboratory of physical wonders and increasingly a subject of research investigations which attempt to link solar activity with not only weather but also with a host of geophysical events.

Solar research will, in the years to come, continue to yield important results. Many fields of human endeavor will be directly affected by this research. In the process of learning more about the Sun, we feel some of the awe experienced by our earliest forebears.

1. What is the basic cause of all weather? Energy from the Sun, pouring earthwards, acts as a trigger mechanism to our ocean of air, the atmosphere. The interplay of this energy between Earth and air results, after many physical and chemical occurrences, in the various processes which we familiarly call weather. A breeze, a tornado, a raindrop—all can be inevitably traced to the effect of the Sun's energy upon our atmosphere. Weather might be considered as a child, as all living things are, born of the union between the Sun and the Earth.

2. Why are meteorologists increasingly interested in solar activity?
First, because of obvious effects of solar activity on weather—seasonal weather changes, for example, or temperature differences between the equator and the poles. But of greater significance, perhaps, are the known and suspected effects of solar activity on the upper limits of the atmosphere where, 50 miles above the Earth, many phenomena occur of fundamental importance. These phenomena involve such problems as the Earth's magnetism, auroral displays, cosmic-ray research, ionospheric physics, long-range communication and weather cycle changes. The upper atmosphere, bombarded with telling effect by the Sun's radiations, is a thoroughfare of rockets and may soon become the gateway for space vehicles.

3. How far is the Sun from the Earth?   The Sun is our nearest star. Its average distance from the Earth is about 93 million miles. In early July, when the Earth is farthest from the sun (aphelion), the distance is about 94½ million miles. The closest distance occurs in early January (perihelion)—about 91½ million miles. This average distance of 93 million miles is used in astronomy as a unit of distance measurement. It is called the Astronomical Unit, abbreviated A.U.

In order to avoid the use of cumbersome astronomical figures, distance may be expressed in terms of the speed of light which is about 186,000 miles per second. At this speed, light from the Sun reaches the Earth in approximately 8⅓ minutes. Therefore, the Earth can be said to be 8⅓ light-minutes away from the Sun. Alpha Centauri, the nearest star beyond our Sun, is about 4⅓ light-years away. Sirius, the Dog Star, is about 8¾ light-years away and the bright star Vega is 26½ light-years away.

4. How hot is the Sun?   Scientists have been able to measure the surface temperatures of the Sun rather closely. The average surface temperature for the entire disk is very near 10,000° F. It is a bit cooler in sunspots—about 7500° F. As for temperatures which may exist in the center of the Sun, guesswork rather than measuring takes over. By laws of physical and chemical inference, and by using theoretical model laboratory suns, scientists believe that the temperature in the center of the Sun is about 40 million degrees F.

5. How big is the Sun?   This blazing hub of our solar system has a diameter of 864,000 miles—a little more than 100 times the Earth's diameter. Its mass is equal to about 330,000 earths. The Sun is gaseous from surface to center, but not in a manner in which we normally think of gas, as something tenuous and lightweight. At the Sun's center, this gas is far denser than any earthly metallic substance. Pressures there are several billions of times more than our Earth's atmospheric pressure.

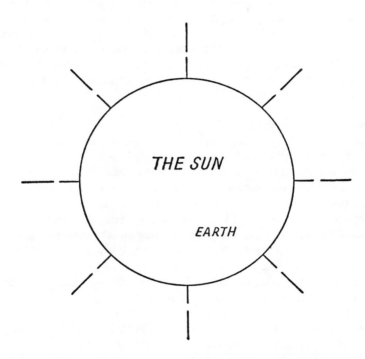

6. How does our Sun compare with other stars? Despite its relative colossal size and importance to the Earth, our Sun is quite an ordinary star. Its brightness, its volume, its density and its diameter can be matched and surpassed by thousands of other stars. Thousands more are not so big, bright or hot. Our Sun is especially average with relation to the mass of other stars. 90% of all stars have a mass not smaller than one tenth or larger than ten times the mass of our Sun.

7. What does the Sun Consist of? The Sun is a mass of incandescent gases. About $\frac{4}{5}$ of the solar atmosphere is hydrogen, a bit less than $\frac{1}{5}$ is helium and about $\frac{3}{100}$ oxygen. Iron, nickel and aluminum traces are present and many earth-found elements like sodium, carbon, silicon and calcium. About $\frac{2}{3}$ of all the elements known on the Earth have been identified as existing in the Sun. The bright surface seen by the casual observer through a filtered field glass or telescope is called the photosphere or *light* sphere. It measures about 500 miles in depth. Beyond that is the chromosphere or *color* sphere; named because of its colored appearance observed during solar eclipses. The chromosphere extends to about 25,000 miles. The next layer is the outer part called the *corona*. It is seen at times of eclipse and sometimes reaches out as far as the diameter of the sun or even more.

8. Does the Sun move? Many people think the Sun stays in one place where it majestically shines. Actually, its movement includes three important points of reference. First, it rotates about its axis, like the Earth, from west to east. It takes about 25 Earth days for the Sun to make one rotation at its equator, but it spins more slowly at the poles. Secondly, with respect to nearby stars, it is moving with an apparent velocity of $12\frac{1}{2}$ miles per second. Incidentally, as it moves through space (in the general direction of the constellation of Hercules), it takes the Earth and other planets with it. Thirdly, the Sun, along with millions of other stars, is revolving around the center of the milky way galaxy. The Sun's orbital velocity during this revolution is an estimated 200 miles per second.

9. How does the Sun create energy? Because of the tremendous heat and fantastically high pressures existing in the Sun's core, atoms of matter are broken down. Electrons are torn from their nuclei. A transmutation of elements occurs following a series of steps of atomic fusion. Helium is converted from hydrogen and in the process of this chain reaction, radiant energy is born. In atomic explosions on Earth, scientists have reproduced this release of energy which takes place on the Sun.

10. Is the Sun burning itself out? Even though the Sun constantly creates energy by the chain reaction described in the preceeding question, about 4 million tons of its mass is destroyed every second. But because the Sun's mass is so enormous, this destruction is hardly significant. Scientists estimate that it will take some 30 billions of years before the Earth's inhabitants notice a lessening of the Sun's energy. Since our planet is only some 4 billion years old, we have a long way to go and will probably face many other kinds of trouble before the cooling of the Sun becomes a problem.

11. How much of the Sun's energy does the Earth receive? Our Earth (about 8,000 miles in diameter), being about 93 million miles from the Sun, intercepts a surprisingly small amount of the total amount of Sun's energy that is being radiated 360 degrees into space. This amount is one/two-billionth—one part in two billion.

12. How much energy is this in terms of horsepower? Horsepower is one way of expressing a measurement of power. It is a unit of force required to raise 33,000 pounds at the rate of one foot per minute. In order to express the total amount of continuous rate of horsepower delivered by the Sun into space from its total surface, it is necessary to write the number 500 followed by 21 zeros! A stream of energy is emitted by the Sun from each square yard of its surface at a 72,000 horsepower rate. The Earth's share, though but an infinitesimal fraction of the Sun's total outpouring of energy, amounts to about 5 million horsepower per square mile. It should be borne in mind that this supply of power strikes the upper limits of the atmosphere. Not more than about 70% reaches the Earth's surface because of reflection and absorption effects of the atmosphere. If we could trap and harness this final ground-striking raw power, we would be free from want.

13. What instruments are used to study the Sun? One of the most important is the *spectroscope*. This device, based on the use of a prism which breaks up sunlight into a series of colors, makes it possible to analyze the light from the Sun (or other celestial bodies), and to determine the various chemical elements of which the Sun is composed. The *spectroscope* permits identification of the individual light-signals of each separate element. The *coronagraph* and *spectroheliograph* are instruments which, in effect, create artificial total eclipse conditions by blanking out the Sun's image in the focal plane of a telescope and allowing the outer portions of the Sun to be photographed. The *coelostat* is a clock-driven set of mirrors which reflect the image of the Sun. These and the large observatory solar tower telescope and camera-scanning equipment have revealed much of the complex make-up and startling characteristics of the Sun.

14. What are sunspots?   They are whirling columns of gas coming from below the Sun's surface. As they reach the surface, they expand and pressure is reduced. They usually are seen as a central dark area surrounded by a lighter area. They appear darker because they are cooler than the surrounding general surface. The diameter of the dark central area or umbra is from about 500 miles in the small ones to about 50,000 miles in the largest ones.

15. How can sunspots be observed?   Almost any telescope or good field glass can be used in sunspot hunting. Or, when spots are unusually big, the unaided eye may be used. It is exceedingly important to remember that the Sun should never be seen directly, either with glass or naked eye, unless proper filters are used. Specially severe damage or blindness may result from looking at the Sun directly through a telescope or binoculars without heavy filters. A possible type of filter for naked eye use in sunspot observing is a number 12 or 14 welder's glass. The Sun should desirably appear as a flat, nonglaring yellow disk.

16. Where on the Sun do sunspots occur?   Sunspots have some of the mannerisms of terrestrial hurricanes. They are whirling storms which form in twin zones, about 30 to 40 degrees on either side of the Sun's equator. They grow irregularly in size and number and move toward the solar equator with the life cycle dying just before they reach the equator. New ones often break out in higher solar latitudes as older ones are dying out. Sunspots usually do not last long, a few days to a week being about the average duration.

17. What is the sunspot cycle?   Years of study have revealed that while a few spots are almost always visible at any time, a maximum number appears at intervals of about 11 years. The high sunspot occurrence or climax period lasts for about two years. This 11-year cycle is by no means a rigid one. Many fluctuations occur so that intervals have ranged from 7 to 17 years.

18. What causes sunspots? Sunspots have been observed and studied longer in man's history than any other solar phenomena. Despite this, their exact cause is unknown. They may result from magnetic disturbances deep in the Sun's interior, according to a Swedish school of thought, where two repelling eddies move toward the Sun's surface in the form of a horseshoe magnet with one pole on each side of the equator.

19. What are solar prominences? They are flamelike clouds of luminous gases, mainly hydrogen, which erupt from the chromosphere (the "color" layer of the Sun's atmosphere). These prominences usually surge swiftly outward in erratic, irregularly shaped bursts to tremendous distances—frequently reaching 250,000 miles beyond the Sun's surface. They then curve back in streamer and knot formations to the surface. They are most common in the vicinity of sunspots and are probably associated with spot activity. It may be inferred that some of the Sun's gas can shoot clear to the Earth from these violent eruptions.

20. How are weather and other events on Earth linked to these solar disturbances? Explosions on the Sun and other types of solar activity are responsible for the aurora borealis. The erupting radiations from the solar disturbances bombard the tenuous gas regions of the upper atmosphere, exciting them to glow. Magnetic storms generally accompany auroral activity. Severe radio communication interruptions may occur with fade-outs and sometimes complete breakdown taking place. Also at times of auroral displays, compass needles frequently show big fluctuations. These disturbances make a study of sunspots economically significant as well as scientifically important. As far as other effects, a great deal is only suspected rather than known. Research meteorologists, for example, have consistently attempted to seek correlations between solar activity and rain forecasts, weather-cycle changes and various weather abnormalities. So many attempts have been made to link sunspots with terrestrial phenomena of different types, that even a brief compilation here will not be attempted. We must bide our time and await results from the field and laboratory. In the meanwhile, sunspots will probably continue to be linked to stock-market fluctuations, wars, births, deaths, hurricanes, animal behaviorism, sports results and psychological moods.

21. How does the Earth move with relation to the Sun?   It spins or rotates on its axis. The axis is an imaginary line running through the Earth from pole to pole. It takes 23 hours, 56 minutes and 4 and a fraction seconds for the Earth to complete one rotation. This motion is, of course, the basis for night and day. As the Earth rotates on its axis, it also travels or revolves around the Sun. The path or orbit of revolution is not circular but is elliptical and accounts for the varying distances between the Earth and Sun (See Question 3). The period of time of the Earth's revolution around the Sun determines the length of the year. Because of many different points of reference, there are actually several kinds of years. The ordinary seasonal year on which the calendar is based is called the tropical year and contains 365 days, 5 hours, 48 minutes and 46 seconds. As the Earth moves in its orbit around the Sun, it moves with a velocity of 18½ miles per second. This means that during the time it takes for this question and answer to be read, the earth has moved about 830 miles!

22. Is the Earth in an upright position with respect to the Sun? No. Its axis is tilted about 23½ degrees from an upright position. By upright we mean perpendicular to the plane of the Earth's orbit. This 23½ degree leaning of the Earth remains the same throughout the daily rotation on its axis and the annual revolution around the Sun.

23.   What is the importance of this position?   It is a major cause of the different seasons. Since the direction of the Earth's tilt remains the same as it revolves around the Sun, it follows that at times the north pole will be leaning toward the Sun and at other times it will lean away from the solar body. When the north pole leans toward the Sun, the rays of the Sun will strike the Northern Hemisphere in a more direct and more concentrated manner per square unit, bringing summer to the northern half of the Earth. At the same time, obviously, the south pole will be angled away from the Sun with opposite, winter conditions prevailing over the Southern Hemisphere.

24. What is meant by winter and summer solstices? The word *solstice* (from the Latin *sun to stand*) refers to a position reached by the Sun in the sky with relation to the Earth's equator. When, in the course of the Earth's revolution around the Sun, the Sun appears farthest south of the equator, it marks the winter solstice or beginning of winter for the Northern Hemisphere (astronomically, not meteorologically, speaking). When the Sun has journeyed to a point farthest north of the equator, it marks the summer solstice for the Northern Hemisphere and the beginning of summer. The Southern Hemisphere experiences exactly opposite solstices.

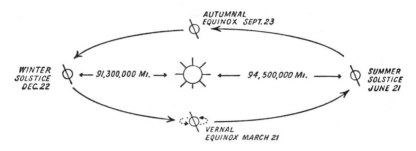

25. What are the equinoxes? The equinox (from Latin meaning *equal night*) represents a midway point reached by the Sun between the two solstices. For the Northern Hemisphere, the vernal equinox occurs at the moment when the Sun crosses the equator on its apparent journey northward. This marks the beginning of spring for the Northern Hemisphere. The autumnal equinox occurs when the Sun crosses the equator on its apparent movement south of the equator. Exactly opposite equinoxes occur, therefore, at simultaneous times for each hemisphere.

26. On what dates do the solstices and equinoxes fall? Usually as follows: winter solstice—December 21; vernal equinox—March 21; summer solstice—June 21; autumnal equinox—September 22. These dates may vary, however, up to about a day either way. This is because the Earth doesn't oblige us by coinciding its rotational and its wobbling motions with our man-made and faulty calendar.

II. THE OCEAN OF AIR—
OUR ATMOSPHERE

Introduction. We are creatures who inhabit the floor of a swirling gaseous ocean, breathing its sustaining mixture by reflex and seldom aware of the vast forces imbedded in its currents. In the bottom layer of the air ocean are the familiar sights, sounds and odors of weather. Indeed, to inhale and exhale is to breathe a piece of weather, an ever-changing yet earth-old panorama of wonders.

As the greater part of an iceberg is hidden from view, so is the endless interplay of physical and chemical processes in the atmosphere which result in the phenomena we call weather.

Only recently in man's history has a reasonable cross-section of our atmosphere emerged. The laboratory microscope; the instrument-bearing kite, balloon, plane and rocket; radio; spectroscope and radar —among other devices—have extended man's sensory perceptions in probing our aerial sphere. The study of the atmosphere is far from complete although a great fund of knowledge has been accumulated. It is the hope of the meteorologist to sound the air ocean's tenuous reaches hundreds of miles above the Earth. He hopes to leave to succeeding generations a complete charting of the atmosphere's dynamic anatomy.

Studies thus far have revealed our atmosphere to be a tremendously complex structure. Some insight into the behavior of its gaseous components must precede an understanding of weather.

27. What is the sky? The sky generally has two references, astronomical and meteorological. Astronomically, the sky is the space and expanse of the heavens beyond our atmosphere, the firmament apparently arching overhead and the realm of celestial bodies. The more popular conception of the sky is the lower and upper atmosphere, especially with reference to its *appearance* as affected by cloudiness, impurities in the air and color effects. Thus, when we say the sky is cloudy, bright, hazy or blue, this would be a weather sky. When we speak of the stars of the southern sky, the concept is astronomical.

28. What is meteorology? Meteorology is the science treating of the atmosphere and its phenomena. The word comes from the Greek *meteora* meaning "things in the sky." Some confusion arises about the word because a meteor usually implies a shooting star (in itself an erroneous term) or meteorite. Therefore, to many people, a meteorologist is one who studies meteors. Actually, there are many different kinds of meteors. They can be grouped as follows: hydrometeors, such as rain or snow; aerial meteors, such as winds and dust-devils; lithometeors, such as pollen and dust; and luminous meteors, such as rainbows and mirages. Igneous meteors which fall under the luminous classification include lightning. All of these meteors are weather phenomena and they are the concern of the meteorologist.

29. Is meteorology considered a separate science or a branch of physics? Meteorology is as closely bound to the laws of physics as engineering is to mathematics. It has, therefore, usually been considered a branch or offshoot of physics. However, its rapid growth, its scope of investigation and its impact certainly warrant its classification as a separate science. But the atmosphere and its phenomena have not been fully contained by the dynamic laws of mathematics. Until this occurs to a very strong degree, meteorology cannot be considered an exact science. Recently, because of the merging tendencies of the Earth, air and space sciences (geology, oceanography, meteorology and astronomy), scientists of each of these disciplines are frequently considered as geophysicists. Meteorology may then be considered as one of the branches of geophysics.

30. What are the basic units of all matter? All matter—animal, mineral or vegetable, animate or inanimate, solid, liquid or gaseous— is composed of submicroscopic particles of positive, negative and uncharged energy called protons (positive), electrons (negative) and neutrons (uncharged).

31. What is an atom? The atom has a structure something like a microcosmic solar system. The nucleus or central sun of this tiny system is positively charged and contains protons and neutrons. Around this nucleus revolve negatively charged electrons. The num-

ber of electrons revolving around the nucleus is equal in number to
the protons in the nucleus.

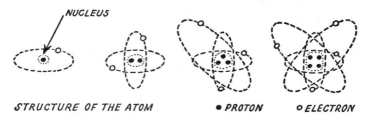

STRUCTURE OF THE ATOM   • *PROTON*   ○ *ELECTRON*

32. What is an element?   An element is any substance whose chemical character is determined by the number of its electrons. These elements have been given numbers. Number 1, for example, is hydrogen.
Number 2 is helium. Number 7 is nitrogen. Magnesium, sulphur,
nickel, carbon, oxygen are all examples of different elements. A total
of 102 elements on Earth have been detected thus far. More are
expected.

33. What is a molecule?   A molecule is the smallest quantity of
any element or compound that can exist in the free state. Molecules
of elements consist of one or more similar atoms. Compounds, on the
other hand, consist of two or more atoms of different elements.

34. What are ions?   Under conditions of very high temperatures,
atoms may lose electrons from their orbits. The atoms then exist with
a diminished number of electrons. If they lose electrons, they have
more positively charged particles remaining. These charged atoms are
called ions. The process of losing electrons is called *ionization*.

35. What is air composed of?   Air is a mixture of gaseous elements
—an incomprehensible number of atoms combined into molecules.
By far the greatest part of air is nitrogen, representing about ⅘ by
volume measurement. Nitrogen is relatively inert and acts more or
less as a mixing medium for other gases. Oxygen is the second main
gas and comprises ⅕ of the atmosphere. Besides these two, perhaps
18 gases are present in far lesser quantities or traces such as water
vapor, neon, argon, carbon dioxide, helium, hydrogen, methane,
krypton, ozone, ammonia and iodine.

36. What are the impurities of the air? Salt spray from the oceans, pollen, industrial smoke, dust, volcanic ash and a continuous fine rain of meteoric sediment are some of the impurities which add to our atmospheric make-up.

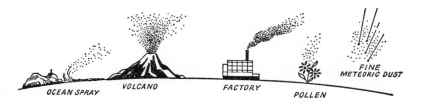

OCEAN SPRAY   VOLCANO   FACTORY   POLLEN   FINE METEORIC DUST

37. What is the most important gas in the atmosphere? Colorless, odorless and tasteless oxygen has been called the *gas of life*. It is essential for respiration of all animals and necessary for most types of combustion.

38. Is the composition of air fairly constant at all places? The chemical composition of dry air is generally uniform from place to place on the Earth's surface. Differences occur in the amount and kind of impurities and also in elements which are called *variables,* like water vapor and ozone. This constant composition is understandable when we consider the continual mixing process caused by surface heating differences and winds which blow around the Earth.

39. What is meant by atmospheric pressure? It is the actual weight or push on the Earth's surface of all the various gases and other elements of which air is composed. It is the force per unit area exerted by the atmosphere. At any given point, it represents the force of a column of air of unit area which extends to the top of the atmosphere.

40. How much does air weigh? Air exerts a tremendous pressure at the earth's surface. At standard sea-level conditions, the human body (or any other object) is subjected to a pressure of $14\frac{7}{10}$ pounds per square inch. At this rate, the average man supports about a ton of air. It causes no discomfort because our internal pressures push outwards and equalize this force so that we are not normally aware of the push of the atmosphere.

41. How much does the entire atmosphere weigh?   About 5¾ quadrillion tons of air clings to the Earth's 197 million square miles of surface.

**A ton of air weight
per person**

SEA LEVEL

42. What is the comparative density of air?   The density of any substance depends upon the amount of matter in a given volume. The more molecules which occupy a unit area, the denser the matter. Density is calculated by dividing the mass by its volume. For example, a cubic foot of water weighs 62⁴⁄₁₀ pounds and a cubic foot of mercury weighs 848.64 pounds. The density of water is 62.4 pounds per cubic foot, of mercury 848.64 pounds per cubic foot. Comparative densities of different elements are normally compared to water which has been given a value of 1.0000. Gold has a comparative density of 19.32; aluminum, 2.65. The comparative density of air is 0.001293 or a little more than ¹⁄₁₀₀₀ the density of water at a standard temperature.

43. How dense is the air at the Earth's surface?   Air molecules are crammed together at the Earth's surface in number almost beyond conception. So dense is the air we breathe at the surface that a tiny microscopic air particle has only to move less than one millionth of an inch before it collides with a neighbor particle.

44. What is the air density and pressure at high altitudes? The density and therefore weight of air diminishes sharply with altitude. If one were standing on a mountaintop 18,000 feet above the surface, only 7½ pounds per square inch of pressure would be exerted by the atmosphere. At 36,000 feet, this pressure would be only 3¼ pounds. For each doubling of the altitude, the pressure divides in half. In this way, fully one half of the entire weight of our air ocean lies in the lower 3½ miles. The pressure at about 65 miles is estimated at only about one millionth of what it is at sea level!

45. How is atmospheric pressure expressed? In terms of pounds per square inch (English system), by millibars (metric system) and by the length of a mercury column in inches. In meteorology, the millibar is used most frequently. It is based on the *dyne,* a unit of force in the metric system. One millibar is equal to 1,000 dynes per square centimeter.

46. What is the average sea-level air pressure in these units? In inches of mercury—29.9212; in millibars—1013.25; in pounds per square inch—14.7; one inch of mercury equals 33.86 millibars or 0.49 pounds per square inch; 1 millibar equals 0.03 inches of mercury or 0.01 pounds per square inch; 1 lb. per square inch equals 2.036 inches of mercury or 68.9 millibars. No matter what the measurement is, it is important to remember that it is an expression of the force exerted by the atmosphere on a given unit. It is a measurement of a piece or column of air, at a split moment, which lies above the instrument providing the measurement.

47. What instruments are used to measure atmospheric pressure? The mercurial barometer, the aneroid barometer and the aneroid barograph.

48. What is the principle behind the mercurial barometer? The mercurial barometer provides the most accurate measurement of atmospheric pressure. It is a device that has a glass tube a little more than 30 inches long closed at one end and filled with mercury. The

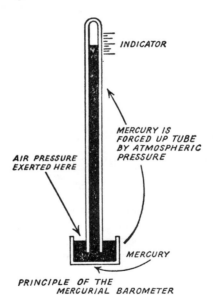

INDICATOR

MERCURY IS FORCED UP TUBE BY ATMOSPHERIC PRESSURE

AIR PRESSURE EXERTED HERE

MERCURY

PRINCIPLE OF THE MERCURIAL BAROMETER

tube is placed open end in a cistern (vessel) partially filled with mercury. Mercury in the tube falls until the weight of the mercury column balances the atmospheric force exerted on the mercury in the open vessel. If colder air should move over the area (and hence over the instrument), the colder, denser, heavier air would exert a stronger push on the mercury in the vessel and force mercury higher in the tube. We would say, then, that the barometer is *rising*. Warmer, lighter air would not press down so hard. Some of the mercury would lower in the tube into the vessel and we would say that the barometer is *lowering* or *falling*. The barometer, therefore, simply indicates the relative weight of constantly changing air of different densities above the instrument.

49. How does an aneroid barometer work? This is a less expensive instrument and one commonly found in the home or office. An aneroid (from the Greek *without liquid*) barometer is based upon the action of a tiny corrugated metal cell from which most of the air has been removed. As the air's weight over the barometer changes, the little accordionlike cell (called a sylphon cell) is pressed together or stretches out. This change is reflected through a system of links and levers to a clock-face type dial. The dial reading scale is calibrated to reflect the atmospheric pressure in terms of inches of mercury or millibars. Usually the aneroid barometer shows both scales simultaneously.

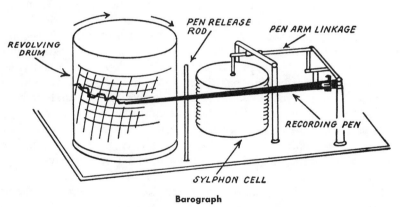

Barograph

50. What is the purpose of a barograph? The barometers described in the two preceding questions provide instant sight readings of atmospheric pressure. Frequently a written record of changing pressure is desired. This is provided by the barograph. It works on the same principle as the aneroid barometer up to the point of using the little corrugated cell to reflect the changing weight of the air. But instead of having this change passed on through linkage to a dial face, it is transmitted to an inked pen arm which records a history of the changing pressure on a graphed chart that rotates on a clock-driven drum.

51. How high is the atmosphere? It extends hundreds of miles above the Earth but thins out so gradually to merge with interplanetary space that it is impossible to say where its exact top is. Scientists have estimated that atmospheric molecules exist as high as several thousands of miles above the Earth. This, however, is an extremely tenuous kind of atmosphere, very nearly a complete vacuum.

52. What are the layers of the atmosphere? Over a period of many years, by observation and by the use of instrument-bearing kites, balloons, planes and rockets, a fairly good profile understanding of our atmosphere has been reached. For convenience, meteorologists have divided the atmosphere into various shells or layers, laid one atop the other from the Earth's surface to the top of the atmosphere. These layers or spheres are classified and named according to important characteristics of each layer. They have no exact division but rather merge into one another. The shell of air that hugs the Earth is called the troposphere. Above it is the stratosphere, followed by the ionosphere and capped by the exosphere. The task of classification is not ended and is always subject to revision depending upon convenient methods of labeling based upon new discoveries. Special high-soaring balloons, rockets and artificial satellites will undoubtedly change our profile understanding of the atmosphere so that we may, as time goes on, refer to six, eight or ten layers instead of four. Recently, for example, two additional layers have been suggested for classification purposes, the chemosphere and the mesophere.

53. How are the divisions between each layer classified? By adding the suffix *pause* in place of *sphere* to the lower layer. Thus, the tropopause divides the troposphere and stratosphere. The stratopause divides the stratosphere and chemosphere, etc.

54. What is the troposphere? This is a relatively shallow layer of the atmosphere which hugs the Earth's surface. In this dense and moist layer, man observes the entire sweep of the weather story. The word troposphere stems from the Greek *tropos* meaning to "turn or

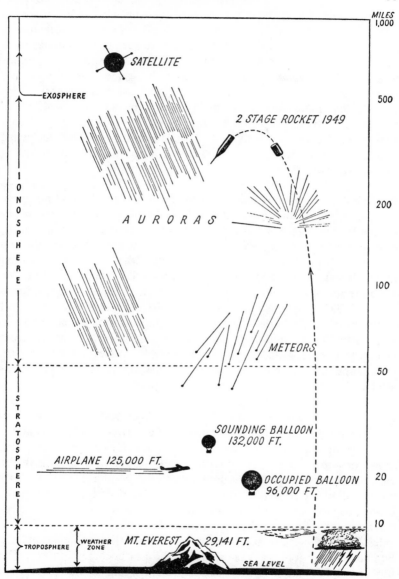

MILES
1,000

SATELLITE

EXOSPHERE

500

2 STAGE ROCKET 1949

I
O
N
O
S
P
H
E
R
E

A U R O R A S

200

100

METEORS

50

S
T
R
A
T
O
S
P
H
E
R
E

SOUNDING BALLOON
132,000 FT.

AIRPLANE 125,000 FT.

OCCUPIED BALLOON   20
96,000 FT.

10

TROPOSPHERE   WEATHER
ZONE   MT. EVEREST   29,141 FT.

SEA LEVEL

Cross section of the atmosphere

mix." Within this lower rung of the atmospheric ladder, vigorous mixing of air takes place. Air moves vertically as well as horizontally. Almost all the water vapor and impurities are here so that clouds, rain, snow, sleet, fog, lightning, thunderstorms—the span of familiar weather phenomena—range throughout this zone. The top of the troposphere extends to about 11 miles over the equator and slopes to about five miles over the poles. Temperature decreases with elevation through the troposphere at a rate, normally, of about 3½ degrees Fahrenheit for every 1,000 feet. Some meteorologists find the word *weathersphere* to be more descriptive than troposphere when referring to this atmospheric layer.

55. What is the stratosphere?   The second layer of the atmosphere presents quite a contrast to the turbulent, mixing vat of the troposphere below. The word stratosphere comes from the Greek *stratos,* or region of stratification or smoothing out. It extends from the top of the troposphere to about 60 miles above the earth. It is marked by an almost complete lack of clouds and an absense of up-and-down air currents. Winds blow very strong and horizontally. Temperature remains fairly constant at about minus 70 degrees Fahrenheit until about 40 miles above the Earth where a concentration of ozone exists. This ozone layer, sometimes called the *ozonosphere,* absorbs a great deal of ultraviolet radiation from the sun which drives the temperature up to about 170 degrees Fahrenheit. Temperatures fall sharply thereafter with altitude to the top of the stratosphere.

56. What is the ionosphere?   This layer extends from about 60 miles to several hundred miles above the Earth. At elevations above 60 miles or so, pressures are extremely low (fewer molecules and atoms of atmospheric gases than at lower altitudes). Electrons of the various gas atoms are more vulnerable to the bombardment of short-wave radiations from the Sun. They can be dislodged more easily from their orbits around the atom nuclei. The stripping away of electrons causes an electrical unbalance in the various atoms and molecules. If a sufficiently large number of electrons is dislodged and the number of free ions increase thereby, the gas becomes ionized. The process of ionization is the basis for the naming of the ionosphere and causes many special phenomena in the high atmosphere (See Question 34).

57. How is the ionosphere linked to radio communication? When a process of ionization takes place as described in the preceding question, an increase in electrical conductivity occurs. Several ionized layers at different altitudes in the ionosphere act as radio wave reflecting shells. These reflecting layers have been identified by measuring the reflections, absorptions and time of travel of radio waves. Four layers have been identified: the D, E_1, F_1 and F_2 layers. The lower the altitude of the layer, the more weakly it is ionized and, in general, the lower layers reflect back longer waves. Higher, more strongly ionized layers, affect shorter waves.

The D layer appears at heights of 40–60 miles. The E_1 layer occurs at heights ranging from 70 to 90 miles. The F_1 merges with the F_2 at night with the height of the F_2 layer averaging about 200 miles above the earth. These layers vary considerably in height and intensity during the night and day. During solar disturbances, they act in extremely erratic manner, changing rapidly with respect to position and intensity within a few seconds. At such times, communications are often disrupted.

58. What is the exosphere? The exosphere is the outermost shell of the atmosphere which merges gradually with interplanetary space. In this critically thin and tenuous zone, so few atoms of gas exist that collisions of air particles are negligible. Their motions are somewhat similar to the bubble and spray effect of water particles at the tops of ocean waves. Some of the air particles, particularly the atoms of the lighter gases, hydrogen and helium, travel fast enough to leap clear of the air ocean. Most, however, fall back after some free flight under the force of gravity. The height of the base of the exosphere has not been exactly determined. Authorities generally place it as being between 500 and 750 miles above the Earth. How high its top is, that is, the top of the atmosphere, is open to conjecture.

59. What temperatures exist in the ionosphere? Speculations about temperatures in the ionosphere have been based on rocket, sound-wave and meteor studies as well as theoretical considerations. It is estimated that temperatures rise through the ionosphere to about 4,000 degrees Fahrenheit at around 400 miles above the earth. This rather startling estimate should not be translated in terms of our usual concept of temperatures. The ionospheric temperature estimation is based on a *kinetic* measurement—a measurement of the average velocity of gas molecules. It is quite true that the air molecules at 400 miles above the Earth move with tremendous velocities and, therefore, result in a high temperature which reflects this speed. But there are so few molecules that despite their individual speed, there are not enough of them to convey any heat sensation such as we normally associate with high temperatures at the Earth's surface.

60. What is the chemosphere?   The chemosphere is a layer of considerable chemical activity involving the breakup of water and carbon dioxide molecules by short-wave radiation. This change in molecular structure causes the molecular components to produce some longer wave length radiation of their own. The lowest of the auroral glows are found here. The chemosphere is that part of the atmosphere between approximately 20 and 50 miles above the Earth.

61. What is the mesophere?   The mesophere extends from about 250 miles to about 620 miles above the Earth. Like the ionosphere, it is characterized by the presence of free ions and electrons, but its effect on radio communication is much less than that of the ionosphere.

62. What is the difference between heat and temperature?   Heat is often confused with temperature. Heat refers to the sum total energy of all the moving molecules in a given substance. Temperature is a proportional measurement of their average speed of motion at a given moment. An object can have a high heat content (much mass, many molecules) but a low temperature (average energy of the individual molecule is less). An example would be an iceberg or a large fish tank of water. Another object could have a low heat content and high temperature. Example: a burning match or a cup of hot coffee.

63. How is heat measured?   Heat is measured by the effect it produces. The units expressing measurement are directly tied to temperature. Basic units used in meteorology are the gram calorie and the British thermal unit (B.T.U.). The gram calorie is the amount of heat necessary to raise one gram of water one degree centigrade. The British thermal unit is the amount of heat required to raise one pound of water one degree Fahrenheit. One B.T.U. is equal to 252 gram calories.

64. What is specific heat? Some substances warm up more quickly than others. This is more than apparent at the seashore during a hot day. The sand will heat up rapidly during the day and cool quickly during the night. Comparatively, the ocean water will hardly be affected. It takes one calorie to warm a gram of water one degree centigrade, but only about $\frac{1}{5}$ of a calorie to warm up a gram of aluminum the same amount. The specific heat of a substance, then, is the number of gram calories required to raise the temperature of a gram of a substance one degree centigrade. It takes one calorie to warm a gram of water one degree centigrade. The specific heat of water, therefore, is listed as 1.00. It takes $\frac{3}{100}$ of a calorie to warm a gram of gold one degree centigrade. The specific heat of gold, then, is 0.03. Some comparative specific heats: water 1.00, rubber 0.44, aluminum 0.21, copper 0.09, ethyl alcohol 0.56, hydrogen 3.389, mercury 0.0331.

65. What is meant by heat transfer? Heat transfer refers to different ways heat and energy can be carried from one place to another. Heat is of vital importance to practically all weather processes. A knowledge of how it is transferred through the air is fundamental to understanding weather. The methods of heat transfer are conduction, radiation, convection and absorption.

66. What is conduction? Conduction is the transfer of heat by direct contact. Heat from a hot teapot handle is transferred to the hand by conduction. The molecules of the substance being heated move at faster paces and jostle their neighbor particles and so transmit heat throughout the substance. Heat conductivity varies considerably with different substances. The denser, more molecule-packed the substance, the better the conduction. Metals are good heat conductors. Glass and plastics are poorer conductors. Liquids are poor and gases the worst conductors. Air, being a poor conductor of heat, is, therefore, a good insulator. A vacuum is the best of all insulators and is the basis for the ordinary thermos bottle. A near vacuum blankets the bottle and does not conduct heat away from the imprisoned liquid. In meteorology, conduction is of secondary importance because air is a poor conductor. Conduction is only important in heating immediate layers of air in direct contact with the ground.

67. What is radiation?   Radiation is a quick, practically instantaneous process of heat transfer. It is quick because radiant energy moves with the speed of light. Radiation is a process by which energy is transferred through space without the aid of a material medium. It is by radiation that the earth receives energy from the Sun.

68. What is convection?   Convection is the most important method of heat transfer in weather processes. It is the transfer of energy by means of vertical currents in liquids or in air. Boiling water is an example of heat being transferred throughout the liquid by currents. Air, too, boils in a sense. A portion of air may be heated by contact

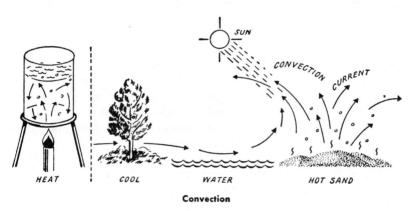

Convection

with warm earth. The air expands and decreases in density. Being relatively lighter than the surrounding cool air, this air becomes buoyant and is forced to rise through the surrounding cool air much as a piece of cork rises through water. This rising air current is called a *thermal* and the process is called *thermal convection*. Sometimes air is forced to rise by moving against objects in its path like hills or mountains. When air is forced aloft in this manner, the process is called *mechanical convection*. Tremendous amounts of heat are carried through the troposphere by air currents which rise convectively.

69. What is insolation? Insolation simply means *in*coming *sol*ar radi*ation*. It is the radiated energy that strikes the earth from the Sun (See Questions 11 and 12).

70. What is absorption? Absorption has been called the Houdini of heat transfer methods in meteorology. It is a process whereby matter accepts radiant energy in one form, absorbs it and then re-radiates it in another form. An example is a portion of ground which absorbs the Sun's short-wave radiation and then returns this energy out in the form of long-wave radiation.

71. What is the electromagnetic scale of energy? The energy streaming earthward from the Sun has an extremely wide range which includes not only familiar light and heat but many other types of energy with widely varying properties. The entire scale is called the electromagnetic scale of energy. The different types of energy on this scale range from cosmic, gamma and x rays of the short-wave radiation type to radio waves and alternating current of the long-wave radiation types. These types of energy all travel in a wave formation, somewhat analogous to ripples through a flag but in wave lengths so small from crest to crest that a special unit is needed to express measurement. This unit is the *angstrom*—about 250 millionths of an inch. The speed of the energy types is the same—about 186,000 miles per second, despite the fact that some have a longer interval of waves and some shorter. The shorter the wave is, the higher the frequency and energy. Long-wave radiations are of lower frequency and lower energy. Thus, gamma waves, at the short-wave end of the scale will penetrate a foot of lead while ultraviolet waves of radiation, even though on the same scale, are highly absorbable by transparent media.

72. How much of this energy can we actually see? By far the greatest amount of energy on the electromagnetic scale is invisible to us. We sense or observe the effects of the different short- and long-wave radiations. X-ray effects in medical usage are well known, for example. Ultraviolet energy causes sunburn. We feel heat, but cannot see it. Actually, a very narrow band on the total scale is visible to us. This is called visible or white light. The colors of visible light range from the violets at the short-wave end to the reds at the long-wave end. To remember the colors in correct order from long wave to short, some people remember the name ROY G. BIV (red, orange, yellow, green, blue, indigo, violet). Many others find it easier to remember the word VIBGYOR. The short-wave band just above visible light is that of ultraviolet. Just below visible light is the longer wave infrared.

73. What different scales are used to measure temperatures? In meteorology, three scales are used; centigrade (° C.), Fahrenheit (° F.) and absolute or Kelvin (° A or ° K). The centigrade and absolute scales are more suited for scientific work while the Fahrenheit scale is used popularly in Great Britain and the United States. Another scale, called the Réaumur, was once used in meteorology but is no longer used except when converting from old European climatological records.

74. What is the Fahrenheit scale? In the winter of 1709, Gabriel Fahrenheit, a German scientist, proposed a temperature scale based on the setting of zero at a point where the mercury sank in his thermometer on a very cold day in Danzig. He proposed that the opposite reference point of the scale be chosen as the temperature of the human body. On this rather illogical basis, the freezing point of water fell at 32° and at sea level, the boiling point of water at 212°. Although not suitable for scientific work, the Fahrenheit scale has the advantage of natural small-scale divisions. Also, in temperate or tropical zones, it is rarely cold enough to require negative readings.

75. What is the centigrade scale? In 1742, the Swedish astronomer Anders Celsius proposed a scale on which the temperature of a mixture of ice and water is zero and that of boiling water 100 degrees. The degree is defined as one hundredth of the difference between the two reference points and hence the term centigrade (100th part). This is a more rational scale than the Fahrenheit and is widely employed.

76. What is the absolute or Kelvin scale? Many meteorological computations, especially involving the upper air, employ the absolute or Kelvin temperature scale. It was suggested by the English scientist William Thomson (Lord Kelvin) as a more exacting method of temperature measurement. On this scale, the zero point or absolute zero represents the lowest temperature that can possibly exist—a theoretical state when the molecules of a substance cease vibrating completely. This value is obtained by theoretical considerations of a sample portion of gas being reduced to a zero pressure. On the centigrade scale, absolute or Kelvin zero would correspond to −273.18° C. or −459.72° F. For practical purposes, therefore, 0° C. may be considered as 273° absolute or 273° Kelvin. In studies of energy and matter, scientists have produced temperatures quite close to absolute zero. Conversely, temperatures in atomic explosions represent about the highest temperatures ever produced on Earth, more than 1,000,000° K.

77. What is the Réaumur scale? In 1731, the French scientist René de Réaumur proposed a temperature scale based on the use of alcohol as the expanding liquid. The zero reference was the freezing point of water. A degree was arbitrarily taken as one thousandth of the volume contained by the thermometer bulb and tube up to the zero mark. Under standard conditions, the boiling point of water on this scale falls at 80 degrees.

78. What are the boiling and freezing points of water on the different scales? At a standard sea-level atmospheric pressure of 14.7 pounds per square inch (1013.25 millibars), the boiling and freezing points of water are as follows:

	Fahrenheit	Centigrade	Absolute or Kelvin	Réaumur
Boiling	212°	100°	373°	80°
Freezing	32°	0°	273°	0°

79. How is centigrade converted to Fahrenheit? Multiply the centigrade temperature by 1.8 and add 32°. For example, 100° C. x 1.8 + 32 = 212° F. As a formula F. = $\frac{9}{5}$ C. + 32.

80. How is Fahrenheit converted to centigrade? Subtract 32° from the Fahrenheit temperature and divide the product by 1.8. For example: 212° F. − 32° = 180°/1.8 = 100° C. If minus temperatures are involved, the addition of 32° must be done algebraically. Thus, $-40°$ F. $-32° = -72/1.8 = -40°$ C. As a formula

$$C. = \frac{5}{9}(F. - 32).$$

81. How is Fahrenheit converted to absolute? Add the Fahrenheit temperature algebraically to 459.4. For example, 30° F. $+ 459.4 =$ 489.4° Fahrenheit absolute. $-30°$ F. $+ 459.4 = 429.4°$ Fahrenheit absolute.

82. How is centigrade converted to absolute? Add the centigrade temperature algebraically to 273. For example: 15° C. $+ 273 =$ 288° absolute. $-20°$ C. $+ 273 = 253°$ absolute.

83. What instruments are used in measuring temperature? The most commonly used is the *liquid* type of thermometer (mercury or alcohol-in-glass or mercury-in-steel). The *deformation* type is another kind of thermometer based on an expansion or contraction of a metal. A third type is the *electrical* which uses resistance or electrical connection.

84. How does a standard, liquid-in-glass air thermometer work? A standard air thermometer essentially consists of a glass tube marked with a scale and partially filled with mercury or alcohol, the balance of the tube containing a vacuum. As temperature increases, the molecules of liquid are agitated to move upwards in the bore of the tube which is narrowly constricted. The top of the liquid will therefore reach a higher position or reading. The lower the air temperature, the less will be the molecular movement of the liquid and therefore the lower position or lower reading.

85. Why is mercury used in some thermometers and alcohol in others? Mercury freezes at around minus 30 degrees Fahrenheit and so is not normally used where extremely low temperatures may be expected regularly. During consistently cold weather periods, for example, at certain Alaskan or central Canadian weather stations, alcohol thermometers are used because alcohol freezes at $-202°$ F. Mercury, however, is a more efficient liquid in terms of exactness of measurement. The ordinary air thermometer used in the home or office is usually alcohol-filled and colored red for easy observation.

86. How are the highest and lowest temperatures of a given period measured? Two specially constructed thermometers called the maximum and minimum thermometers are used. They are simple but ingenious in concept. The maximum thermometer has a constriction in the bore made so as to allow mercury to rise past it when temperature rises but to prevent the mercury from falling back when temperature lowers. The minimum thermometer contains a tiny glass

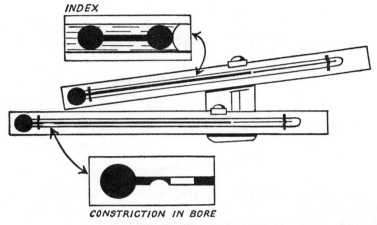

INDEX

CONSTRICTION IN BORE

Maximum and minimum thermometers

index which rides down the bore with the alcohol when temperature falls, but doesn't slide up in the bore when temperature rises. In the maximum thermometer the top of the mercury column will always show the highest point reached. In the minimum thermometer the index will be at the lowest reading. At the end of any particular given period, usually 24 hours, the instruments are reset to normal conditions.

87. How is a written record of temperature changes obtained? A continuous trace of temperature changes is obtained by the use of a thermograph. One type, the exposed thermograph, consists of a slightly curved metal tube which is filled with alcohol. It is called a Bourdon tube. The temperature changes affect the alcohol which causes the curvature of the tube to change. One end of the tube is rigidly fixed and one end free to reflect this change through a linkage system on to an inked pen arm which records a trace on a graphed paper. The chart is attached to a clock-driven drum. Another type uses a spiral bi-metallic strip. Changes in temperature cause the coil to wind or unwind. This change is communicated to a recording paper by linkage, and a continuous record is traced on a chart which is attached to a clock-driven drum.

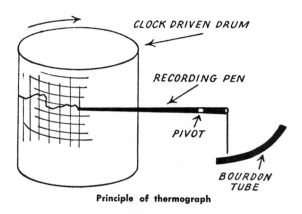

CLOCK DRIVEN DRUM

RECORDING PEN

PIVOT

BOURDON TUBE

Principle of thermograph

88. How are distant temperature recordings made? Electrical thermometers are used to obtain distant readings. The unit of the thermometer is exposed in a remote area and electrically connected so that the changing temperatures are indicated on sight-reading dials in an observatory. The change can also be converted to a written trace on a chart.

89. What is meant by an official temperature? It is obvious that at a given time during any day temperature readings will vary considerably, depending upon the actual location of the thermometer and its exposure. A thermometer in the shade exposed to a strong sea breeze will read several degrees lower than a thermometer in wind-blocked direct sunlight, even though the thermometers are just a few yards apart. Standard conditions, therefore, are required so that a truly representative temperature reading can be taken. If an official weather report announces that at 1:00 P.M. yesterday the temperature was 76 degrees Fahrenheit, it means that the reading was taken from a standard, mercury-in-glass thermometer which is kept in a specially constructed weather instrument shelter. The shelter is a wooden box painted white with louvered sides. It is freely exposed to sun and wind and yet the thermometer will not be in direct sunlight or be affected by direct wind motion. The shelter is raised about 4 feet from the ground so as to avoid heat effects from the ground or rooftop.

90. What is a degree-day? It is a measurement which indicates deviations of temperatures from normal. It is a useful guide or index wherever temperature is a critical weather element—in the heating industry, for example, or in fruit and vegetable crop production.

91. How is the degree-day calculated? As an example of its use to heating engineers, start with an outside air temperature of 65° F. This is considered a critical heating temperature or standard. When air temperature is above it, requirements for heating are not significant. If outside air temperatures are below 65° F., heating is generally required. The next figure to consider is the mean temperature of the day. If, for example, the high is 50° F. and the low is 40° F., then the mean temperature is 45° F. This number, 45, is deducted from the standard number 65. The balance of 20 is the final calculation. That particular day is said to have 20 degree-days. A continuous record of degree-days is kept from September through April with each day's total added to the preceding grand total. This mounting number is watched carefully by heating engineers with respect to its deviation from normal or average conditions. If the mean temperature of the day is equal to or higher than the standard of 65, then the day is called a zero degree-day and so does not add to the total.

92. What is water vapor? Water isn't always a liquid. Sometimes it becomes solid matter, as when it freezes to ice. It also has an invisible form and exists everywhere around the world in the lower levels of the atmosphere—even over the driest deserts. When water is invisible, it is called water vapor. It is water in a gaseous state.

93. Why is water vapor so important? Water vapor is always present in the lower air layers. It can vary in amount from a mere trace in arid regions to about 4% by volume in moist areas. This represents a very small fraction of the air, but its importance as a weather-producing factor far outstrips its relative quantity. It is the gas embryo for precious rain, billowing clouds, fog, ice and delicate snow crystals.

94. What is the water cycle?   A constant water cycle exists around our planet—a three-way action which allows the total replenishment of water to sky, sea and earth. It starts with a process whereby water vapor rises into the air from oceans, lakes, forests, fields, animals and

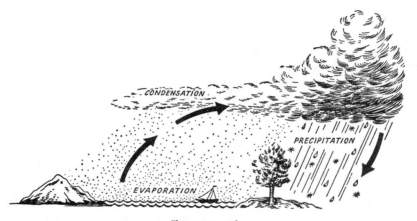

The water cycle

plants the world over (evaporation). This water vapor then develops into visible moisture in the form of clouds or fog by cooling (condensation) and, finally, the cycle of replenishing is completed by the return of the water to the land and seas in liquid or solid form (precipitation).

95. What is vaporization?   When liquids are turned into a gaseous state, the process is called vaporization. The product that results is called a vapor.

96. How does evaporation take place?   In the process of evaporation, swift-moving water vapor molecules dart clear of the surface of a liquid to join other transparent gases of the air. The fastest-moving molecules vaporize first. This leaves the rest of the liquid with a lower energy content. This makes evaporation a *cooling* process.

An example of evaporation is the drying of a city street after a rain-shower. Evaporation can occur from a liquid at temperatures between the freezing and boiling points.

97. When does boiling occur?   Boiling, or ebullition, is vaporization that takes place not only at the surface of a liquid but throughout the liquid when sufficient heat is applied.

98. What is condensation?   The amount of water vapor that can be held in any portion of air is limited to about 4% by volume. This is a kind of quota system. The warmer the air is, the more water vapor it can hold. The cooler the air gets, the less water vapor it can hold. When air is cooled to a point at which the air's capacity to hold the water vapor is zero, saturation is reached. The water vapor becomes visible, as in a cloud or fog. When this occurs, the process is called condensation. Cooling is not always necessary in order to bring about condensation. Sometimes an excess of water vapor is pumped into a portion of air which is already carrying its full quota of water vapor. The excess of water vapor condenses into visible moisture. Steam over a hot cup of coffee is an example.

99. What is the heat of condensation?   Evaporation is a cooling process because the escaping water vapor molecules leave the liquid with a lower heat content (See Question 96). In condensation heat or energy is liberated to surrounding air because the heat content is increased by the addition of water molecules. The heat released by copious condensation often becomes a major supply of energy fuel for large-scale storms like hurricanes.

100. What is precipitation?   Precipitation is a *product* rather than a *process*. It refers to various types of condensed water vapor which fall to the Earth's surface such as rain, snow, drizzle, hail, etc. Precipitation may be liquid, frozen or partly frozen, depending upon the state in which it arrives at the surface of the Earth.

101. What is sublimation?   Most solids are turned into liquid when enough heat is added. Some substances like dry ice or musk turn directly from a solid to a gas without going through a liquid state. The commonly used closet camphor is a perfect example of this process which is called sublimation. In weather, when water vapor cools directly into a solid (ice) form without going through a liquid state, the water vapor is said to sublime and the process is called sublimation. Frost on a window pane is an example (See Question 185).

102. What is the dew point? The dew point is a temperature to which, if an air parcel cools, saturation will occur. When this temperature is reached, the water vapor will change from its invisible state to condense into visible moisture. The bigger the difference which exists between the air temperature and its dew point, the drier the air is and the less chance there is for condensation to occur. When the temperature and the dew point are the same, fog or clouds will form.

103. What is meant by humidity? Humidity is a reference to the amount of water vapor in the air. Humidity measurement is a way of expressing the amount of this water vapor. The amount of moisture which can be contained in a given volume of air is known as *capacity*. It is extremely important for the meteorologist to know the degree to which that capacity is filled. This concern is echoed, though in a different way, by the housewife with hanging wash, the surgeon who doesn't want his eyeglasses to steam during some operations and many people who cannot seem to stop perspiring on certain days.

104. What is relative humidity? It is one of several methods of expressing the water vapor content of the air. It is a ratio between the amount of water vapor the air is actually holding at a certain temperature and the amount it could hold if saturated at the same temperature. This is usually expressed as a percentage. When the water vapor remains the same, increasing the temperature will decrease the relative humidity. If the temperature is lowered, even though the water vapor present is the same, the relative humidity will be higher. This is why, for example, a relative humidity of 65% conveys a different sensation at 85° F. and at 45° F. During the winter season, we often create the relative humidities of the fiercest and driest desert regions in our homes by merely heating dry cold air up to comfortable temperatures.

105. What is humiture?   Humiture is a nonscientific term of fairly recent origin. It apparently originated as an effort to express humidity and temperature in such a way as to provide an instant guide for judging comfort reaction to these two weather elements. It is defined as the mean between the temperature and humidity and is expressed in *humits*. Thus, if the temperature is 70° F. and the relative humidity 58%, then the humiture would be 64 humits. Supposedly, the range of comfort lies between 60 and 70 humits.

106. Is humiture used in meteorology?   As a method of expressing the amount of water vapor in the air, the term has no application to meteorological observations. Meteorologists are concerned with the need for a numerical guide for comfort reaction to the various weather elements and are seeking ways of solving this difficult problem. Humiture might be considered as a step in the right direction, but, unfortunately, it has two immediate shortcomings. First, it breaks down considerably as a useful index when applied to many extremes of temperature and humidity experienced in many different places at many different times. Secondly, it does not reflect the comfort reaction to many other kinds of weather elements which effect our sensory perceptions and which occur simultaneously with temperature and humidity such as winds, pressure, glare, pollen, precipitation, etc. Because of these variables and the way each individual reacts to separate weather elements, a solution in a quick and easy guide poses a complicated problem.

107. When it rains is the relative humidity always 100%?   No. Rain occurs under many circumstances. Frequently, for example, warm moist air may overrun or ride on top of a cooler somewhat drier mass of air which is on the Earth's surface. Rain may fall from the warm, condensed layers of air through the cooler, drier air at the surface. The relative humidity value at the surface reflects the actual moisture content of the lower layer of air which is simply being invaded by rain from above.

108. Can relative humidity be 100% without rain?   Yes, as in the presence of a light fog which is simply a mass of air cooled to condensation but without any precipitation forming.

109. What causes early morning dew on grass? During a calm clear night following a rather warm clear day, typical of late spring or early autumn, the ground will radiate its stored daytime heat and become cool. If the air is calm enough and moist enough, its cooling by contact with the ground will cause the water vapor to be condensed on exposed objects like blades of grass or flowers. Dew forms; it doesn't fall from clouds.

110. What instruments are used to obtain humidity measurements? The two instruments generally used for humidity measurements are the psychrometer and the hair hygrometer. A written record of humidity changes is provided by the hygrograph. These instruments, like the standard air thermometer, are placed in an instrument shelter so as to obtain free-moving air standard values (See Question 89).

111. How does the psychrometer work? The psychrometer is the least expensive and most accurate of instruments for measuring moisture content of the air. It consists of two thermometers attached to a common frame. One thermometer is called the dry bulb. The second is called the wet bulb and its bulb is encased in a muslin or gauze jacket. The wet bulb is dipped in distilled water and the instrument is whirled or ventilated by a fan. Evaporation of water takes place from the wet bulb. Since evaporation is a cooling process (See Question 96), the wet bulb thermometer will show a lower temperature reading than the dry bulb. The difference or spread between these two readings is checked against a series of predetermined calculation tables and provides relative humidity and dew-point temperature values. The larger the spread between the readings the drier the air is. It is obvious that if the air is already saturated, the wet bulb will read the same as the dry bulb and 100% relative humidity will be its value.

112. How does a hygrometer work? The action of a hygrometer is based upon the reaction of strands of human hair contained in the instrument to changing moisture values in the air. It has been found that hair strands lengthen when humidity rises and shorten when the air is drier. At the same time, the strands are not affected by any changing temperatures. In the hygrometer, the expansion or contraction of the hair lengths is communicated through a linkage system to a properly calibrated clocklike dial face. In the hygrograph, instead of the changes showing on a direct-reading dial face, they are linked to a pen arm which records a trace of the changing values on a chart revolving on a clock-driven drum. These hair instruments lack fine precision mainly because of the lag of the hair response to rapid humidity fluctuations.

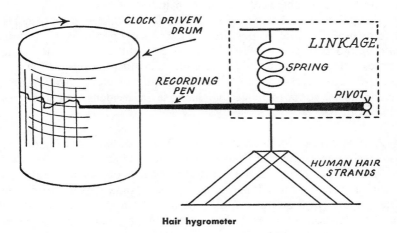

Hair hygrometer

113. How do clouds form? Any cloud, from delicate wispy ice-crystal clouds six miles above the earth to a pea-soup London fog, is a portion of air that has condensed into visible moisture. Cooling of such a portion of air to its dew point is by far the main reason for most cloud formations. The chief cause of cooling lies in some process which causes air to rise or be moved upwards. Large warm air masses may sometimes slide up the slope of a colder heavier mass of air. As the warmer air overrides the cooler, the warm air expands and is cooled itself by the expansion of air molecules. It may finally reach a dew-point temperature with widespread cloudiness forming. Air can also rise by being pushed against a mountain. Forced aloft, the rising air cools by expansion and clouds result. Clouds like these are typical of the windward sides of large mountains which face a direction from which prevailing moist winds blow. A third way air can rise is in vertical convection currents, heated from below by a warm surface. In all cases a cloud in the sky or a fog on the ground is just a larger version of many familiar everyday forms of condensation—the expelled breath smoke on a frosty morning or the steam from a teakettle.

114. What exists in the air that helps water vapor to condense? Our atmosphere contains substances which attract water vapor by a chemical magnetism. Sea salt, spuming into the air from the oceans, is one example. Chemical wastes from industrial processes, like sulphur dioxide, have an affinity for water. These different microscopic bits of lure for water vapor are called *hygroscopic* or *condensation nuclei,* and they play a most important behind-the-scenes role in the process of most forms of condensation. Artificial rain-making experiments are based, mainly, on the seeding of clouds with hygroscopic nuclei so as to induce an increase in condensation and precipitation.

115. How large are cloud droplets? When water vapor condenses into visible moisture, the droplets are much too small to be seen by the naked eye. They gather around the myriads of tiny condensation nuclei in the air—the salt spray, dust, pollen—and join up in incredible numbers as a group to form a cloud. The average cloud droplet of water is about $\frac{1}{2500}$ inch in diameter. It takes almost a million to form a moderate-sized raindrop! It would take about 7 billion fog droplets to make a tablespoon of water.

116. What are clouds composed of? All clouds consist of an assemblage of minute water droplets, ice crystals or a combination of both. The free-floating clouds at very high altitudes (where below-freezing temperatures exist) consist of tiny hexagonal prisms of ice formed by the process of sublimation. In this process water vapor turns directly into ice without going through a liquid stage (See Question 101). Clouds frequently contain water droplets which may exist as water even though at levels where temperatures are below freezing. These water droplets are said to be *supercooled.* Surface tension on the minute droplets keeps them in a water form. They are very unstable and rupture quickly into ice.

117. How are clouds classified? Clouds are identified and grouped according to three basic classifications: their form and appearance, their composition and their heights above the surface.

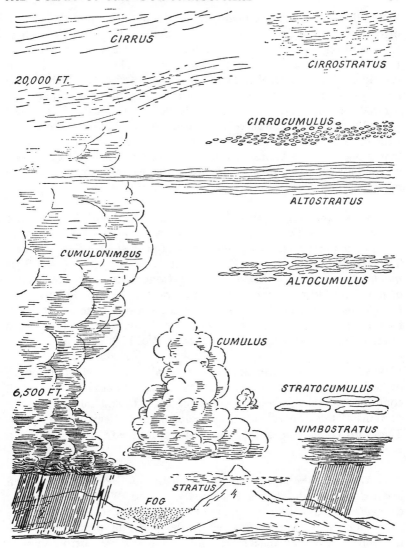

118. What are the basic forms of all clouds? Despite their seemingly endless shapes, clouds can come under two very broad classifications with regard to appearance or form. They are either stratus type (stratiform) or cumulus type (cumuliform). The stratus cloud is sheetlike, fibrous, layerlike. Think of the word *straight* which sounds like stratus or the Latin word *stratum* meaning layer or sheet. It is a rather formless type with very little definitive markings. In some ways a stratiform cloud is very much like a wisp of cigarette smoke hanging in still air. Cumulus, on the other hand, is a type of cloud that is piled up or ac*cumulated* in appearance. It has a rolled, cottony or cauliflower appearance. The word cumulus comes from the Latin meaning mound or heap.

119. How are clouds classified according to height? They are broken into *families* of clouds. Family A or *high* clouds, exist at heights having a mean lower level of 20,000 feet above the surface. They range through the top levels of the troposphere. Family B or *middle* clouds exist at heights ranging from an average lower level of 6,500 feet to an average upper level of 20,000 feet. Family C or *low* clouds are found from close to the Earth's surface to about 6,500 feet. Family D or *clouds of vertical development* have an average lower level of 1,500 feet and an upper level of high clouds.

120. Which are the high clouds (Family A)? The high-cloud group include cirrus, cirrocumulus and cirrostratus clouds. They are all composed of microscopic six-sided ice crystals and inhabit the top portions of the troposphere where below-freezing temperatures exist. The word cirrus comes from the Latin translated, rather loosely, a curl.

121. What are the characteristics of cirrus clouds? They are often isolated, detached, fibrous filaments of cloud. They are generally delicate, thin and silky in appearance, sometimes in feathery plume shapes or tufted streaks. Icy white, the edges of these clouds are indefinite. They may reflect the light of the sun well before sunrise or after sunset and appear brilliantly colored at those times. Cirrus clouds consist of ice crystals and produce no precipitation.

122. What are the characteristics of cirrocumulus clouds? They are usually arranged in very small ripples, layers, patches or tiny-appearing globular forms without shading. When arranged in somewhat parallel rippled patterns, they are familiarly known as a mackerel sky. They do not last long, usually appearing in company with other, more fibrous cirriform clouds. Cirrocumulus is an ice crystal cloud and gives no precipitation.

123. What are the characteristics of cirrostratus clouds? Cirrostratus are veil-like fibrous sheets of cloud, often with a shredded appearance. They usually cover large portions of the sky and when forming an overcast impart to the sky a whitish or milky look. They do not block out the sun or prevent shadows from forming. This cloud causes the ring around the sun or moon, usually a 23-degree halo caused by the refraction of sunlight or moonlight through the minute ice prisms of the cloud (See Questions 582–585).

124. Which are the middle clouds (Family B)? Middle clouds include altocumulus and altostratus clouds. These clouds range from about 6,500 feet to 20,000 feet above the surface. They are usually composed of water droplets, although they may sometimes contain ice crystals, particularly in the upper portions of altostratus. The prefix *alto* comes from the Latin word *altus* meaning high. However, it is probably better to think of the musical concept of alto—lower than soprano and not the highest range.

125. What are the characteristics of atlocumulus clouds? They may appear as small, isolated globular patches, parallel bands or in flattened, slightly rolled masses. They usually do not completely obscure the sun. Open sky patches are often seen through an altocumulus deck. These breaks in the cloud are called *interstices*. Sometimes a small rainbowlike ring seems to hug the sun or moon. The ring is called a corona and is red on the outside with blue coloration on the inside. This phenomenon is caused by diffraction of sunlight or moonlight through the water droplets of the cloud. Altocumulus clouds may produce some occasional light rain or snow.

126. What are the characteristics of altostratus clouds? This fibrous, shredded-looking cloud, when forming an overcast, is usually a relatively unrelieved medium gray in color. When thickening, altostratus becomes a darker gray and seems to impart an iron grayness to the entire landscape. When seen through this cloud, the sun may appear as though it is shining through ground glass. No shadows are cast. Altostratus clouds may produce a fairly steady light rain or snow, especially when the upper portions are icy.

127. Which are the low clouds (Family C)?   The low-cloud group includes nimbostratus, stratocumulus and stratus. Stratocumulus and stratus are water droplet clouds. Nimbostratus may contain ice crystals and water droplets. They appear at heights ranging from near the surface to about 6,500 feet.

128. What are the characteristics of nimbostratus clouds?   They are low, thick, dark gray masses of clouds, often threatening in appearance. Their bases are usually ragged and wet-looking. This cloud produces steady rain or snow. It actually is a lowering or thickening of an altostratus cloud and usually has a mixture of water droplets and ice crystals.

129. What are the characteristics of stratus clouds?   Stratus is a uniform, thinly fibrous cloud of indefinite shape, very much like a light fog lifted a few hundred feet off the ground. It gives the sky a light gray or medium gray tone. Stratus is usually not very thick and dissipates quickly under the sun's heat. It does not produce any appreciable amounts of rain or snow but may sometimes give a fine drizzle. Stratus is a water-droplet cloud.

130. What are the clouds of vertical development (Family D)? They are the most easily recognized and the most familiar forms and include the various cumuliform clouds. They are piled-up, wool pack or cauliflower in appearance, ranging from the small puffy fine weather cumulus to the towering cumulonimbus or thunderhead. These clouds, because of their vertical structure, often exist throughout all cloud levels, reaching from the low cloud heights to the very top of the troposphere.

131. What are the characteristics of fair-weather cumulus clouds? They usually appear as isolated, puffy and dense masses with characteristically flattened bases, rounded sides and domed tops. They often have a whipped, cottony or woolly look. Depending upon the position of the sun, cumulus clouds may have a bright white portion and a darker gray side or base. This water-droplet cloud is not sufficiently developed to produce rain or snow.

132. What are the characteristics of heavy cumulus clouds? These are heavy swelling clouds with considerable vertical development, usually in isolated piled-up masses. They have a typical *boiling* appearance with generally flattened bottoms and rounded bulging sides. Heavy or congested cumulus consist mainly of water droplets and sometimes build high enough to produce brief showers.

133. What are the characteristics of cumulonimbus clouds? These are massive, cauliflower-shaped clouds with tremendous vertical development. They extend to icy levels in heavily rolled mountainous turrets of cloud. The cumulonimbus usually have mantles or tops of sprayed-out fibrous cirriform clouds. These ice-crystal caps are called *anvils* which shape they generally represent. The cumulonimbus, popularly called thunderhead, has a dark and menacing-appearing base, usually ragged and sometimes extending to a few hundred feet from the Earth's surface. The base of this cloud often seems to boil downwards in pendulous forms. This cloud is sometimes called a factory of clouds or the grandfather cloud. When a massive and isolated cumulonimbus cloud dissipates, it frequently leaves patches of clouds which may be identified as belonging to all other cloud families. Cumulonimbus clouds contain a mixture of water droplets and ice crystals and generally produce heavy showers of rain or snow, sometimes of hail and frequent thunderstorm activity.

134. Is fog a cloud? Fog is simply a cloud whose base rests upon the surface. Because of its importance as a hazard to different types of transportation, it is usually considered separate from clouds. Like any cloud, a fog is a mass of air whose water vapor has condensed into visible moisture, usually from some cooling process. It consists of minute water droplets suspended in the air and reduces the horizontal range of visibility in different degrees, depending upon the fog's intensity.

135. Does fog always consist of water droplets? No. Fog can consist of minute ice crystals rather than water droplets. This type of fog is called ice fog and forms by sublimation whereby water vapor turns into ice without going through a liquid stage. In this respect it is like a cirriform cloud on the ground. Ice fog usually occurs in higher latitudes where low temperatures prevail.

136. How are different types of fog named or classified? Fogs have been classified in many ways. Sometimes they are identified with a particular locality—California fog, Grand Bank fog, London fog, etc. Sometimes fogs are intermingled with a general meteorological term like monsoon fog, summer fog, sea-breeze fog, etc. The best method for a more scientific understanding of fogs is to classify them according to the basic processes which are involved in their formation. There are many classifications and sub-divisions of fog-forming processes, but only three general categories will be listed here; radiation, advection and evaporation fogs.

137. What causes radiation fog? The basic ingredients favorable for radiation fog formation are (1) a long clear night to allow the earth to radiate away its day-stored heat, (2) a considerable amount of moisture in the air, (3) a light wind of about three to six miles

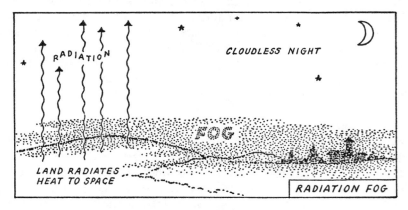

per hour and (4) a type of terrain which readily permits pooling or collection of air such as valleys, depressions, etc. The type of fog usually developing as the result of these factors is familiarly known as ground or summer fog. Under such circumstances, the layer of air in immediate contact with the surface is cooled by contact to condensation. A relatively shallow and patchy fog builds up from the ground from a few feet in depth to perhaps a hundred feet in well-developed ground fogs. They usually dissipate (burn off) upon being exposed to the sun's heat during the day.

138. What causes advection fog?   Ground fogs are local in character and shallow in depth, occurring mostly in late summer or fall. Advection fogs, in contrast, may cover very large areas at one time, may be some one or two thousand feet in depth and may exist at any

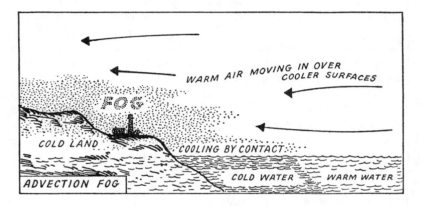

time of the year. The word advection stems from the Latin *advectio* or conveying. Advection fogs are caused by the transport of warm moist air over colder surfaces. When this occurs, the lower levels of the warm air are cooled by contact with the cold surface to the dew point with fog forming in persistent blankets.

139. What causes the thick fogs off the Grand Bank?   The vicinity off the Grand Bank near Newfoundland is a typical advection fog source region. Warm moist air, traveling over the Gulf Stream northward, blows over the Labrador Current, a cold flow of water from high latitudes. At the junction off the banks, the warm Gulf air cools quickly by contact with the Labrador Current air and condenses into persistent walls of fog.

140. What is evaporation fog?   In the radiation and advection types of fog, cooling of air to saturation is the keynote. In the case of evaporation fog, the formation does not depend upon cooling. It occurs when an excess of water vapor is pumped into air which has a limited capacity for more water vapor.

141. How does a typical evaporation fog form?   An example of evaporation fog formation occurs almost daily in a kitchen in the form of steam arising from boiling water. Water vapor molecules escape the liquid to join the air over the pot. The air has a specific capacity for accommodating additional water vapor, and if an excess amount is pumped into the air by the process of evaporation from the pot, then the excess vapor will condense into visible moisture. Notice how this differs from the fog types in Questions 137 and 138 where a *cooler* surface was required to touch fog off. In evaporation fog, a *warmer* surface is an important factor. This situation may frequently occur, for example, over an inland lake or river in the late fall season. The air is often quite cold and the water still retains its heat. Evaporation of water vapor takes place with water vapor being supplied from the lake or river to the overlying air. Saturation is reached and fog forms over the body of water. The more extreme the differences in temperature are between the water and the air, the more favorable for this type of evaporation fog (steam fog).

142. What makes a fog dissipate?   Fogs will dissipate, generally, under conditions opposite to those which aided in their formation. Cooling of air to condensation causes many types of fog. Increase of heat will increase the air's capacity to hold water vapor so that a fog will tend to dissipate when heat is increased. Shallow ground fogs will dissipate rather quickly under a bright sun. Thick layers of advection fog may persist for days. Fogs can also dissipate by being literally broken away by a change of air mass from a different region. Fogs may thin and break by being carried over warm surfaces (if they formed from lower air layer cooling). An increase in wind may also thin off a fog.

143. How is the amount of clouds (sky cover) expressed?   The familiar weather report words like overcast, clear, scattered, etc., refer to the portion of the sky covered by clouds. The words have specific reference to a scale from 0 to 10 which is given in tenths.

For example, a sky which is completely covered by clouds, with no blue sky to be seen, is described as overcast or $^{10}/_{10}$ cloudiness. This terminology is used by the U.S. Weather Bureau for airway weather reports.

144. What does clear signify?   Clear means that either no clouds are in the sky or that if there are any, less than $\frac{1}{10}$ of the sky is covered by clouds.

145. What does scattered mean?   When clouds cover from $\frac{1}{10}$ to $\frac{5}{10}$ of the sky dome, a scattered condition exists.

146. What is a broken sky?   When more than $\frac{5}{10}$ of the sky, but not more than $\frac{9}{10}$, is covered by clouds, a broken condition exists.

147. What is an overcast?   An overcast condition exists when clouds cover more than $\frac{9}{10}$ of the sky. If the blue sky can be seen in places, then the condition is described as breaks in overcast.

148. What is a ceiling?   This term is mainly used with relation to aviation weather reports by the U.S. Weather Bureau. It is defined as being the height above the ground of the lowest clouds which cover more than half the sky. For example, if a scattered (less than $\frac{5}{10}$) cloud deck existed at 4,000 feet and a broken (more than $\frac{5}{10}$, not more than $\frac{9}{10}$) cloud stratum was at 7,500 feet, the ceiling would be 7,500 feet.

149. How is a cloud height defined?   The height of a cloud is considered to be the distance from the Earth's surface to the base of the cloud.

150. What is an unlimited ceiling?   When no broken or overcast layer of clouds is reported, the ceiling is said to be unlimited. This does not mean an absence of clouds. Scattered clouds could exist at several levels, but as long as each cloud level does not cover more than half the sky, there is no ceiling. If one of the scattered levels should become broken, then the height of that cloud layer would become the ceiling.

151. What is ceiling zero? This term is used to report a condition when the ceiling is 50 feet or less.

152. What instruments are used to measure cloud heights? To measure the distance from the surface to the base of a cloud, the following instruments are used: ceiling balloon, ceiling light projector, clinometer, ceilometer and radar indicator. Weather observers can also estimate cloud heights with reasonable accuracy. Estimations are made only when ceiling measurements are not critical or marginal or when mechanical failure prevents instrument measurement.

153. How is a ceiling balloon used? This is not the most exact method and is usually not used with ceilings much above 2,500 feet. Small balloons weighing about 10 grams and colored so as to be easily seen are inflated with lighter-than-air helium and released. They are filled with enough gas so as to lift a weight of 45 grams. This *free lift* establishes a close estimate of the average rate of ascension for the balloon which rises the first 100 feet in about 8 seconds and takes about 15 seconds for each additional 100 feet. The balloon ascent is timed from the moment of release until its disappearance in a cloud layer. Since its rate of rise is predetermined, the base of the cloud becomes the product of the time multiplied by the ascension rate. Vertical wind motions in the lower air make this method only fairly accurate.

154. How are a clinometer and light projector used? These devices are used in combination at night. A powerful searchlight, usually placed 500 or 1,000 feet from the observer, projects a beam of light vertically on the cloud base. The observer sights a tube called a clinometer on the spot of light on the cloud. He centers the light spot on cross-hairs in the tube. The angle formed by the line of sight to the light spot and the ground is measured by means of a pendant which hangs vertically when free on the clinometer but which is tightened against an angle elevation scale marked on the side of the instrument. The observer then knows the distance from the light projector to his position and the elevation angle of the projected light spot. He has established a right triangle problem and can quickly determine the distance from the surface to the cloud base by reference to a trigonometry table.

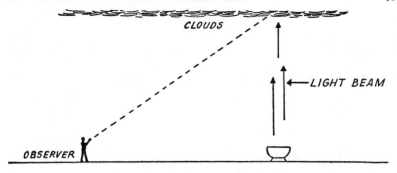

155. How does a ceilometer work? The ceilometer is an electronic improvement over the clinometer method described in the preceding question. It employs the same principle of creating a right triangle and solving the cloud height by determining the angle of a projected

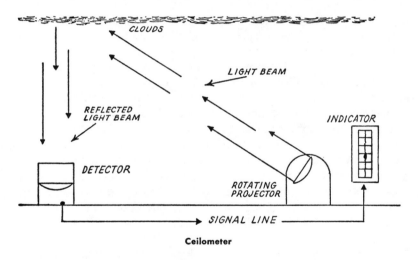

Ceilometer

light beam on a cloud base, but the equipment is more elaborate and effective. It can operate during the day as well as the night, can provide a continuous record of the ceiling and can measure cloud heights with more accuracy and greater distances. The ceilometer includes a rotating beam projector which sweeps the cloud base at brief intervals with a light modulated to a known frequency. Connected by a signal line at a known distance away is a fixed-position detector containing a selective electronic unit. This unit will intercept the light beam as it reflects on a cloud base overhead. The automatic signals created by this interception are calibrated directly into ceiling height.

156. How is radar used to measure cloud heights?   A radar-operated device can be used to indicate cloud outlines from bases to tops. A short wave length radar beam is thrown vertically. As the beam strikes a layer of clouds, part of the energy involved is reflected back to the surface. From this reflection and from electronic calculation based on the radar pulse length and repetition rate, a good profile of the clouds above may be obtained.

157. What instruments are used to determine direction of cloud movement?   An instrument called the *nephoscope* is most widely used. There are two types of nephoscopes, direct vision and mirror-vision. The mirror type consists of a horizontal mirror of dark glass, circular in shape, which can reflect the clouds in the sky. This reflection is seen through an eyepiece attached to the mirror support. The eyepiece is adjustable for height and is mounted in such a way that it can be moved around the mirror's rim. A directional scale is etched on the perimeter of the mirror itself so that, in effect, a miniature reflected sky can be observed in the mirror with convenient markers to point the direction of cloud movement. The direct vision device is called a comb nephoscope and essentially consists of a long vertical rod with a shorter horizontal piece on top. Along the horizontal cross piece, vertical spikes are placed at equal distances. The vertical rod may be rotated and is marked with directional points. The observer lines up the cloud to be observed with the central vertical spike and, as it appears to travel along the vertical side spikes, its direction is indicated by the direction disk attached.

158. Why do only some clouds produce rain?   It might seem that rain is simply a continuation of a cloud formation process with cloud droplets growing larger until they fall to earth as rain. Actually, the processes of rainbirth are complicated and difficult to observe. In the following question, we will examine one theory of rain formation based on the composition of a cloud and the interplay of its water and ice mixture.

Comb nephoscope

159. What kind of clouds generally produce large-scale rain?
Clouds are composed either entirely of water droplets, entirely of ice
crystals or a combination of both (See Question 116). A law of
physics states that objects of like electrical particles repel each other
and objects of unlike qualities attract. It is reasoned that if a cloud
consisted entirely of microscopic water droplets and that if they were
relatively undisturbed, each tiny droplet being composed of aggre-
gates of molecules with like charges residing on each particle, they
would repel one another. Each droplet would keep its distance from
its neighbor. Not joining together to get bigger, they would continue
to float along in cloud form. A similar nonmingling situation would
occur in a cloud consisting entirely of ice crystals. The crystals hang

suspended in the air and make up a cirriform cloud in the aggregate, but do not get together. What force causes a coming together so that the particles grow larger and fall? The answer lies in a cloud deep enough to contain both water droplets and ice crystals so that attractions of particles can take place.

160. How does rain actually start in such clouds? Water droplets can exist at below-freezing levels (See Question 114). These supercooled droplets often exist at temperature well blow freezing, to —10° F. or even —20° F. At a critical temperature ranging from about 5° to —15° F., if a supercooled water droplet and an ice crystal are side by side, difference in vapor pressure exists, stronger from the water to the ice. The water droplet will join or sublime directly on to the ice crystal because of this difference. The ice crystal will grow larger at the expense of neighboring supercooled droplets. Finally, it will grow large enough to fall earthward under gravity. If it passes through warm layers of air, it may melt and become a raindrop. If it parachutes to earth still retaining its frozen state, it becomes a snowflake.

161. Does all rain start from ice-and-water clouds? No. Until quite recently, the ice crystal process described in the preceding question was thought to be the basis for all significant rainfall. Its importance is still unquestioned. However, other factors should be recognized. For one thing, a considerable amount of rain falls in the tropics from clouds which contain no ice crystals. Also, the ice-crystal process seems to require an additional process for the crystal to grow larger. It is self-restricted as far as its ability to grow is concerned. We should, therefore, consider a second rainbirth factor.

162. How does rain form in clouds containing only water droplets? It is a fairly simple process as opposed to supercooled water latching on to ice crystals at below freezing temperatures. It is based on a collision or merging factor. Some cloud droplets fall or are jostled into collision with others. This happens in clouds with strong internal air movement and water droplets of many different sizes. When this jostling effect takes place, the large water droplets grow bigger at the expense of the smaller ones. As growth increases, the large droplets start to fall, sweeping smaller ones into their bodies as they fall earthward. In this type of coalescence rain formation, the droplets of water are literally forced to mingle and merge with one another. In water droplet clouds containing no or very little vertical movements (stratiform clouds), rain formation would be prevented by the absence of any force to drive the droplets together.

163. How big are raindrops? Raindrops, or, to identify them scientifically, precipitation of drops of liquid water, are generally less than $\frac{2}{100}$ inch in diameter. They range from about $\frac{1}{100}$ to $\frac{1}{4}$ inch. This may be about a million times larger than a cloud droplet. Raindrops can get just so big and then they are cut down to size by the inexorable equalizing rules of nature. Any drop larger than about $\frac{1}{4}$ inch will, when meeting up with air friction induced by the rate of fall of the droplet, break up into smaller drops.

164. How fast do raindrops fall? Raindrops fall a bit faster than 600 feet a minute through still air (about 7 miles per hour). The size of the drop, interference, movement of air, etc. will affect this speed. The largest drops (about $\frac{1}{4}$ inch) will break up if they reach a falling speed of about 1,650 feet per minute or approximately 18 miles per hour.

165. What is the shape of a raindrop? The popular image of a raindrop is usually presented as the shape of a miniature liquid pear, symmetrically bulging at the bottom and coming to a point at the top. Another popular conception is the completely spherical raindrop. Actually, high-speed photographs of raindrops nearing the earth show them to be rather flattened at the bottom—a sort of mushroom-top shape. This is quite understandable when we consider that the raindrop is meeting air resistance which exerts force on the underside of the drop as it descends. This air resistance flattens the bottom of the drop somewhat and makes the droplet bulge on top. The smaller the droplet and the more quiescent it is (as when it is in a cloud state), the more spherical it is.

166. What instruments are used to measure rainfall? An instrument called the standard rain gauge is most commonly used. To obtain a continuous written record of the rainfall amount, a gauge called a tipping bucket is used. The basis for measurement is to determine the depth of water that falls on a level surface within a certain period of time.

167. How does the standard rain gauge work?   The basic simplicity of this instrument reflects the fact that of all the weather elements, rainfall was probably the first to be accurately measured. It consists of a receiving funnel, a measuring tube, an overflow can and

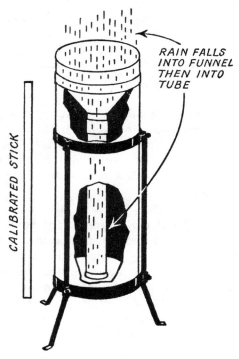

a measuring stick. The gauge is supported vertically by a type of tripod. The area of the receiver is ten times the area of the measuring tube. Therefore, as water collects in the tube, its depth is exaggerated tenfold, permitting a more precise measurement of rainfall. Since the inner measuring tube is usually 20 inches high, it actually can only contain two inches of water. The depth is determined by inserting a special hardwood measuring stick through the receiving funnel into the measuring tube. When withdrawn, the moist portion of the stick indicates the depth of collected water. The stick is calibrated to allow for the vertical exaggeration of the gauge. Thus, 10 inches on the stick means one inch of rain, one inch means $\frac{1}{10}$, etc. Should more than two inches of rain fall between measurements, the excess water falls into the overflow can. After the water in the inner tube is disposed of, the excess is then measured and its depth added to the total amount.

168. How does the tipping bucket work?   As the rain falls, it is caught by a little rocking bucket divided into two openings which tip on a common spindle. When $\frac{1}{100}$ inch of rain falls, its weight tips the bucket over and that amount of rain spills over into a funnel below while rain continues to be caught in the other half of the tipping bucket. This movement of the bucket is connected electrically to a more distant point and a contact is made when each $\frac{1}{100}$ inch of rain tips over. This electrical contact can be synchronized and registered as a recording trace on a chart which is attached to a clock-driven drum.

169. What is the difference between a trace, light, moderate and heavy rain?   The rate of rainfall within a given period of time determines the intensity of the rain. A trace of rain describes a condition when the amount is too slight to be measurable. Light rain refers to a rate of rainfall from a trace per hour to $\frac{1}{10}$ inch per hour ($\frac{1}{100}$ inch or less in 6 minutes). Moderate rain is described as a rate of rainfall within the limits of $\frac{11}{100}$ inch per hour to $\frac{30}{100}$ inch per hour (more than $\frac{1}{100}$ to $\frac{3}{100}$ inch in 6 minutes). Heavy rain is a term used to describe rain which is falling in excess of $\frac{3}{10}$ inch per hour (over $\frac{3}{100}$ inch in 6 minutes). These intensity descriptions apply to the rainfall time rate when the observation is actually made. They do not apply to accumulated water depths.

170. What is the difference between drizzle and rain?   The minute and numerous droplets of water termed as drizzle have diameters of less than $\frac{2}{100}$ inch. Sizes range from about $\frac{1}{500}$ to $\frac{1}{50}$ inch. They are therefore much smaller than raindrops (See Question 163). Drizzle droplets are rather uniformly dispersed and appear to float along with the air currents. They appear, in their aggregate, to be a thick fog, but unlike fog, drizzle falls to the ground. It generally falls from low stratus clouds and is usually attended by poor visibility and fog conditions. Drizzle never appears in showery form. The intensity of drizzle fall is rarely over $\frac{2}{100}$ inch per hour. This compares to light rain intensity (See preceding question), but its smaller droplet size and its association with fog separate drizzle from rain. It is also separated from rain by the way in which its intensity is described. Rain intensity is judged by its rate of fall in a period of time. Drizzle intensity is judged by the range of horizontal visibility when drizzle is present.

171. What is the difference between light, moderate and heavy drizzle? Light drizzle describes a condition when, in its presence, visibility is $\frac{5}{8}$ mile (3,300 ft.) or more. Moderate; when visibility is $\frac{5}{16}$ mile (1,650 ft.) to, but not including, $\frac{5}{8}$ mile. Heavy; when drizzle is present and visibility is less than $\frac{5}{16}$ mile. In the last classification, fog would probably be present.

172. What is the difference between showers and rain? A rain shower is a type of precipitation as opposed to continuous rain. It is the type of rain usually falling from a cumuliform cloud containing a large degree of vertical air movement. Showery rain is sporadic in nature. Its intensity may vary from light to heavy in a brief time. Also, it usually affects limited areas. A brief rainshower may fall, for example, on a few thousand square feet from a moving isolated cloud. Blue sky may be seen at such times and hence the popular term sun shower. Showers can also be heavy as in the torrential downpours in the tropics or as associated with well-developed thunderstorms. Continuous rain, on the other hand, is manifested by its steady rate of fall and usually drops from a thick stratiform or sheet-type of cloud. At such times the sky is usually an unrelieved gray.

173. How does snow start? Snow formation starts like rain—in clouds deep enough to contain minute water droplets and ice crystals (See Question 160). At low temperatures (about 5° F. to —15° F.), water droplets in supercooled form sublime (turn from water to ice) directly on nearby ice crystals. The crystal grows larger and heavier and falls under gravity force, in the meanwhile coalescing with other subcooled minute water droplets and finally reaching the Earth. If it holds its frozen state throughout its journey to the surface, it is termed as snow. If these snowflakes pass through an air layer with above-freezing temperatures, it may simply melt and reach the Earth as rain.

174. What does snow consist of? The delicate featherlike snowflakes that appear to float rather than fall are actually pieces of ice. They are most intricately patterned and consist of a great variety of hexagonal crystal forms. Snowflake images revealed in enlargements made from microphotographs are subjects of jewel-like beauty. An American photographer named Bentley was so fascinated by snow crystals that he amassed a collection of thousands of snow crystal photographs over a period of many years. He claimed that he never found two crystals exactly alike.

175. Why are snow crystals six-sided? The beginning of the six-sided ice crystal can be traced to the construction of the water molecule which consists of two atoms of hydrogen and one atom of oxygen (H_2O). This construction is formed in the shape of a triangle with three equal sides. When crystallization takes place, each new ice-crystal bud is formed at an angle of 60 degrees from the hub or apex of the triangle. Continuing this process, the hexagon is formed when six of these molecular triangles are completed. As the crystal falls and is made larger by further sublimation and coalescence, its six-sided and six-angled form becomes the latticed framework for further growth and extension into several different basic crystal forms. These include columns, needles, flat plates, capped columns, stars and irregular groupings.

176. Why are some snowflakes small and powdery and others large and moist? The flake is a historical reflection of its own path as it descended, of its reaction to the temperatures and water content of the air layers it moved through and of its collision with other flakes. Some soggy flakes, measuring about an inch in diameter when they reach the earth, are conglomerations of matted-together flakes which have passed through a low and moist segment of air. Small-flaked dry snow, on the other hand, shows a history of fall through dry cold-air layers.

177. Why is snow such good insulation against sound and heat?
Snow crystals contain tremendous amounts of tiny openings because of their delicate and fibrous structure. By the capturing of air in these openings or cells, a good deal of heat transfer is prevented. In this respect, snow crystals are miniature cousins to industrial products like rock wool, fiber glass or certain plastics which imprison air in a multitude of tiny cells. Sound can be dampened in a similar way, the tiny hollows in snow crystals trapping and muting sound waves to an astonishing degree. A few inches of snow can muffle a very high percentage of noise energy.

178. How is snow depth measured on the ground? Depth of snow is measured with an ordinary yardstick or ruler at three or more representative areas of snowfall. The recorded snowfall is an average of several measurements. Drifted snow poses a special problem and sometimes ten measurements are made to get a representative value. A weighing type of snow gauge is also used. It weighs the snow as it falls and automatically registers the depth according to its weight.

179. What is the water equivalent of snow? Snow can be moisture-crammed or very dry. Normally, the water equivalent of snow to water is ten to one. Thus, ten inches of snow would provide a water equivalent of one inch. This ratio can vary considerably. Very wet snow ratio to water is about six to one. Ten inches of extremely fine powdery-type snow may melt down to about only ⅓ inch of water.

180. What instrument is used to measure the water equivalent of snow? The same instrument used in measuring rain is used, the standard rain gauge (See Question 168). The exception is that when snow is to be measured, the collector funnel and the measuring tube are removed. The outside or the receiving area of the gauge is used. When ready to measure, the measuring tube is filled with hot water and poured into the gauge. When the snow has melted, the tube is filled with the liquid and thrown away. The remaining liquid is then measured.

181. What is sleet? Like mist, sleet has several different meanings. In England, sleet refers to a mixture of rain and snow. In the United States, some people think of sleet in the same manner. U.S. Weather

Bureau observers record sleet as the name applied to *ice pellets*. These are tiny hard pieces of ice, transparent or translucent and irregular in shape. They average about $\frac{4}{100}$ to $\frac{16}{100}$ of an inch in diameter. They form when raindrops or partially melted snowflakes freeze while falling through a surface layer of cold air which is almost always below freezing. These pellets are hard and rebound when they hit hard ground.

182. What are snow pellets?   They differ from ice pellets in that they are opaque and white in appearance. They are about $\frac{2}{100}$ to $\frac{20}{100}$ inch in diameter and are granular in shape. They easily burst to the touch or when they strike hard ground. Snow pellets are sometimes called *graupel*.

183. What is freezing rain?   When rain falls as liquid but freezes on striking the ground or different objects on the ground, it is called freezing rain. This is a particularly dangerous form of precipitation since it coats power lines, aircraft and highways with tenacious ice. It occurs generally when supercooled water droplets (See Question 116) fall and burst on impact with a cold object and turn directly into ice. It may also occur when regular liquid raindrops from above-freezing air levels falls to the surface where temperatures are below freezing.

184. What is the difference between glaze and rime ice?   Glaze, or clear ice, is a transparent or translucent coating of ice. It is glasslike in appearance. Being nonporous, glaze ice is relatively heavy and tenacious. Rime ice is an opaque, white ice with a granular texture. Unlike glaze, which usually develops by a rupture and freezing of water droplets, rime is formed when tiny droplets of water freeze without rupture. It is therefore porous in structure and not as heavy or as tenacious as glaze. Rime ice is frequently seen on the linings of a freezer compartment in a refrigerator.

185. What causes frost?   Frost forms under similar conditions to dew (See Question 109). Unlike times when dew forms, the temperature must be at or below freezing at the surface area where frost forms. Frost consists of small thin ice crystals which develop, generally, when a layer of air cools to condensation by contact with a cold surface, the temperature of which is below freezing. In this process, the water vapor in the air condenses directly into ice crystal form without liquifying (sublimation). These crystals are deposited on various objects on the surface. Frost forms when nights are clear and calm and the air next to the surface relatively moist. If surface temperatures stay above freezing under these general conditions, dew is likely to form. Frost and dew are not forms of precipitation. They form and do not fall.

186. How is the intensity of frost determined? Because of the importance of frost formation to agriculture, its intensity is judged by the degree of its destructive effect upon different types of vegetation. Frost is described as being light, heavy or killing.

187. What is black frost? When low temperatures cause destruction of vegetation without the existence of frost crystals, it is sometimes called black frost.

188. What is hail? Hail is almost exclusively a product of thunderstorms. It consists of small concentric-layered ice balls or broken ice-ball fragments, largely transparent but sometimes in layers of alternating clear and opaque ice. It is formed in the violent updrafts of cumulonimbus or thunderstorm clouds. Liquid droplets are carried to below-freezing levels where they freeze into small ice fragments. They then drop only to be picked up again by other upsurging air currents. This process of transporting the fragments back and forth between freezing and nonfreezing strata of air results in the onionlike layers of ice which are built up on the original pieces. When they finally become heavy enough to overcome the force of the vertical air currents, they fall to the surface as hailstones.

189. How large are hailstones? Fragments or balls of hail range from about $\frac{2}{10}$ inch to 2 inches, and are sometimes larger. Hail the size of golf balls or oranges may fall from particularly violent thunderstorm clouds. The largest hailstone definitely recorded fell on a farm in Nebraska in the summer of 1928. It was weighed, measured and photographed immediately after falling. The weight was $1\frac{1}{2}$ pounds, with a diameter of about 5.4 inches. Many reports describe much larger hailstones and it seems more than probable that larger ones have fallen. But these reports probably refer to masses of ice which result from the freezing together of two or more separate hailstones coming in contact with one another on the ground after falling.

190. What is dry haze? The atmosphere contains many impurities which mingle in minute form with the atmospheric gases (See Questions 36 and 35). Some of these impurities, like dust or salt particles, are dry and so small in individual size that they cannot be felt or seen by the unaided eye. Frequently, when the air is stable (little or no vertical air currents), these particles concentrate in number in a layer of air next to the earth. They restrict visibility and impart to the air a slightly veiled appearance. Against a dark background, dry haze appears bluish, like smoke. Against a bright background, dry haze gives a saffron tint to the atmosphere.

191. What is damp haze? Damp haze lies somewhere in the scale between dry haze and light fog. It consists of microscopically small water droplets or particles which attract water (hygroscopic nuclei) suspended in the air. It differs from dry haze in that it has a grayish appearance and occurs with high relative humidities. It differs from light fog since its droplets are most widely dispersed and smaller in size. Damp haze frequently is observed in windy weather at seashores, for example, where salt particles are carried up to fairly high levels. Light fog, on the other hand, usually occurs in air which has very little movement.

192. What is smog? Smog is a mixture of smoke and fog. Every once in a while, two factors occur simultaneously to produce smog. The first is the existence of an area where large amounts of industrial smoke are pumped into the atmosphere. This smoke may contain a wide variety of types of chemical waste products from refineries, smelters, factories, railroad engines, steamboats, automobiles, etc. Under normal atmospheric conditions, the smoke is mostly dispersed by vertical air movement (rising warm air currents) or by a change of air masses which sweep in frequent waves over temperate zones. Occasionally (the second factor) a huge air mass will stagnate over such an industrial region for several days. The air mass sinks slowly or subsides and is characterized by little or no rising air currents. This air may be conducive, simultaneously, to fog formation. The industrial smoke then is imprisoned in the lower layers of air and mingles with the fog. The worst features of each will be compounded into a thickening murky pollution of different degrees of intensity and irritant effect depending upon how long the air mass dominates the area and the amount and nature of the chemical products in the air.

193. What is meant by visibility? Visibility may be defined as the maximum horizontal distance at which prominent objects can be seen and identified by an observer of normal eyesight under existing atmospheric conditions. In standard weather reports, only horizontal surface ranges of visibility are included. Another kind of visibility is also important, the slant visibility from an airplane to the ground.

194. How is visibility determined? The observer first marks down various known objects like trees, houses, towers, etc., on a map with their known corresponding distances from the observing point. These objects are spread around so that they extend to the cardinal points of the compass. The observer then divides the horizontal circle into a number of equal sectors and estimates the visual range in each sector by identification of known objects contained therein. The visibility value in each sector is arranged in increasing numerical order and the value in the middle represents the *prevailing visibility.*

195. What is sound? Sound, as a physical phenomenon, is a disturbance occurring in the air or some other elastic medium caused by vibration or shock. The particles of air or matter vibrate and propagate pressure waves capable of being heard by the human ear. All sound vibrations must be transmitted through some elastic medium (solid, liquid or gas) before they can be heard. Without particles of matter to vibrate and pass along the pressure waves, sound ceases to exist.

196. What is the speed of sound through the air? At the surface, sound travels at a rate of about 1,100 feet per second, the same rate in every direction from the source. This velocity is effected by such factors as temperature and humidity of the air, as well as altitude. All of these factors are tied in with changes of density of the air. Another way, therefore, of expressing conditions which affect the speed of sound is to say that it varies with density of the air.

197. How do temperature and humidity affect sound velocity? In warm air at the surface, molecules are less densely packed and will transmit sound somewhat faster than if the air were colder. Sound velocity increases slightly over one foot per second for each degree of increase in temperature (Fahrenheit). At 32° F., velocity of sound in the air at the surface has been determined to be 1,090 feet per second. An increase in humidity also tends to result in an increase in sound velocity. In humid air, the density of the air is lessened and therefore sound will travel faster through it. Humidity, however, is not too important a factor.

198. Can sound be heard in the upper atmosphere? At very high altitudes, air thins out to a point at which there are not enough particles to transmit sound. It is estimated that audible sounds are not propogated at altitudes of 80 miles and higher above the earth.

199. What limits the distances at which sound can be heard? Many atmospheric factors which exist normally seldom permit sound waves to be carried to full range. Wind, for example, is an important factor. When wind blows strong and steadily or with gusty tendencies, it often distorts the shape of the sound wave and may move it completely away from hearing range in very short distances from the sound source. Air also moves vertically (turbulence) and can dent or break the sound wave. These vertical and horizontal air motions force the sound wave to lose unity of strength and direction and so it becomes disorganized and fades. Part of the energy of sound is also lost by reflection when a sound wave enters air of different density.

200. Under what conditions can sound be heard better and farther? In weather conditions when the atmosphere is in a state of relative rest, undisturbed by vertical air currents or gusty, strong winds. This condition, for example, usually exists in the early morning hours following a calm and clear fall or winter night, especially over level land. The layer of air resting on the earth at such times is usualy cooler than the air immediately above. This stable condition, called an inversion, prevents sound waves from being broken up by vertical air movement. Sound then moves outwards in a narrow horizontal disk shape to surprisingly large distances. On damp and foggy days, uniform, stable air conditions also allow sound to travel further.

201. What causes an echo? Like light striking a mirror and returning so that an image can be seen, sound waves also can be reflected. These reflections of sound are known as echoes. We hear echoes as repetition of sound returning to us from various surfaces like the faces of cliffs, high river banks, walls and forest edges. A single echo is produced by sound reflecting from a solitary vertical wall. Multiple echoes occur from sounds which rebound back and forth between parallel walls. Overlapping multiple echoes are produced from a number of distant reflecting surfaces which are not spaced equally, such as the walls of a canyon.

202. What is the minimum echo distance?   The human ear cannot distinguish separate sounds when they are $1/10$ of a second apart. Since sound travels about 110 feet in $1/10$ of a second, then a vertical plane must be more than 55 feet away in order for an echo to be heard. In an ordinary room, sound rebounds too quickly to be heard as an echo. It merely re-enforces the original sound.

203. What is the Doppler effect?   The Doppler effect is based on the approach or retreat of a sound source. A common observation is the piercing whistle of a rapidly approaching train. As the train rapidly nears, a greater number of sound waves per second is received by the ear and therefore the pitch seems higher. When the sound source retreats from the ear, relatively fewer sound waves reach the ear per second and the pitch appears to lower. The Doppler principle is of extreme importance in astronomy as applied to measurements of distances of celestial bodies in which cases light emissions are analyzed rather than sound waves.

204. What is the general purposes of weather balloons?   The balloon is a handy and effective means of gathering information about the upper air. Balloons filled with lighter-than-air helium (noninflammable) or hydrogen (inflammable) can perform a multitude of useful tasks. Their ascent and path of travel can be tracked by the unaided eye, telescopic devices, radio signal and radar, resulting in information about cloud heights and wind direction and speeds aloft. Most important, they can carry sensitive weather measuring instruments and radio transmitting equipment through the weather zone of the atmosphere.

205. What is a radiosonde?   The term radiosonde (radio-sound) or raob for short, describes a method of sounding the upper air by means of transmitting instruments borne aloft by balloon. The purpose of the sounding is to obtain a fairly accurate record of temperature, pressure and humidity values to as great an elevation as possible.

206. How does the radiosonde work? The radiosonde is a lightweight box containing small elements which react to changes in atmospheric pressure, temperature and humidity. The box also contains a radio transmitter which emits a radio signal to a receiving station on the ground. The unit is carried aloft by a large helium-filled balloon. As the radiosonde box is borne upwards, the small elements respond to changes in pressure, temperature and humidity. These changes are transmitted by radio impulse to a ground receiving station where they are automatically recorded. When the balloon bursts because of expansion of the gas in the bag, the radiosonde box is carried to the surface by means of a parachute. The box is sometimes recovered, although repair costs are usually more than it is worth.

207. What is a drop-sonde? A drop-sonde is somewhat similar to a radiosonde in that it emits signals to a receiving station which provide a record of temperature, pressure and humidity values in the upper air. But instead of being borne aloft by a balloon, this instrument is *dropped* from a high-altitude airplane and floats down on a parachute. Is is especially useful for getting upper-air information over normally inaccessible areas. It falls at a rate of about 2,500 feet per minute.

208. How high has man probed into the atmosphere? Man himself has attained an altitude of over 120,000 feet, more than 22 miles above the Earth, in a rocket-powered airplane launched at an altitude of over 30,000 feet (1956). In 1957, a U.S. Air Force Major was carried aloft in a hermetically-sealed enclosure by a huge plastic balloon which soared to 100,000 feet (nearly 19 miles) over Minnesota. Where men have not been able to go themselves, they are sending vehicles and instruments. In 1949, a two-stage rocket soared to an altitude of 250 miles above the Earth. These altitude records will undoubtedly be surpassed in the near future, as the technical assault on the upper atmosphere increases.

209. How does our atmosphere protect us? Our aerial blanket protects us from a variety of potential dangers. If not for the absorbing qualities of an ozone layer at about 40 miles above the earth, lethal doses of ultraviolet radiation would consume us with burning energy. Mysterious cosmic rays, emanating from interstellar space, pierce the top of our atmosphere with enormous energy. These rays do not reach the surface of the Earth in their primary form. A chain reaction is formed in the upper atmosphere, and a shower of secondary rays is released which penetrate to the Earth. We are all pierced by these invisible rays which apparently are harmless. The effects of primary cosmic rays on the human body are unknown. We are also protected from a daily bombardment of millions of small meteoric fragments which tear into the upper air. As these fragments slant lower into the atmospheric depths, they meet with increased air molecule resistance and are mostly dissipated out of existence by frictional effects many miles above the Earth's surface.

210. Do other planets have atmospheres? Most of our neighbor worlds have atmospheres, some of which have been observed in surprising detail. Of all the other atmospheres in the solar system, ours is unique in the sense that it is apparently the only one which can support life forms as we know them. The Martian atmosphere most closely resembles the Earth's in a meteorological sense. But its thin atmosphere, depleted of oxygen and water vapor, cannot sustain our kind of life with the possible exception of certain microbes which require no oxygen. Consider some of the other solar system members; the airless, bleak Moon; incredible Mercury, sunbaked to lead-melting temperatures on one side and near absolute zero on the other; carbon dioxide-laden Venus; or the frozen methane and ammonia surfaces of the outer planets. If man is to set foot on other planet surfaces bound with atmospheres of such life-damaging properties, he must take with him a sizeable portion of his own planet's air or manufacture it at some distant interplanetary terminal point. As far as atmospheres beyond our solar system are concerned, the sheer number of existing stars would indicate that other planets also exist. If they do, then an extension of any law of mathematical chance would argue for an atmosphere similar to ours in the most far-flung parts of our universe. At present, we can only speculate.

III. OUR AIR-CONDITIONED EARTH—
WIND AND STORM

Introduction. To an observer studying our earth from a vantage point in interplanetary space, our atmosphere might be noted simply as a gaseous envelope. The observer might also note that it transfers energy around the earth in more or less large and rhythmic patterns. He might even say, "Those earthlings possess a remarkable global air-conditioning system!"

Because we are at the bottom of our air ocean, we are unfortunately prevented from enjoying this global view. A refreshing sea breeze, the biting cold wind of winter or the punishing gales of a hurricane are usually considered against an immediate weather background. But all the winds of the Earth, from the zephyr to the globe-encircling trade winds, are smaller inlays of a vast atmospheric mosaic. As these bodies of air flow, mix and clash above and around us, we call them winds and storms and weather.

211. How is wind defined? Wind may be defined as air in motion. It usually describes air moving as a body, large or small, in a gen-

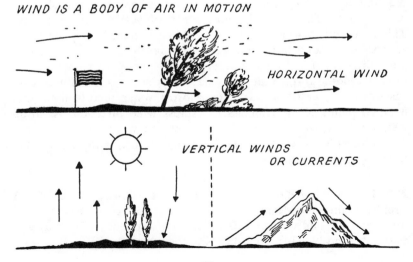

WIND IS A BODY OF AIR IN MOTION

HORIZONTAL WIND

VERTICAL WINDS
OR CURRENTS

erally horizontal direction. When air moves vertically, it is called a *current*. Heated air will be buoyed up by the surrounding denser air and will rise. Such a rising air current (thermal) reduces the pressure at a given locality. The surrounding denser, cooler heavier air (high pressure) will move horizontally to replace the rising warmer, lighter air (low pressure). By this process, a flow of air develops (wind), from higher pressure to low, caused by differences in temperature.

212. How is a wind described? To describe a wind at a given time, its direction and velocity must be specified.

213. How is wind direction defined? In meteorology, wind direction is a direction *from which* the wind is moving. For example, if the wind is blowing from south to north it is called a south wind. A wind blowing from the northwest to the southeast is regarded as northwest. A wind vane is designed to point into the wind. Therefore the arrow is pointing in the direction from which the wind is coming. A wind vane pointing northward indicates a north wind.

214. How is wind direction expressed by points of a compass? The compass may be divided into as many reference points as are required, up to 360°. The compass circle can be divided into 4, 8, 16 or 32 points. The cardinal points are north, east, south and west. This represents a 4-point division. Adding intermediate points such as northeast, southeast, southwest, and northwest provides an 8-point division. If still greater accuracy is required, 16 points may be used by adding further intermediate points such as north-northeast, east-northeast, east-southeast, etc. Most weather reports use either 8 or 16 points of the compass to express wind direction. A northwest wind is a wind blowing from directly between north and west. A south-southeast wind is a wind blowing from a point halfway between south and southeast.

215. How is wind direction expressed by a 360-degree scale? A circle contains 360 degrees. North corresponds to 0°, east to 90°, south to 180° and west to 270°. North corresponds to 360° as well as 0°. A wind blowing from 45° is a northeast wind.

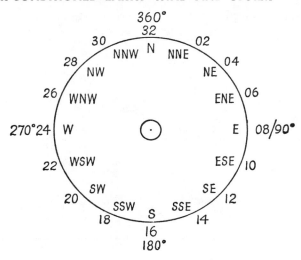

216. How is wind speed defined and expressed? Wind speed refers to the rate of motion of air and represents the speed of air past a given point on the surface of the Earth. It is expressed in several ways. In the United States, wind speed has been usually expressed in statute miles per hour (one mile equals 5,280 feet). There is an increased tendency, however, to express it in knots, a term common in nautical and aviation navigation problems. A knot is equal to one nautical mile per hour. It is incorrect, therefore, to use the expression *knots per hour.* A nautical mile is equal to 6,076 feet. In countries where the metric system is used, wind speed may be expressed in kilometers per hour or meters per second. A kilometer is equal to 3,281 feet and the meter to 39.37 inches.

217. What instruments indicate surface wind direction? Wind direction is indicated by an anemoscope or wind vane which is designed to point into the wind. The wind arrow points to the direction from which the wind is blowing. The way the vane responds to the wind depends upon the mass of the vane and friction in the bearing. Most wind vanes have fins large enough to offer reasonable resistance to the wind. The fin is counterbalanced by a small but relatively heavy arrow. The vane moves freely on a vertical axis. The wind direction is indicated directly by reference to a compass point direction marker near the vane. This direction marker shows the cardinal points of the compass and is oriented to true north.

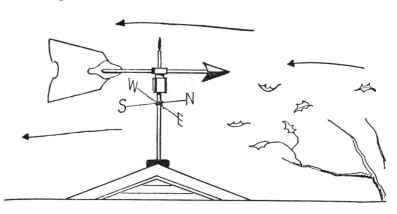

218. How is surface wind direction indicated to a remote point? If the wind vane is on top of an observatory, the main shaft can be extended down through the supporting pipe to a room below. An indicator can be erected in the room with a pointer which is geared to the movements of the wind vane through the shaft. A more practical and effective method is to use an electrical connection so that changes in the wind vane close an electrical circuit connected with a series of electric bulbs located at some distance away. The bulbs correspond to various points of the compass. Another electrical type and the best for direct reading of wind direction uses two motors. One is activated by wind vane movements and causes a corresponding movement to a second motor which, in turn, is connected to a pointer on a clock-type dial containing compass point directions.

219. How is wind direction automatically recorded? The electrical connections described in the preceding question can be extended to a device called a *register*. As the wind changes its movement, the vane touches electromagnets which are connected to the cardinal contacts of the vane. At each contact, a circuit is closed which corresponds to that particular direction. This is transmitted to an inked pen which marks a series of dots automatically on the chart corresponding to the wind direction.

220. What instrument is used to measure surface wind speed?
Anemometers are the most universally used instruments for measuring wind speed. There are four principal types used in meteorological work: the rotating cups, bridled, pressure tube and a self-synchronous motor type called an aerovane. Recently, the U.S. Army has developed a *shooting sphere* anemometer which employs a gun that fires a small steel ball upward into the wind. By checking the bullet's return and the gun's angle, the operator can estimate low-altitude winds to within two miles an hour.

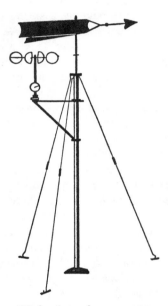

Wind vane and anemometer

221. How does the rotating cup anemometer work? This is the most widely used although it will probably give way gradually to one of the newer types. The Robinson cup anemometer consists of three or four aluminum cups, 5 inches each in diameter, extended about a vertical axis. The cups revolve as the wind blows. An electrical contact is made for each $\frac{1}{60}$ of a mile of wind that passes the instrument. This measurement of the wind speed may be indicated at a distance by causing a light to flash, or may operate a buzzer at $\frac{1}{60}$ of a mile or one-mile intervals. The number of $\frac{1}{60}$ of a mile buzzes or flashes in one minute indicates the wind speed in miles per hour. An automatic record of the wind speed can also be electrically transmitted to a magnetically operated pen which leaves a written trace of average wind speed for short intervals on a clock-driven drum.

222. How does the bridled anemometer work?   This is a modification of the cup anemometer described in the preceding question. The difference is that, in the bridled type, there are 32 or more cups which do not revolve. The cups are held by a spring on the vertical shaft around which the wheel of cups is extended. As the wind blows, the cups are displaced somewhat or move. This movement by the cups against the spring represents the wind force which can be translated in terms of wind speed on an indicating dial or recorder.

223. How does the pressure tube anemometer work?   It is devised so as to measure pressure changes produced by wind blowing into a tube placed in a vane heading into the wind. The vane is attached by tube to a float in water which rises or lowers according to the increase or decrease in pressure. This motion, in turn, is linked to a pen arm which can record a trace on a revolving chart. Pressure type anemometers are used in airplanes to indicate air speeds.

224. What is an aerovane?   The aerovane transmitter designed by Bendix-Friez, is a combined anemometer and wind vane in one unit. It can indicate wind direction and speed, and record these elements. A three-bladed plastic rotor drives a magneto which generates a voltage directly proportional to wind speed. A dual recorder makes a permanent record of each wind gust or lull and every variation in wind direction on a continuous strip chart with a normal rate recording period of 14 days.

225. How are upper winds measured?   Upper wind measurements are usually made by tracking the flight of a balloon which is filled with helium and is free to rise and move with the air. Its position during the flight is observed at regular intervals. By plotting these observations, computations of the direction and speed of the wind can be made. There are three methods employed to observe the balloon: visual or optical, radió and radar.

226. What is the optical method? A balloon, called a pilot balloon, is filled with helium. Such balloons usually are either 30 grams or 100 grams in weight, made of rubber or neoprene. The 30-gram balloon is given a free lift of about 5 ounces and the 100-gram balloon about 18 ounces, so as to provide necessary buoyancy. Balloons inflated with this helium and this free lift, under average atmospheric conditions, will rise at a fairly uniform rate of about 600 feet per minute. The balloon is released at a specified time. At the point of release, its path of flight is followed by means of a sextant-like telescopic device called a *theodolite* which is capable of measuring elevation and azimuth angles to a free balloon at great distances. Readings of these two angles are taken at regular intervals and are plotted on a disc chart. The result of this plotting will provide wind speed and direction to such heights as are limited by the disappearance of the balloon in a cloud, the bursting of the balloon, or the loss in tracking. At night, the balloon can be tracked by its carrying a small battery and electric light bulb.

227. How is radio used to measure upper winds? In Questions 217 and 218 the radiosonde is described as a method of using a large balloon to carry sensitive elements through the air which react to changing pressure, humidity and temperature. These changes are sent to a ground receiving station by a radio transmitter carried by the balloon. At the same time, the balloon's progress can be followed by a separate sensitive direction finding radio receiver. This is called a rawinsonde (radio-wind-sounding).

228. How is radar used to measure upper winds? A balloon carries an attached target that is tracked by a precision radar on the ground in a manner similar to a theodolite. As the balloon rises and is carried by the moving air, the short-wave bursts from the radar intercept the balloon and are echoed back from it. The time difference between the sending of the burst and the echo received provides the basis for computing the wind's speed and direction to elevations of 50,000 to 100,000 feet. The advantages of radio and radar are much greater than with the theodolite method because the balloon can be followed to greater heights and in all kinds of weather.

229. What is the Beaufort scale of winds?   It is a numerical scale for indicating wind speed devised in 1805 by Admiral Sir Francis Beaufort of the British Royal Navy. Beaufort numbers originally ranged from 0, corresponding to calm, to 12, corresponding to a hurricane wind. Each numerical value corresponded to a proportionately stronger wind force. The scale of numbers has now been extended to 17.

230. What is the Beaufort scale in terms of wind force?   The following scale is used by the U.S. Weather Bureau:

Beaufort Number	Miles per hour	Knots
0	Less than 1	Less than 1
1	1–3	1–3
2	4–7	4–6
3	8–12	7–10
4	13–18	11–16
5	19–24	17–21
6	25–31	22–27
7	32–38	28–33
8	39–46	34–40
9	47–54	41–47
10	55–63	48–55
11	64–72	56–63
12	73 or more	64 or more

231. What is the difference between light, gentle, moderate, strong, etc. winds?

Light	—1–7 m.p.h.	Strong	25–38 m.p.h.
Gentle	8–12 m.p.h.	Gale	39–54 m.p.h.
Moderate	13–18 m.p.h.	Whole gale	55–72 m.p.h.
Fresh	19–24 m.p.h.	Hurricane	over 72 m.p.h.

232. How can surface wind speed be estimated by its effect on a flag?   The ordinary flag makes a good wind speed indicator. In a flat calm, it hangs limply along the staff. When the wind is about 10 miles per hour, the average flag will wave out at about ⅓ from the perpendicular staff to a horizontal position. At about 20 miles per hour, the flag is about ⅔ up. At 30 miles an hour or more, the flag streams out in full horizontal position.

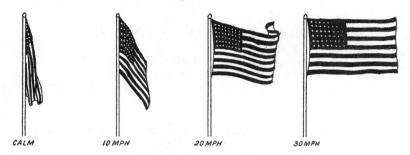

CALM   10 MPH   20 MPH   30 MPH

233. How can surface wind speed be estimated by effects on familiar objects on land? A careful observer can judge the wind, not by wetting his finger (a dubious method at best), but by observing how the wind affects common objects. The scale below may be helpful:

Wind in M.P.H.	*Description of Effects*
Under 1	Smoke rises straight up. No perceptible motion of anything.
1–3	Smoke drift shows direction. Tree leaves barely move. Wind vane shows no direction.
4–7	Leaves rustle slightly. Wind felt on face. Ordinary vane moved by wind.
8–12	Leaves and twigs move. Loose paper and dust raised from ground.
13–18	Small branches are moved. Dust and paper raised and driven along.
19–24	Small trees sway. Large branches in motion. Dust clouds raised.
25–31	Large branches move continuously. Wind begins to whistle. Umbrellas used with difficulty.
32–38	Whole trees in motion. Walking difficult.
39–46	Tree twigs break. Walking progress slow.
47–54	Slight structural damage.
55–63	Exposed trees uprooted. Heavy structural damage.
64–75	Widespread damage.
Above 75	Severe damage and destruction.

234. How are wind speeds estimated by appearance of ocean or sizeable lakes?

Wind force (M.P.H.)	*Wind Effects observed at sea*
Less than 1	Sea like a mirror.
1–3	Ripples with scale appearance, no foam crests.
4–7	Small wavelets, short but pronounced crests appear glassy, do not break.
8–12	Large wavelets with crests beginning to break. Foam appears glassy. Scattered white foam crests.
13–18	Small waves, becoming larger. More frequent white horses (white foam crests).
19–24	Moderate waves; many white horses, possibly spray.
25–31	Large waves begin to form. White foam crests more extensive everywhere.
32–38	Sea heaps up. White foam blows from breaking waves in streaks along wind direction.
39–46	Moderately high waves. Well-marked streaks of foam blow with wind direction.
47–54	High waves. Dense streaks of foam along direction of wind. Spray may affect visibility.
55–63	Very high waves with long overhanging crests; great patches of foam blown in dense white streaks along direction of wind. White appearance to sea surface. Visibility affected.
64–72	Exceptionally high waves. Sea completely covered with long white patches of foam lying along direction of wind. Wave crest edges blown into froth. Visibility affected.
73 or more	Air filled with foam and spray. Sea completely white with driving spray. Visibility very seriously affected.

235. What pressure force is exerted by the wind?   Wind moving directly against any surface creates pressure proportional to the square of its velocity. This is what makes hurricanes dangerous—the rapidly mounting wind pressure with increasing velocities. It means that a 100-mile-an-hour wind is not ten times as punishing as a 10-mile-an-hour breeze, but hits 100 times as hard. A light breeze (4–7 miles per hour) exerts approximately $\frac{8}{100}$ pound of pressure on one square foot. A fresh breeze exerts a force of about $2\frac{1}{3}$ pounds and a whole gale (55–72 miles per hour) strikes directly against a surface with a pressure force of about 12 pounds per square foot.

236. What is a prevailing wind?   A prevailing wind is the average or characteristic wind at any given place.

237. What is a wind rose?   A wind rose is a diagram designed to provide wind information at a glance. It usually consists of eight lines extending from the center of a compass rose outwards along the cardinal and intermediate points of the compass. The length of each

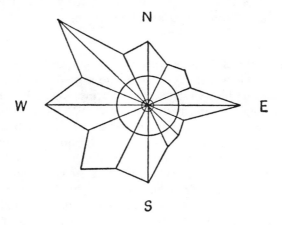

line is drawn to show the relative frequency and/or the average speed of the winds blowing from different directions in a specified region. These wind roses or *stars* were widely used on nautical charts to show prevailing wind conditions along different sailing routes.

238. What is the difference between a backing and veering wind?
A backing wind is a wind which is changing direction in a counter-clockwise direction in the Northern Hemisphere and a clockwise

direction in the Southern Hemsphere. If, for example, a wind is blowing in Boston harbor from the east, then from the northeast, then north and finally northwest, it is called a backing wind. Change in the opposite direction is called a veer.

239. What is a wind shift?   A wind shift is a change in the direction of the wind, usually more than 90 degrees. It differs from a veering or backing wind in that it is a relatively sudden change rather than gradual, occurring most often when a new air mass moves in to replace another behind a front.

240. What is a wind shear?   A shear line is a boundary between winds of different speeds at different altitudes. This shear can occur with winds moving in one direction but with levels of different speeds. For example, a rapidly moving mass of air may drag more slowly at the Earth's surface because of friction than at a few thousand feet higher where the air surges forward more freely. The boundary between the stronger winds aloft and the dragging winds below is called a shear line.

241. What is the difference between a true and apparent wind? A true wind is one which blows relative to a fixed point on the Earth. An apparent wind blows relative to a moving object. For example, if a vessel is heading in a northerly direction at 10 knots in a dead calm, its motion would induce an apparent wind from the north at 10 knots. Even though a calm existed, the ship's anemometer would measure 10 knots and the wind vane point to the north. If, instead of a calm, a true wind was blowing from the north at 10 knots under the same conditions of the ship's direction and speed, then the apparent wind would be reinforced by an additional 10 knots. The wind then experienced by the ship would be called a *resultant* wind of 20 knots. A resultant wind takes into account the speed and direction of a moving object and the true wind force and direction.

242. What is the difference between windward and leeward? Windward describes the side facing the direction from which the wind is blowing. Leeward describes the side in the general direction toward which the wind is blowing. In many areas of the world, particularly mountainous regions which are in the path of prevailing moist winds (Hawaii, Ceylon, the Appalachians, etc.), the windward sides of the mountains show a great contrast of weather to the leeward. Considerable precipitation and cloudiness may characterize the windward slopes as against clear and dry weather to leeward.

243. What is a wind wave?   Undulations or ridge patterns on the surface of a liquid area are called a wave. A wind wave is generated by the friction between the wind and the fluid surface. Most ocean waves are caused by this frictional effect, although some are caused by differences in atmospheric pressures and differences in ocean-bottom contour. Some sand-dune patterns are similarly molded by the effect of wind. The ripple appearance of the dune is shaped in long, more or less parallel lines which are generally perpendicular to the wind direction. The height and width of each dune are shaped by the general land contour and by the nature of the wind as it carries sand particles up the gentle windward side and down the leeward side.

244. What is the basic cause of wind?   The basic cause of all winds can be traced to contrasts of temperature. These differences occur because air is not heated at all points with equal intensity. The differences also occur on scales of varying magnitude. A coal-stove fire, for example, causes differences of heating in a small cabin. At the seashore on a summer afternoon, differences in temperature exist between the hot sand and cool water. The island of Ceylon is warmer than the Indian Ocean waters in the summer. On a planetary scale, the equatorial belt is warmer than the Temperate Zones.

245. How do temperature differences result in wind?   When air is heated, its molecules are agitated and their movement accelerated. They tend to draw away from one another and the air expands. As the molecules expand, they occupy a greater volume and the density of the heated air parcel is decreased. Like a huge invisible bubble, the heated air starts to rise. Surrounding cooler air flows in to replace the rising air. This movement of air, from cooler (higher pressure) to warmer (lower pressure) areas is a *wind*.

246. What makes the atmosphere move around the Earth?   The initial forces or trigger mechanism are the differences in terrestrial heating which lead, in turn, to differences of pressure causing air to move from high to low pressure. This process is described in the preceding two questions, and can be applied to the broad waistline of Earth's equator which accepts more direct heat throughout the year than any other region. Air over the equator, being relatively lighter and warmer, rises in great currents to the top of the troposphere. Global motion of the atmosphere starts with cooler air from either side of the equator moving toward the equator to replace the rising warm air. This gigantic coal-stove effect is modified and shaped by a variety of factors.

247. What is the primary circulation?   The Earth's atmosphere at the surface moves in a series of permanent, though fluctuating, wind belts which circulate continuously around the Earth. This primary circulation of large-scale wind belts results, in a basic way, in distributing heat and energy throughout the planet's surface. Warm air is transported poleward and colder air infiltrates southward.

248. What is the secondary circulation?   Superimposed on the primary wind belts of the world are secondary or more localized wind patterns. These secondary and local winds range in size from circulations which may be large enough to affect an entire continent to the immediate and more familiar local winds of mountains, valleys or seashores.

249. What forces shape the primary wind belts of the earth?   As the atmosphere starts its basic motion to equalize pressure differences caused by differences in heating (See Question 245), the winds are influenced in their direction over the Earth's surface by four factors: the Earth's rotation, centrifugal force, frictional and topographical effects, and the secondary circulation.

250. How does the rotation of the Earth affect the motion of the atmosphere?   The atmosphere, as a gaseous sphere, does not cling to the Earth as a building or a person does. The Earth rotates west to east and all solid things attached to it move with the Earth. Air being relatively free to move, continues straight onward by momentum while the Earth slips under it. For example, if a mass of air starts to move southward in the Northern Hemisphere, it will move directly from north to south with reference to space. But the Earth rotates eastward under the air so that with respect to the Earth's parallels and meridians, the air gradually assumes a motion from the northeast to the southwest. This force, called the Coriolis force, may be summed up as follows: a large body of air tending to move in a given direction in the Northern Hemisphere will tend to move to the right of the path of this direction.

251. What are the wind belts of the primary circulation? Acting under the influences of temperature and pressure differences, rotational effects of the Earth, centrifugal and frictional effects on moving

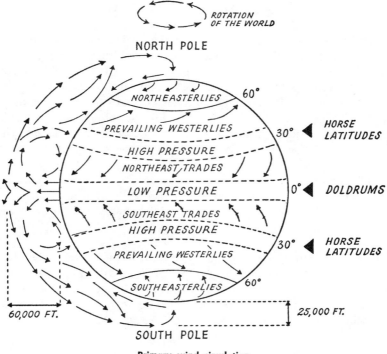

Primary wind circulation

air, the atmosphere moves in large-scale wind belts in different areas of the Earth. These belts are the doldrums, the trades, the prevailing westerlies and the polar easterlies.

252. What are the doldrums?   The greatest amount of heating takes place at the equator. Air rises there creating a zone of low pressure at about 10 degrees each side of the thermal equator. This zone is called the *doldrums*. As air rises over the doldrums, it cools by expansion. Since the air is characteristically humid, very little cooling is required before condensation takes place. Hot, stagnant and moist winds prevail at the surface, frequently dead calms. Conditions are favorable for widespread vertical cloudiness and heavy rain, often in the form of thunderstorms. Huge columns of air rise to the top of the troposphere over the doldrums where the air spins northward and southward toward the poles.

253. What are the horse latitudes?   As the great columns of air rise over the doldrums, they cool by expansion. The cooling air becomes heavier and tends to sink down again when they reach the top of the troposphere about 11 miles over the doldrums. They are prevented from this by the upward surges of the air beneath. The result is that the air is pushed out from the equator and starts to flow poleward in a spiral fashion at high altitudes. Some of the air dams up at about 35 degrees north and south and sinks or subsides to the Earth, causing belts of high pressure at those latitudes around the Earth. The sinking or subsidence causes a warming of the air by compression and has a cloud-dissipating effect. The result is that these two belts of high pressures are regions of calms or variable winds and generally clear skies. The term horse latitudes is generally applied only to the northern of these two regions in the North Atlantic Ocean or to the portion of it near Bermuda.

254. How did the name of horse latitudes originate?   The term might have started in the early days when windjammers sailed the Atlantic. The calms and heat of the area killed cargoes of horses which were frequently transported. Coleridge captures the feeling of this calm and clear weather belt in his "Ancient Mariner":

> Day after day, day after day,
> We stuck, nor breath nor motion,
> As idle as a painted ship
> Upon a painted ocean.

255. What are the trade winds?   As air subsides at the high-pressure belts about 35 degrees north and south latitudes, it flows toward the equatorial zones. The rotation effect of the Earth causes the wind direction to veer somewhat to the right of its path of motion. Thus, in the Northern Hemisphere the trade winds blow as northeast winds and in the Southern Hemisphere as southeast winds.

256. What are the prevailing westerlies?   From the high-pressure zone of the horse latitudes, winds blow in a poleward direction. The Earth's rotation causes these winds to veer to the right in the Northern Hemisphere and to the left in the Southern Hemisphere. From about 35 to 55 degrees, then, in both hemispheres, belts of westerly winds prevail. In the Southern Hemisphere, where there is relatively more ocean and the wind is not influenced by too many surface irregularities, the winds are of higher velocities. They blow especially strong around the southern tip of South America where the region is called the Roaring Forties.

The westerlies dominate most of the Temperate Zones and represent the meeting ground of warm semitropical air from the horse latitudes and cold polar air from high latitudes. During the winter, this clashing of opposed air becomes intense and the westerly belt reaches southward. In the summer, the circulation is less intense and the westerly belt is pushed northward.

257. What are the polar easterlies?   Some of the air which rises from the equator moves at high altitudes all the way to the poles where it arrives cold and dense to form a dome of high pressure at each pole. This air breaks out irregularly in bulges and moves southward toward the prevailing westerlies. The Earth's rotation causes these polar winds to deflect somewhat to the right so that in the Northern Hemisphere the winds blow from a northeasterly direction in a zone from the pole to about 60 degrees north. In the Southern Hemisphere, the winds at corresponding latitudes are generally from the southeast. Where the polar air meets the semitropical air at about 60 degrees latitude, a belt of low pressure exists around the Earth. This zone of clash between the polar easterlies and the prevailing westerlies is called the Polar Front in the Northern Hemisphere and is extremely important in creating the changeable Temperate Zone weather.

258. What is a center of action? The simplified, ideal circulation of air around the Earth described in preceding questions is modified by *centers of action*. These main generators of air movement are situated in various locations around the Earth and, because of their persistency, are termed *semipermanent cells*. They generally coincide

Centers of action

with the broad belts of high and low pressure as described in the preceding questions. These cells are either high- or low-pressure systems, adding to or supplementing the primary circulation, and are named according to the regions in which they are found.

259. Which are the important centers of action in the Northern Hemisphere? The Aleutian Low, the Icelandic Low, the Azores or

Bermuda High, the Pacific High, the North American High, and the Siberian High.

260. What is the Aleutian Low?   In the northern Pacific, this large low-pressure area circulates counterclockwise. Its outer edges brush the Alaskan coast and feed air across the northern Pacific down to the northeast coast of Siberia (See Question 318).

261. What is the Icelandic Low?   Like its Pacific counterpart, the Aleutian Low, the Icelandic Low is strong in the winter. It is found between Greenland and the Scandinavian coasts, spinning counterclockwise to carry outbreaks of cold winds from the North Atlantic to eastern Canada and across to the Scandinavian coasts, the British Isles and northwest Europe.

262. What is the Azores or Bermuda High?   With its center in the general vicinity of the horse latitudes and at a point about midway between Florida and Spain (the Sargasso Sea area), this giant wheel of air spins clockwise. It drives wind from the Atlantic to the eastern coast of the United States, then across to England and the western European coast, to the northwest coast of Africa and curves westward to start the cycle again.

The Bermuda High expands in the summer. It provides the east coast of the United States with summertime prevailing winds from a southerly quadrant; pumping warm moist air from the Atlantic and Gulf of Mexico to the entire coastline and frequently to the eastern third of the nation. It acts as a block frequently to cooler drier air from the west and north. It has an important bearing on Atlantic hurricanes, spinning young storms westward toward the West Indies and southeast United States and often affecting the path of a hurricane as the storm moves toward the horse latitudes.

The Bermuda High drives the Gulf Stream along its western edge and generates warm winds to England and the French coast, preventing a colder climate in those coastal sections.

263. What is the Pacific High?   It is a clockwise rotating cell of air lying off the coast of Southern California during the summer and moving inland during the winter. It drives air from the Japanese coast looping across the Pacific to the California coast and dominates the weather pattern for most of that area.

264. What is the North American High? In the winter months, this clockwise-blowing cell provides most of the push for the prevailing northwest winds which cross the central and eastern portions of the United States. It is not so dominant in the summer, giving ground to the summertime-expanded Bermuda High. North central Canada is the spawning grounds for this center of action.

265. What is the Siberian High? Like its North American counterpart, the Siberian High becomes strong in the winter. It sprawls over the heartland of Siberia and rotates clockwise to pour cold dry air from the polar zones over large portions of the Asiatic continent and eastern Russia.

266. What is an air mass? An air mass is a portion of the Earth's atmosphere with fairly uniform horizontal characteristics. It is a body of air that has acquired the characteristic thermal and moisture properties of a large uniform surface over which it has remained.

267. What is a source region? Source regions are those sections of the Earth's surface where air tends to remain and acquire the characteristic thermal and moisture properties of the surface.

268. What is meant by air-mass analysis? Even after air masses move far from their original breeding-ground source, they may still be recognized by their characteristics. Meteorologists are able to chart the progress of air masses over large sections of the Earth's surface. They can study the paths over which the air segments are moving and attempt to anticipate the weather it tends to produce along its path and the weather which may result from one air mass colliding with another.

269. What factors change the characteristics of an air mass? Forced by the general circulation, large air masses (covering many hundreds of square miles across and several miles deep) may migrate over long distances of the earth's surface. While traveling, they still retain in some degree their original source region characteristics. But these characteristics modify as the mass moves over new regions. They change, either slightly or considerably, depending on how far the air has traveled, how long it has been from its "feed" grounds, and the type of terrain over which it is moving. The changes usually occur first at the surface, later aloft.

270. What conditions favor the existence of a source region?
Since uniformity is the essential feature of an air mass, it can be assumed that the principal source regions of the world will have reasonably uniform surfaces. Another important condition is an area where large air scale currents *diverge* or spread outwards from descending currents. *Converging* air currents causes air masses of different properties to be thrust against one another, hence to change rapidly.

271. Where are some primary source regions?   Some areas on the Earth are uniform enough and situated favorably so that air currents subside over them, stagnate, arrive at an equilibrium with their surfaces and carry their imprint to other areas. Some of these source regions are the Atlantic Ocean around Bermuda, the tropical Pacific around Hawaii, the Sahara Desert region, the interior of Siberia and central Canada.

272. What is a secondary source region?   There are some large areas where surfaces are uniform but over which air does not tend to stagnate. Air that flows over such a region may take on enough of the surface characteristics in transit to justify it being called a distinct air mass. Some examples of secondary source regions are the dry Southwest of the United States and interior of Mexico during the summer, the North Atlantic Ocean between Canada and northwest Europe and the North Pacific Ocean between Siberia and Canada.

273. What is equilibrium of an air mass?   When an air mass reaches a balance with the surface of a source region—when it has been warmed or cooled, made dry or moist from the surface so that these properties will change only slowly with the passage of time—it is said to have reached equilibrium.

274. How long does it take for an air mass to reach equilibrium?
The type of surface in most cases determines the time it takes for an air mass to acquire the characteristics of their sources. Air masses reach equilibrium much more rapidly when the surface is warmer than the air above it. When heated from below, the air mass modifies more rapidly because vertical currents are set up (unstable air) so that heat and moisture acquired at the surface are carried aloft. In such a warmer-than-air surface condition, an air mass may reach equilibrium in two or three days. Colder-than-air surface, on the other hand, prevents mixing vertical currents. The air is said to be *stable* and modification to equilibrium may take a week or two.

275. What is meant by a stable or unstable air mass? Stability and instability refer to the relationship between the vertical temperature distribution of an air mass and a vertically moving air parcel. If the temperatures within an air mass decrease sharply with altitude, then conditions will be generally favorable for air currents to rise vertically through it. The air mass is said, then, to have unstable characteristics. An air mass which is generally cooler at the surface than aloft has stable characteristics and will have very little vertical motion.

276. What is the lapse rate of an air mass? The change of temperature per unit vertical distance as one changes altitude is called the lapse rate. When changes are strong, when temperatures decrease sharply with altitude, the lapse rate is *steep*. When temperature only decreases slightly with altitude, it is a *weak* lapse rate. When temperature remains the same with an increase in altitude, the condition is called *isothermal*.

277. When is an air mass stable or unstable? The stability or instability of an air mass depends upon the lapse rates of its air. Air is unstable when a parcel (an individual parcel of heated air) continues to rise through the free air because of its comparative higher temperature and lessened density than the surrounding air. Thus, if there is a strong lapse rate in the free air of the mass (sharp temperature drop with altitude) and a portion of air is heated at the surface to a higher temperature than the surrounding air, it will rise because of its being less dense and will continue to rise just as long as its temperature remains higher than the surrounding air. This is an unstable condition. If a portion of air is forced upward and remains cooler than the surrounding air, it will have a tendency to sink down to its own density level. This is a stable condition.

278. How are air masses classified? Air masses are classified according to their source regions and their temperature and moisture properties. There are actually just four basic types of air masses to be found at one time or another everywhere in temperate latitudes. Air is either moist or dry, cold or warm. Combining these the classification shows the following types: Cold-dry air masses, cold-moist air masses, warm-dry air masses and warm-moist air masses.

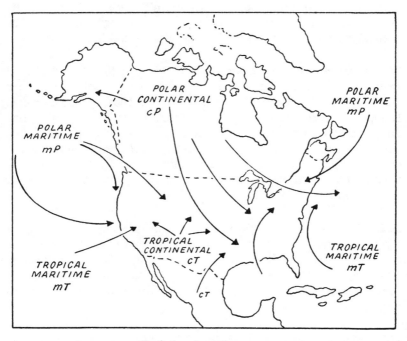

North American air masses

279. How are these air mass types named? For purposes of identification on weather maps and for general discussion, the names of the air masses in the preceding question are:

Polar Continental (symbol cP)—corresponds to cold-dry
Polar Maritime (symbol mP)—corresponds to cold-moist
Tropical Continental (symbol cT)—corresponds to warm-dry
Tropical Maritime (symbol mT)—corresponds to warm-moist

If the lower part of the air mass is *colder* than the surface over which it is moving, the letter *k* is added to the air mass symbol. If the lower part of the air mass is *warmer* than the surface, the letter *w* is added. For example, Tropical Maritime air (mT) may move from the Gulf of Mexico over the Mississippi Valley in the winter with the lower layer of air warmer than the surface. The symbol would be mTw.

280. Where does Polar Continental (cP) air originate in the Northern Hemisphere? It originates over the arctic region of the Earth, particularly over north central Canada, Russia and Siberia.

281. What are the characteristics of cP air in the United States? In the summer it is characterized by moderate to cool temperatures and low humidity. Outbreaks of this air, propelled by the Canadian High, one of the North American centers of action (See Question 264), usually sweep southward into the central and eastern United States. In the summer, these outbreaks bring northwesterly winds, clear skies and relief from the heat. In the winter, more violent Polar Continental air invasions may reduce temperatures by 30° F. or more within a few hours and are called *cold waves*.

282. Where does Polar Maritime (mP) air originate in the Northern Hemisphere? Polar maritime air originates over the Bering Sea and the northern Pacific Ocean and may be considered as a separate classification, Polar Pacific (pP). Polar Maritime air also originates over the colder portions of the Atlantic Ocean and is classified as Polar Atlantic air (aP).

283. What are the characteristics of Polar Pacific air in the United States? Polar Pacific air (pP) has moderately low temperatures and moderate humidity. It modifies rapidly over the western Pacific and the North American continent. East of the Rockies, Polar Pacific air becomes much drier and somewhat warmer, particularly at lower levels. It closely resembles Polar Continental air when it reaches central United States in the summer.

284. What are the characteristics of Polar Atlantic (aP) air in the United States? Polar Atlantic air frequently spreads over the northeast United States, particularly in the fall and winter. It is characterized by low to moderate temperatures and high humidity as well as considerable cloudiness. Polar Atlantic air seldom extends far south or west of New England because its invasion depends on an east-to-west movement, whereas in the temperate latitudes, winds from the west predominate.

285. Where does Tropical Continental Air (cT) originate in the Northern Hemisphere? It originates over southwestern United States and over Mexico in the summer. The broad regions of the Sahara Desert in Africa furnish an ideal source region for Tropical Continental air.

286. What are the characteristics of Tropical Continental air in the United States? It is warm and dry air, sometimes excessively so in the summer and may bring on drought conditions.

287. Where does Tropical Maritime air originate in the Northern Hemisphere? This moist and warm air originates over the Gulf of Mexico, the Caribbean Sea and over the Sargasso Sea area of the Atlantic. Air from the Gulf and the Caribbean can be classified as Tropical Gulf (gT). The Atlantic air can be classified as Tropical Atlantic (aT). Tropical Maritime air also originates over the North Pacific Ocean trade-wind belt northeast of Hawaii and is classified as Tropical Pacific (pT).

288. What are the characteristics of Tropical Maritime air in the United States? In the summer, Tropical Maritime air has high temperatures and very high humidity. It is the prevailing type of air over most of the eastern third of the United States and its coastlines in the summer. It flows from the warm-water regions northward usually brought in by the circulation of the Bermuda High (See Question 262) which circulates air clockwise from northwest Africa to the West Indies and then to the eastern United States. This air is responsible for the typical warm, muggy and thundery weather of the southern and eastern United States in the summer. In the winter, Tropical Maritime air frequently is forced aloft over cooler Polar Continental or Polar Maritime air at the surface and produces considerable cloudiness or rain.

The Tropical Maritime air which originates over the Pacific is somewhat cooler and less moist than the Atlantic version and has moderate temperatures and fairly high humidity. This air is usually found along the Pacific coast, but sometimes is forced east of the Rockies at high altitudes. Tropical Pacific air produces heavy cloudiness and rain when it meets up with mountains and is the cause for most of the frequent rains of the Pacific Northwest.

289. Where are the air mass source regions in the Southern Hemisphere? Polar air masses surge from the antarctic regions. Tropical Maritime air masses are found in the South Pacific off central South America; the South Atlantic between eastern South America and southwest Africa; and in the tropical waters on a latitude between the central eastern coast of Africa and western Australia. The interior of Australia is an ideal source region for Tropical Continental air.

290. What is a front? A front is a boundary between two different air masses. This may be a sharp or very narrow zone that marks a sudden change from one air mass to another or it may be a wider, diffuse separation zone, more accurately described as a frontal zone. The surface of discontinuity between two meeting air masses of different properties is called a frontal surface. The boundary or imaginary line on the Earth's surface that separates the colder air and warmer air is called a front.

291. How are large frontal zones on the Earth formed? The primary circulation tends to produce large air masses in certain source regions. The same circulation tends to produce areas where the various types of air masses meet or converge. Where such interaction takes place more or less continuously, a primary frontal zone exists.

292. Where is the Equatorial Front? This frontal zone is situated in the doldrums. It is not a pronounced front or an important one because the air masses which meet are not too different as far as temperatures are concerned. Equatorial air masses and the subtropical air masses brought toward the equator by the trade winds are the components.

293. Where is the Arctic Frontal Zone?   This zone separates Polar Maritime air from true arctic air to the north.

294. Where is the Polar Front?   The most important frontal zone in the formation of North American weather is the Polar Front which has an average position between 50 degrees and 60 degrees north around the globe. During the winter, the Polar Front extends

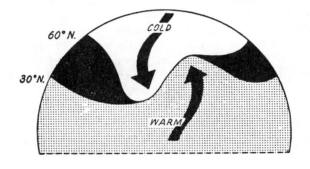

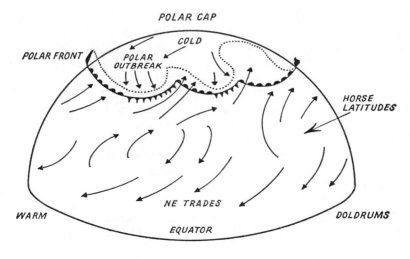

well southward over the continental areas of North America and Eurasia. It can, at times, push south to the tropics in winter. During the summer, the line backs up northward and frequently is near or north of the arctic circle. Over the oceans, this seasonal migration of the Polar Front is not as well marked as that over the land areas.

295. Why is the Polar Front so important?   Along this broad earth-circling zone at about 50 to 60 degrees north latitude the greatest amount of opposite-type air masses collide and cause the familiar changeable weather patterns of the middle latitudes.

296. Which air masses converge at the Polar Front?   North of the Polar Front, cold air pours southward from a general northeast direction (the polar easterlies—See Question 257). South of the Polar Front a different kind of circulation exists. Streams of warmer air, part of the belt of prevailing westerly winds (See Question 256) flow generally from a southwesterly to northeasterly direction. The Polar Front thus becomes a zone of impact with air masses of different characteristics meeting head long.

297. What happens when opposite-type air masses converge at the Polar Front?   Along the primary front, two air masses may sometimes continue to flow side by side with no underrunning or overrunning action between them. But for a variety of reasons, based on topographical effects or thermal irregularities, the warm and cold air masses start to swirl into each other. This produces a *wave* on the original Polar Front line. The cold air bulges southward and eastward into the warm and the warm air retreats northward into the cold air. This wave deepens. The opposite motion effects of the converging air masses takes on a counterclockwise rotation. It develops into a *system* which moves and produces different kinds of weather according to the differences of the converging and interlocking air masses. The central area around which the winds spin counterclockwise is called the *vortex* or area of lowest pressure. The lines or surfaces in the system which separate the interaction of the air masses are called active *fronts*.

298. What is this converging system of air masses called?   This development along the Polar Front is called a *low-pressure area* or

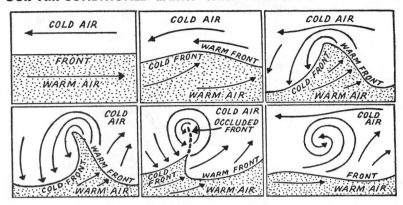

Life cycle of a Low

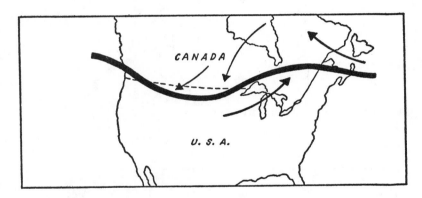

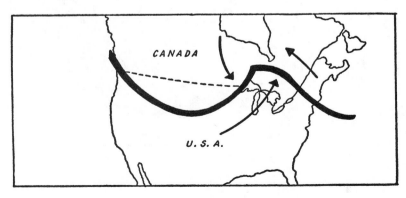

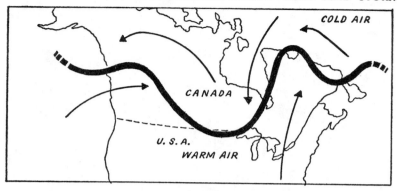

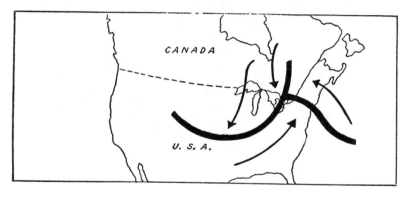

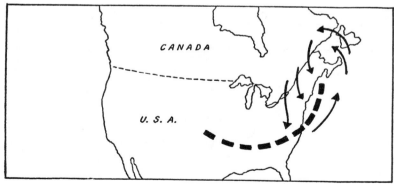

revolving storm. More technically, a low-pressure area formed along the Polar Front is called a *frontal depression* or *extra-tropical cyclone.* In the Southern Hemisphere winds blow clockwise in a low.

299. What are the similarities of cyclones, extra-tropical cyclones, tropical cyclones, tornadoes, hurricanes, typhoons, etc.? Starting from one common denominator, they are all areas of relatively low atmospheric pressure. They all are systems of winds which revolve about a center of low pressure, counterclockwise in the Northern Hemisphere and clockwise in the Southern Hemisphere. They are all lows.

300. What are the differences? There are differences of classification based on areas of formation, areas of activity, and different sizes, characteristics and effects. A *cyclone* may be considered as an approximately circular portion of the atmosphere having winds that blow counterclockwise around the center in the Northern Hemisphere and clockwise around the center in the Southern Hemisphere. An *extra-tropical cyclone* is a cyclone occurring outside the tropics (See Question 319). A *tropical cyclone* is a violent cyclone of large dimensions originating in the tropics. Tropical cyclones are called by different names in different parts of the world. In the Atlantic they are called *hurricanes.* In the Pacific they are called *typhoons,* etc. (See Questions 321–346). A *tornado* is an extremely violent revolving storm of relatively small diameter which travels over land and produces great damage along a narrow path (See Questions 348–356).

301. What are the characteristics of the Temperate Zone cyclone? The diameter of the extra-tropical cyclone (depression, low, disturbance) varies from about 100 miles to 2,000 miles. It usually covers an area averaging about 1,000 miles. In the Northern Hemisphere winds blow counterclockwise and slightly inward around the center or vortex. The different air masses which make up the components of the system are separated at the surface by lines of discontinuity called *fronts.* The portion of the cyclone with the warm air mass at the surface is called the *warm sector.* This sector is in the southern or southeast quadrant of a low and the winds in the warm sector are generally from the south. The northern, northeastern and western portions are cooler. Generally, cloudiness and precipitation are associated with lows, especially around the frontal zones.

302. What is a cold front?   The leading or advancing edge of a cold air mass is called a cold front. As a bulge of cold air penetrates into the middle latitudes from polar regions, its front or leading edge burrows under the warmer air in advance. The cold air hugs the

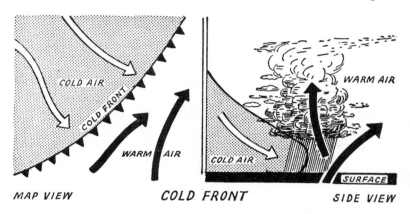

MAP VIEW   **COLD FRONT**   SIDE VIEW

ground because it is relatively denser and heavier in weight than the warm air it is displacing at the surface. As the cold-air wedge continues to move under the force of momentum of the piled up cold air behind it, it forces the warm air in advance aloft. The warm air is pushed upwards by the "snowplow" effect of the cold air.

303. What kind of weather is associated with a cold front?   The area of stormy weather at the cold front is generally limited in extent. The kind of weather which will occur with a cold front depends considerably upon its momentum, the nature of the air mass it is displacing and the differences in characteristics between the two clashing air masses. When a fast-moving cold front displaces moist unstable air, the weather will probably be squally with sudden gusty winds and heavy but brief precipitation. Rapid clearing occurs. The wind shifts and blows from the cold region. Temperatures drop, pressure rises and humidity lessens after the cold front has passed. A slow-moving front displacing stable warm air may be attended by very slow-clearing weather. In general, the drier and colder the air behind the cold front is and the warmer and more moist the air in advance is, then the more turbulent and stormy will the weather be at the cold front; also the more sudden will be the change from warm and moist to cold and dry.

304. What is a warm front?   A warm front is the leading edge of a mass of warm air which is displacing colder air at the surface. Unlike a cold front which burrows under warm air, a warm front edge overrides the cold air in advance. Its lighter weight prevents it from piercing a cold air mass from underneath. It slides over or overruns the cold air, instead.

305. What kind of weather attends a warm front?   Warm-front weather is much more widespread and persistent than cold-front weather. Unlike cold air which displaces warm air suddenly at the surface, warm air rides up over a wedge of cold air with a very gentle slope. As the air cools in its slow ascent over the long slope,

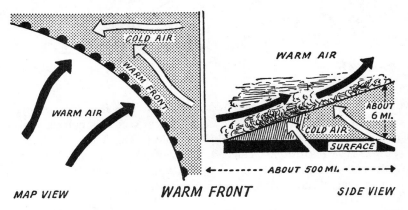

MAP VIEW   **WARM FRONT**   SIDE VIEW

it condenses into clouds which may exist as much as a thousand miles or so in advance of the surface position of the warm front. Its forward movement, then, is heralded long in advance by gradually thickening and lowering clouds and development of rain or fog near the front. As the warm front passes, the temperature rises, the wind shifts, precipitation that has been falling usually stops and the cloudiness decreases.

306. What is a stationary front? If the frontal separation between two air masses remains in approximately the same position, it is called a stationary front.

307. What is an occluded front? When a cold front overtakes a warm front, the air in the warm sector of the low pressure system is lifted aloft. Cold air from the rear of the cold front will then meet with cold air in advance of the warm front. The chances are that a thermal difference exists between the two. If the air mass in the rear of the cold front is colder, it will proceed to displace the rela-

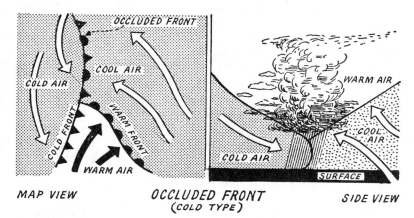

MAP VIEW   OCCLUDED FRONT   SIDE VIEW
(COLD TYPE)

tively warmer air by lifting the warm air aloft. This is called *cold-type occlusion*. If the air in advance of the warm sector is colder, it will lift the air in the rear of the cold front aloft. The front will then be a *warm-type occlusion*. Occluded fronts usually mark the beginning of the end of the life cycle of a Temperate Zone cyclone. They usually bring stagnant and poor weather conditions.

308. How are lows classified? Not all lows are formed along a primary front like the extra-tropical cyclone. Some lows are characteristic of regions where considerable seasonal heating takes place. The desert region in the southwest United States or the interior of India in the summer are examples of *nonfrontal* or nontraveling lows. The equatorial belt of low pressure (doldrums) is another example of a broad area of nonfrontal low pressure which is manifested by the *heat lows* over South America, South Africa and Australia.

309. What is the difference between a cold low and a warm low? In recent years, meteorologists have made distinctions between *cold-core* and *warm-core* lows. If the temperature is lowest at the center, the low is cold-core. It is a warm-core low if temperatures at the center are highest. Cold lows increase in intensity aloft. Warm lows decrease.

310. What is a high? It is an approximately circular portion of the atmosphere having relatively high atmospheric pressure. Winds in a high blow clockwise and slightly outward around the center in the Northern Hemisphere and counterclockwise in the Southern Hemisphere. It is technically called an *anticyclone*.

311. What are the characteristics of a high? Highs are generally classified as being migratory or permanent; cold or warm. The average diameter of a high is somewhat similar to a low, about 1,000 miles, but similarities do not exist much further. Winds in a low converge. Air in a high settles downward (subsides) and diverges outwards from the center. This settling or subsidence heats the air by compression and has a cloud-dissipating effect. Fine weather is therefore characteristic of a high as against a low where converging and upward-moving air currents favor cloud formation. Highs do not always bring fair weather. Frontal weather from adjacent low pressures can often penetrate highs. All in all, the good weather in a high, especially in its central portions, far outweighs the bad weather it can have.

312. What is the difference between a migratory and a permanent high? A permanent or stationary high is usually characteristic of regions where air is accumulating at high altitudes, subsiding or sinking downwards and flowing outwards from the center in a divergent manner. Two principle regions of the Northern Hemisphere favor this kind of air collection at 30 degrees north latitude (horse latitudes—See Question 253), and at the polar cap. Large semipermanent cells of high pressure, therefore, coincide with the 30 to 35 degree north latitude belt. (These are some of the centers of action

described in Questions 258–264). Migrating or *transitory* highs originate in the arctic and antarctic regions and invade the Temperate
Zones as outbreaks of cold air.

313.. What is the difference between a cold and warm high?   Cold-
core highs, typical of the cold air masses which develop in arctic
regions in the winter, are rather shallow in vertical development.
They decrease in intensity aloft. Warm-core highs (warmest temperature at the center) extend to much higher altitudes and increase in
intensity at higher altitudes.

314. How do traveling highs and lows move in the United States?
In general, the circulation in the United States is made up of alternate movements of lows traveling east and northward and highs
traveling generally east and southward. About 60% of lows come
into the United States in the extreme northwest and large proportion
of them move eastward along the northern border, across the Great
Lakes and finally pass off to the North Atlantic coast. Some of the
disturbances move far to the southward in the central United States
and then recurve northeastward. Some lows form over the southern
plains or come into the Southern States from the Gulf of Mexico,
then generally drift northeastward and pass off the North Atlantic
coast. A typical high enters the United States from Canada just east
of the Rockies and then proceeds southeastward into the Mid-Atlantic
States.

315. How fast do lows and highs move in the United States?
There is no great regularity about the direction and rate of movement
of pressure systems, although they vary somewhat with the seasons.
Only in the averages do they show a seasonal effect. Some move
with great rapidity and others move slowly or remain stationary. Cyclones and anticyclones move on an average from 250 to 500 miles
a day in the summer and about 650 miles a day in the winter in
their more or less defined paths. Lows generally move faster than
highs. For his estimate of the future motion of a pressure system,
the forecaster watches the upper air currents, the previous rate and
direction of travel of the system (called *historical sequence*) and
the. changes in pressure in the region in which it is located. (See section on the weather map and forecasting.)

316. How can the location of a high or low center be judged by wind direction? Winds blow clockwise around a high and counterclockwise around a low in the Northern Hemisphere. With this in mind, if an observer stands with his back to the wind, lower pressure will be to his left and higher pressure to his right.

317. What is the range of pressure changes in highs and lows? Barometers fluctuate continuously with the passage of highs and lows across the country and the extreme range of change is less than might be supposed. For example, in Philadelphia, one of the cities along typical high and low tracks, the highest barometer was 31.02 inches (1050.5 millibars) and the lowest was 28.54 inches (966.5 millibars) in a 66-year period. The extreme range is, therefore, 2.48 inches (84 millibars). This is a difference in pressure of little more than one pound per square inch. It corresponds to a change in pressure such as one would experience in rising from sea level to an altitude of 2350 feet. The lowest sea level pressure *range* at any station in the United States is 1.07 inches at San Diego, California, and the highest is 3.02 inches at Hartford, Connecticut. (See Question 795).

318. What causes the characteristic clockwise or counterclockwise winds in highs and lows? The winds in any high or low in the Temperate Zone are governed by a balance of forces all working on air particles in motion. Air originally starts to move over the Earth's surface because of pressure differences. The air tends to move from high pressure directly to low. The rate of change per unit distance in the horizontal direction is known as the *pressure gradient*. When the pressure gradient is strong (greater contrast between high and low), the stronger will the winds be.

Once the air is set into motion directly from high to low pressure, other forces start to work. Because of the Earth's rotation, the moving air is turned to the right of its path of motion because the Earth *slips* under the atmosphere. This air is deflected to the right in the Northern Hemisphere and to the left in the Southern Hemisphere (See Questions 250 and 1131). We now have air starting a movement directly from a high pressure center to a low center but deflected to the right of its path so that a clockwise circular pattern of air flow tends to start around the high. Another force is added at this point— centrifugal force—which pulls the air *outwards* toward the same direction of the pressure gradient and opposite to the Earth's rotation deflective force. When these three forces are balanced among each other, the air reaches a circulation which flows out of a high in a clockwise fashion and into a low where it assumes a counterclockwise rotation. The opposite motions take place in highs and lows of the Southern Hemisphere.

319. How are these storms defined in meteorology? A *cyclone* is generally considered to be any approximately circular portion of the atmosphere having relatively low pressure with winds which blow counterclockwise around the center in the Northern Hemisphere and clockwise around the center in the Southern Hemisphere. An *extra-tropical* cyclone is a cyclone or low pressure area formed along a primary front and occurs outside the tropics. In meteorology it is also called a *Temperate Zone cyclone* or *storm*. A *tropical cyclone* is a violent cyclone originating over tropical waters. Tropical cyclones are called by different names in different parts of the world; hurricanes in the West Indies, typhoons in the North Pacific, etc. (See following questions on tropical cyclones.) A *tornado* is a violent revolving storm of small diameter which travels over land and causes great damage along a narrow path. (See tornado section.)

320. What are tropial cyclones called in different parts of the world? In the western portions of the North Pacific, it is known as a *typhoon;* in the Philippines, a *baguio;* in the Bay of Bengal and Indian Ocean, a *cyclone;* in the Timor Sea and northwest Australia, a *willy-willy*. Tropical cyclones affecting areas south and south-westward of Mexico and Central America are sometimes called *cordonazos*. In the South Pacific, east North Pacific, North Atlantic, Gulf of Mexico and the Caribbean Sea, the tropical cyclone is known as a *hurricane*. A weak cyclone occurring in the tropics is called a *tropical disturbance*.

321. What is the origin of the words cyclone, tornado, hurricane and typhoon? Cyclone comes from the Greek word *kyklon* which, according to language experts, referred to "coil of a snake." The word hurricane does not have a clear origin. It seems to have originated with the natives of the West Indies or Central America being based on a word for "great wind," *huracan*. According to one theory, the origin of hurricane was based on *Hunrakan*, a god of stormy weather to the Indians of Guatemala. Typhoon stems from the Chinese combination of *ty* (great) and *fung* (wind). Tornado apparently originates from a merging of the Latin *tornare*, "to turn," and the Spanish *tronada* (thunderstorm, lightning) and *tornar* (to turn).

322. Where do tropical cyclones originate? They originate in limited portions of all tropical oceans except the South Atlantic. They occur mostly in the western parts of the warm waters and generally affect the eastern coasts of the great continents as they travel and develop. The most characteristic place of origin is in the doldrums, about 10 to 20 degrees from the equator. They do not occur closer to the equator than about 5 degrees. They do not form over the continents.

323. What are typical tracks of tropical cyclones in the Northern Hemisphere? In the Northern Hemisphere, the usual movement for a tropical cyclone is toward the west and northwest and then, as it reaches about 25 degrees latitude, it moves north and then

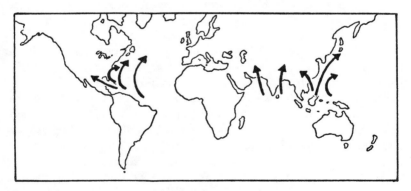

Hurricane tracks

northeastward. The point where it changes its track from northwest to northeast is called the *point of recurvature*. This represents a broad average of tracks. There are many variations from one month to another.

324. What is the average tropical cyclone track in the Southern Hemisphre? In the Southern Hemisphere, tropical cyclones move southwestward from their point of origin and start recurving to the southeast at about 25 degrees south latitude.

325. Where do most hurricanes originate? Atlantic or West Indian hurricanes originate, in most cases, within limited areas. One of these source regions is the southeastern portions of the Atlantic in the area around the Cape Verde Islands. Another area of origination is the western Caribbean Sea and the Gulf of Mexico.

326. How many West Indian-Atlantic hurricanes occur in a year? From 1885 to 1955, an average of 8 hurricanes have occurred each year. The average has been 10 hurricanes per year for the past 20 years. The largest number of hurricanes in any one year was 21 in 1933. Only one hurricane was recorded in 1890. In 1893 and again in 1950, 4 hurricanes were in progress at the same time in either the Gulf of Mexico, the Caribbean Sea or the Atlantic Ocean. The rise in the average amount of hurricanes per year may reflect more thorough reporting and evaluating techniques. On the average, only two tropical storms each year bring hurricane force winds to United States coastal areas.

327. When do most hurricanes occur? The principal hurricane months are August, September and October with September the key month for most hurricanes. Early season hurricanes which start in May, June or sometimes July originate in the western Caribbean Sea. Nearly all of them move northwestward into the Gulf of Mexico, crossing the coastline into Mexico or the Gulf states. During August and September, and less frequently in July and October, most of the hurricanes develop over the eastern North Atlantic near the Cape Verde Islands. Nearly all of them move in a westerly direction across the Atlantic, some reaching the coastal areas of the United States and then recurving to the north or east.

In late September, in October and in November, hurricanes again originate in the western Caribbean Sea. But unlike the early-season Caribbean hurricanes, most of the late ones turn northward and northeastward in lower latitudes, brushing Florida or the Greater Antilles.

328. How did the system of calling Atlantic hurricanes by girls' names originate? The practice probably first started with novelist George R. Stewart in his book *Storm* written in 1941. In this book, one of Mr. Stewart's characters is a Weather Bureau meteorologist who has developed the habit of applying girls' names to storms which he tracks as they move cross country. During World War II, this became a widespread practice, especially by Air Force and Navy meteorologists who plotted the movements of storms over the wide expanse of the Pacific. It quickly became evident that the use of girls' names, in written as well as in spoken communications, was shorter, quicker and less confusing than older cumbersome ways.

329. How are hurricane names actually selected? Names used in alphabetical succession for identifying tropical storms were selected in advance by U.S. Weather Bureau, Air Force and Navy meteorologists at their Hurricane services Co-ordination Conference. In the Gulf of Mexico, Caribbean Sea and Atlantic Ocean area the first hurricane of the season is identified by a girl's name starting with the letter A; the second with a B, etc. Names are selected which are short, easily pronounced, quickly recognized and easily remembered.

330. Is this hurricane-naming system used in the Pacific? In the Pacific Ocean a somewhat similar but separate naming system is used to identify typhoons. Because the Pacific has a much larger number of tropical storms each year, four sets of girls' names are used without regard to the calendar year or season. The first typhoon of each season picks up the name directly following the last name of the previous season. When all 84 names have been used, the Pacific list is repeated from the beginning.

331. How do tropical cyclones start? The exact reasons for the creation of an embryo tropical cyclone are not understood. The weather conditions around an incipient storm are known. They can be tracked, once formed, but the precise set of factors which touch off a tropical storm must await large-scale studies of the atmosphere to great heights over the source regions. Rockets, balloons, planes, radar and radio are closing in on the secrets of the hurricane. In the meanwhile, a few plausible theories have been offered.

332. What is the convectional theory of tropical cyclone formation?
This theory is that when heating takes place over a relatively large area, the rising warm air is displaced by cooler air from the sides and above. A low pressure area is formed. If this large-scale convection of moist air occurs sufficiently far from the equator, the Coriolis force takes effect (See Questions 250, 318) so that a counterclockwise vortex is started. Energy is supplied to the embryo storm by the release of condensation taking place, and the young hurricane is picked up by the trade winds and continues a process of growth and development.

333. What is the equatorial front theory?   Many meteorologists do not hold to the theory that hurricanes are formed within a homogeneous air mass. Some suggest that importation of different air from higher latitudes is necessary to start a cyclonic circulation such as develop into tropical cyclones, that this condition might exist at the Equatorial Front (See Question 292) which separates the trade winds from the equatorial air of the doldrums. This theory of frontal interaction between air masses of different characteristics has not been fully accepted by some meteorologists who point out that no significant temperature differences and no density discontinuity exist at the Equatorial Front.

334. What are some recent theories about hurricane formation?
Considerable thought is given to the development of *easterly waves* which are wave disturbances in the broad easterly current near the equator. They are areas of cloudiness and rain which occur near the boundary of the trade winds and the doldrums. The wave is an area of convergence. Some believe that this convergence is started not by different air masses but by a *shearing* effect caused by irregular *surges* of the trade winds. The stronger winds of the surge are caused by higher latitude disturbances. The boundary line between the stronger winds in the trade winds and the weaker doldrum winds is the shear area which, by its relatively opposite motions, may start the hurricane. Exactly how a stable wave becomes the unstable wave leading to a hurricane is still not completely understood. It is possible that the sheer effect is deepened by topography effects when the wave is moving over land. At Panama, for example, a downslope flow from the doldrums may exist east of the wave and flow into the trades

with an opposite motion caused by upslope winds west of the waves. Another interesting theory is the possibility of upper level low pressure areas which are related to polar fronts and which represent the last remnants of a high latitude disturbance which reach the fringes of the tropics.

335. What basic fact underlies all theories of tropical cyclone formation? Whatever the actual reason, the original vortex or counterclockwise spin of air must occur sufficiently north or south of the geographical equator. At the equator itself, if convergence occurs, it usually takes on a straight-up chimney effect. The Earth's spin (Coriolis effect) does not operate on moving air in the geographical equator. It starts to take effect from about 6 to 15 degrees latitude, north or south.

336. What are the initial stages of a tropical cyclone? Once a hurricane is born, convectional showers are prevalent. Unsettled squally weather develops over a considerable area usually involving thousands of square miles. At first there is no marked center of low pressure. The barometer falls over the whole region. A gradual inflow of air freshens and eventually takes on a cyclonic or counterclockwise rotation. Because of the effect of the Earth's rotation, the winds, which would otherwise blow directly toward the storm center at the geographical equator, are deflected to the right in the Northern Hemisphere. A cyclonic system with winds directed counterclockwise and slightly inwards around the center is established. Some of these disturbances develop into full-fledged hurricanes. Others never quite make it or may simply exist as areas of unsettled weather. The force or energy which drives and intensifies the wind circulation comes from the energy of heat which is released during precipitation (latent heat of condensation).

337. What is the eye of the hurricane? At the center of a tropical cyclone or vortex, there is an area known as the "eye of the storm." It varies in diameter from about 7 to 20 miles. Whereas the rest of the storm is violent, the eye has little or no wind and sometimes is clear enough so that sunshine or stars can be seen. The lowest pressures, highest temperatures and lowest relative humidities exist in the eye. There are indications of some descending air motion. At sea the waves in the eye are often described as mountainous and confused. Around the calm eye is the encircling wall of hurricane winds and clouds. A roaring sound of wind can be heard. The eye is preceded by violent winds from one direction. After the eye passes, winds blow violently from the opposite direction. Death or injury has occurred to many people who left places of protection believing the storm was over when the calm center passed over the region.

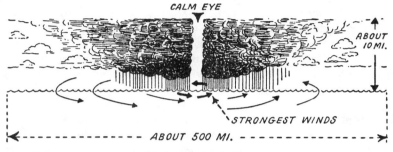

Hurricane cross section

338. What are the barometric pressure characteristics of the West Indian hurricane?
The mercurial barometer in the regions frequented by West Indian hurricanes normally reads about 30 inches at sea level. As the hurricane approaches, the barometer falls slowly at first and then rapidly as the center approaches. In fully developed hurricanes, the barometer nearly always falls below 29 inches. Many readings are below 28 inches. A few records show pressure readings of about 27 inches at the center of tropical cyclones. A reading of 26.19 inches (886.8 millibars) was noted on board a ship in a typhoon. The lowest barometric pressure ever recorded at a weather station in the United States was 26.35 inches (92 millibars) at Long Key, Florida in September, 1926, during a hurricane.

339. What is the forward movement of a hurricane?
Despite the tremendous speed of winds rotating about the center of a hurricane, the forward movement of the entire storm averages only about 12 miles an hour, especially during its early stages in the tropical water regions. This might be compared to a merry-go-round on a truck. The truck may move forward slowly or remain stationary while the merry-go-round rotates rapidly. When a hurricane moves

northward out of tropical waters, its forward movement usually increases. It slows down normally at the point of recurvature around 25 degrees north latitude and may even remain stationary for a brief spell and then may pick up speed to a forward movement at a rate of 25 to 40 or even 50 miles an hour in higher latitudes.

340. What factors influence the movement of a hurricane? West Indian-Atlantic hurricanes move westward at first because they are carried by the easterly trade winds. As they reach a higher latitude near about 25 degrees north, they come under the influence of the prevailing westerlies which cause them to recurve to the north and then generally to the northeast. This characteristic or average path can be changed by pressure systems which pass northward of the hurricane belt. For example, if the Bermuda High (See Question 262) is well developed, its clockwise rotation will guide the hurricane around its edges and deflect it from an anticipated course. Also, hurricanes seem to be attracted to move toward areas of low pressure to the northward. These influences—pressure systems to the north of the hurricane—cause a number of loops, abrupt turns and unusual movements of hurricanes.

341. What are the advance signs of a hurricane? One of the first definite signs is the sea swell. It first appears at sea as a long unbroken wave. The time interval between crests is considerably longer than in waves ordinarily observed. As the storm approaches, the sea becomes heavy and rough and the tide rises above normal. In land areas, one of the first indications is the appearance of high feathery-appearing cirrus clouds which seem to converge at a point on the horizon. This convergence is taken by some observers to indicate the direction in which the storm center lies. At sunset and sunrise the clouds on the outer edges of the hurricane are highly colored, and therefore a red or brassy-looking sky is often an advance sign.

342. What are the tidal effects of a hurricane? The tide begins to rise along the coast toward which a hurricane is moving while the storm center is a long distance away. As the storm advances, the tide rises slowly at first, then more rapidly. Along certain areas of the Gulf and South Atlantic coasts of the United States, high tides caused by hurricane winds have ranged from 10 to 16 feet above normal, causing severe inundation and subsequent great loss of life and property. Tidal effects may be compounded by two factors; one, when a normal gravitational tide is occuring and two, when a funneling action occurs due to the shape of the ocean bed on the coast. High water in the form of storm waves causes more damage to life and property than winds.

343. When do hurricanes die out?   When hurricanes move over large bodies of land, their source of moisture energy is cut off and they begin to die. They maintain a more active life span if they travel only over water. Most hurricanes moving northward into Temperate Zones become involved in extra-tropical storms and acquire fronts and distinct air masses. They lose true hurricane characteristics but can still produce damaging winds and heavy rainfall. The average life span of a hurricane is about 9 days. August hurricanes last, on an average, about 12 days. July and November hurricanes last about 8 days.

344. What is the Hurricane Warning Service?   The U.S. Weather Bureau has an elaborate system called the Hurricane Warning Service which has three main functions: to collect reports of hurricane activity, to issue forecasts and warnings and to distribute reports, forecasts and warnings to the public. The aim of the Hurricane Warning Service is to furnish advance alerts and warnings in sufficient time to all individuals and interests in storm affected areas so that all practicable preparations for the safeguarding of lives and property can be taken.

345. What are some of the most damaging tropical cyclones on record?   At Santa Cruz del Sur, Cuba, in November, 1932, approximately 2,500 lives were lost out of a population of about 4,000. Although the winds of this storm reached tremendous velocities, estimated at 210 miles an hour, the destruction was principally caused by the sea and not directly by the wind force. The rise of the sea virtually swept everything before it. Perhaps the greatest disaster occurring from a storm wave occurred on October 7, 1737, at the mouth of the Hooghly River, on the Bay of Bengal. It is recorded that over a quarter of a million persons perished in lower Bengal or in the bay. A storm wave of 40 feet rose and destroyed about 20,000 craft of all descriptions. One of the worst hurricane tides ever experienced in the United States was at Galveston, Texas, on September 8, 1900. A sudden rise of water of about 4 feet occurred in a *few seconds* causing a sweeping salt-water flood in the city. Six thousand persons were drowned.

346. What are precautionary measures on land? In badly exposed locations, evacuation to higher ground of the entire population and livestock is sometimes ordered. Houses should have storm shutters or at least windows on windward sides should be boarded up. The water supply may become brackish through overflow or spray of the sea and should be carefully checked. Electricity failure frequently occurs so that emergency lighting (oil lanterns, candles, flashlights) should be provided. Removal of all loose gear like cans, outside furniture, signs should be accomplished. Be careful of many kinds of debris which can be blown by the wind, including tree limbs. Above all, people should keep alert through all news communications regarding the storm's movement and characteristics.

347. How does the U.S. Weather Bureau use flags to warn of storm winds? Flags by day and lanterns by night are used at many coastal points to indicate the approach of storm or hurricane winds and the direction from which a storm of marked violence is approaching. *Northeast storm warning* is indicated by a red pennant above a square red flag with black center during the day or two red lanterns, one above the other, displayed at night. *Southeast storm warning* is indicated by a red pennant below a square red flag with black center displayed by day or one red lantern at night. *Southwest storm warning* consists of a white pennant below a square red flag with black center displayed by day, or a white lantern below a red lantern at night. *Northwest storm warning* is indicated by a white pennant above a square red flag with black center displayed by day or a white lantern above a red lantern at night. *Hurricane warnings* are shown by two square red flags with black centers, one above the other, during the day or two red lanterns with a white one in between at night. Beginning in 1958, the U.S. Weather Bureau will employ a new simplified system of storm warnings along the coasts of the United States, the Great Lakes, the Hawaiian Islands and Puerto Rico.

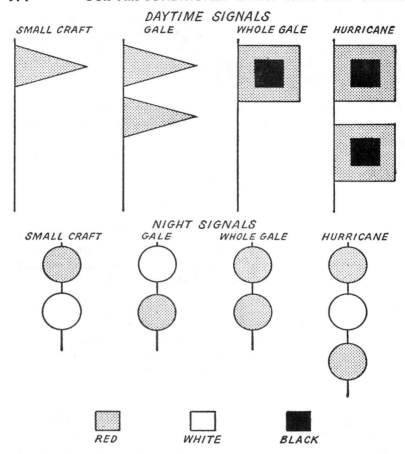

348. What are the general characteristics of a tornado? The tornado is the most violent of all storms. Essentially it is a vortex of destructive whirling winds with a funnel-shaped cloud in which winds blow at tremendous speeds and spiral upwards around a center of low atmospheric pressure.

349. Where can tornadoes occur in the United States? At any place in the United States at any time of the year. They occur most frequently in the Midwestern, Southern, and Central States from March through September. Sixty-eight per cent of all tornadoes occur from March through June. Only 21 per cent occur during late summer and early autumn, July through October; and only 8 per cent in winter months.

350. What region in the world has the most tornadoes?   In all the world no place is more favorable for tornado formation than the relatively flat region which is east of the Rocky Mountains in the United States. The zone of maximum occurrence is in the central plains region.

351. Why do most tornadoes occur in this region?   The topography of the central plains region in the United States is favorable for the collision of warm moist air from the Gulf of Mexico or the Caribbean Sea and Polar Maritime air from the Pacific. There are no mountain barriers to modify the tropical air and in the spring or summer this air arrives in the United States in its most warm and moist state. The air from the northern Pacific crosses the Rockies to reach the interior valleys. As it descends the eastern slopes, it is warmed by compression in its lower levels. It becomes very dry and often develops a steep lapse rate (temperature drops sharply with altitude). In this flat plains region, the action at the meeting places of these dissimilar air invasions results in the severe convergence which sometimes brings on tornado action.

352. How does a tornado actually start?   Like most violent storms in the temperate zones, tornadoes are born where air currents of different temperature, density, and moisture collide along a frontal zone. In most cases, the colder air behind a cold front wedges underneath the warm air with the warm air being forced upwards. In the case of tornadoes, an exception to this seems to occur. Sometimes Polar Maritime air, as described in the preceding question, sweeps eastward into the central plains states with strong force. It moves so fast that friction holds back the air at the surface while a tongue of the cold air juts forward of the surface front at about one or two thousand feet above the ground. An unusual situation is therefore started—a cap of cold dry air on top of warm moist air several miles in advance of the cold front. This is a top-heavy, unstable condition. If upper winds are strong, the cold air tongue and the moist warm air converge together sharply to form a swirling vortex of low pressure which grows and deepens into a furious twisting motion. Resultant cooling starts the formation of a corkscrewlike cloud which builds downward to the surface. The same general characteristics of air mass collision are favorable for thunderstorm activity so that thunderstorms and tornadoes are frequent storm companions.

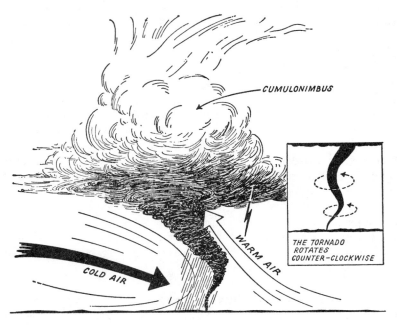

353. Why is the funnel cloud so dark?   The violently whirling cloud which seems to resemble a careening elephant's trunk has a vacuumcleaning effect. It sucks up a great variety of elements into its body such as dust, soil and debris of all kinds. Its cloud and rain composition added to these solid elements makes it appear blackish.

354. What kind of weather precedes a tornado?   Hot sticky days with southerly winds and a threatening, ominous sky prevail. Thunderstorm clouds are present. An hour or two before a tornado, the bases of dark clouds appear to bulge downward in pendulus forms like huge grapes. The clouds often have a greenish-black color. Rain and frequently hail precede the tornado. A heavy downpour of rain usually follows the storm.

355. How do tornadoes travel?   In most cases they move from a southwesterly direction to the northeast, parallel to the cold front line. They usually have a path length of from 10 to 40 miles with an average length of 16 miles. Some tornadoes may move forward for 300 miles. The average width of a tornado path is about 400 yards, but they sometimes cut swaths over a mile in width. They move forward on an average of 25 to 40 miles an hour but this varies greatly. They sometimes move very slowly—about 5 miles an hour— or may rip across the country at over 125 miles an hour.

356. Do tornadoes occur at any particular time of day?   Although they can occur at all hours, tornadoes develop mostly between three and seven P. M. when maximum heating takes place.

357. What are the wind speeds in a tornado?   Tornadoes produce the strongest winds of any storm that occurs on the earth's surface. The vertical and horizontal wind speeds in the vortex of a tornado have never been measured. Horizontal wind speed has been estimated as high as 500 miles an hour. The sound occurring in a "twister" has been described as a roaring, rushing noise similar in a sense to that made by several trains speeding through a tunnel or over a trestle. The noise also seems to sound like a giant blowtorch and can be heard for a distance of several miles. When the tornado is aloft, the noise is faint. As the tornado lowers toward the ground, the noise becomes louder.

358. What are the causes of destruction in a tornado?   The great difference in air pressure which is created as the tornado moves over a region is one major cause for destruction. When this sudden reduction in pressure occurs, air in many enclosures rushes outwards to equalize the pressure. This causes shattering of windows, collapse of buildings and lifting of automobiles. The violent battering effect of tornadic winds is another major cause of destruction. Large trees are uprooted and the bark stripped off. People and farm animals are whirled through the air and then dashed to earth. Bridges are wrenched from foundations. Straws and slivers of wood are driven deep into boards, and posts and large pieces of wood are hurled like javelins deep into the earth.

359. How many tornadoes occur in the United States each year? According to a U.S. Weather Bureau tabulation from 1916 through 1954, about 179 tornadoes are reported on an average each year throughout the country. In recent years, this average has been increasing, but it does not mean that there are necessarily more tornadoes but rather that reporting, observation and tracking techniques are greatly improved. In 1919, 65 tornadoes were reported. In 1954, 690 were recorded.

360. Which states have the most tornadoes? Although no state is immune, the four states which have the greatest average number of tornadoes each year are Kansas (23.05), Texas (17.08), Oklahoma (15.49) and Iowa (14.26). These figures reflect averages from records of tornadoes in the United States each year from 1916 through 1954. The states with the lowest average amount of tornadoes each year are Rhode Island (0.03) and Utah (0.10). Delaware, Nevada, Oregon and Washington report an average figure of 0.15.

361. How many deaths occur from tornadoes in the United States? From 1916 through 1954, a total of 8,776 lives were lost in tornadoes making an average of 225 deaths each year. During this same period estimations of property damages amounted to a total of approximately 790 million dollars or an average of over 20 million dollars per year. Some tornadoes are historically outstanding because of their devasting characteristics and terrible losses of life and property which resulted. These outstanding tornadoes are not particularly more violent than others. Some, in fact, may have had less tornadic force than others, but they happen to strike larger cities or heavily populated areas. There is no doubt that storms of equal or greater violence have occurred in open country with only slight damage to life or property.

362. Why is it difficult to pinpoint tornado forecasts? Severe local storms such as tornadoes and dangerous thunderstorms cover such a small area and develop so suddenly that their exact point of occurrence cannot be forecast. Forecasts of tornadoes, therefore, include a general area where conditions seem favorable for tornado formation.

363. What tornado warning systems exist? The best warning system consists of volunteer community networks which number several hundred and which are increasing. Observers, mostly on the out-skirts of a city or town, are alerted whenever an area tornado fore-cast has been issued by the Weather Bureau. They usually take up stations on hilltops or high buildings to the south and west of the community. When a tornado is spotted, the observer notifies a central Weather Bureau office regarding the storm's location, intensity, and movement. This information is plotted carefully and warnings are issued to radio and television stations and other public outlets reach-ing people in the danger area. Radar is also being used more fre-quently to identify squall lines. A suspicious echo is checked by state police cars in the area.

364. What are some safety rules for tornadoes?
1. When time permits, go to a tornado cellar, cave or underground excavation which should have an air outlet. Keep it fit for use, free from water, gas or debris and supplied with pick and shovel.
2. *If in open country,* move at right angles to the tornado's path. If there is no time to escape, lie flat in the nearest depression such as a ditch or ravine.
3. *If in a city or town,* seek shelter in a reinforced building. *Stay away from windows!* In homes, stay in the southwest corner of the basement. Electricity and fuel lines should be shut off. Open doors and windows on the north and east side of the house. In an office building, stand against the inside wall on a lower floor.
4. Keep tuned to the radio or television for latest tornado advisory in advance. Telephones should not be used to the Weather Bureau because lines are needed for other services.

365. What is a waterspout? Waterspouts are tornadoes which occur over sea. There are two types; the tornado waterspout and the fair weather waterspout.

366. What are the characteristics of a tornado waterspout? It has the same characteristics as a tornado except that it is over water. The lower portion of the funnel cloud is made up of spray instead of dust and debris. A tornado may pass from land to water or water to land without materially changing its appearance or intensity.

367. What are the characteristics of a fair weather waterspout?

It is a more slender whirling column of water droplets and may be seen over water when no threatening clouds are present. Similar to the dust devil over land, it forms at the surface and develops upward, frequently under a clear sky. Because of the moisture present, a small cloud sometimes forms over the fair weather waterspout. They seldom develop into dangerous storms and diminish rapidly when moving over land.

368. Are tornadoes confined to the United States?   The most frequent and most violent ones seem to be, although they occur in many parts of the world, especially in the Temperate Zones. Tornadoes have been reported in England, Canada, France, Germany, The Netherlands, Hungary, Italy, India, Australia, Russia, China and Japan. They have also been reported in the Bermuda Islands and the Fiji Islands.

369. How many tornadoes occur in England?   The Royal Meteorological Society of England lists a total of 50 tornadoes which have occurred during an 82-year period, 1868–1949. They have occurred mostly over the English lowlands and river basins in every month of the year except December. More have occurred in October than in any other month. On the average, throughout the year, tornadoes are relatively frequent in the late spring and late summer. Some of the most destructive ones occurred on October 19, 1870; October 27, 1913; and May 21, 1950.

370. What is a dust devil?   It is a rapidly rotating column of air about 100 to 300 feet in height which picks up dust, straw, leaves or other light material. It usually develops on calm hot afternoons with clear skies, mostly in desert regions and has no relationship to a large-scale dust storm. In the southwestern part of the United States it is called a *dancing devil;* in India, a *devil;* and in South Africa, a *desert devil.* In Death Valley, California, a dust devil may be called a *sand auger.* It is also called a *dust whirl.*

371. What causes a dust devil?   It develops from superheated air over the ground, especially flat, bare surfaces. A small current of warm air starts to rise. Surrounding air from diverse currents flows in to equalize the pressure difference and an eddy of whirling winds is started which ascends in spiral form. They may rotate either clockwise or counterclockwise and rarely cause serious damage although, if the surface air is exceptionally hot and the air aloft quite cool, they can sometimes rise to a 1000 feet and last several hours. Most are short-lived, limited in size and intensity, and usually cause only nuisance damage.

372. What conditions favor a dust storm? A large dust storm may occur over deserts or plains where vegetation is generally scanty and winds generally high in late winter and early spring. At such times and places, two rapidly moving air masses, relatively dry although differing in temperatures, may converge to form a vigorous cyclone which stirs dust in thick clouds. The dust storm starts most often in the strong winds behind the cold front. Dust storms also are caused in warm-dry air masses which are very unstable (warm in lower layers and quite cool aloft). Dust is transported upwards by the turbulence and gustiness of the air. Dust storms become more frequent and deadly over regions which have been made arid by prolonged drought conditions. Drought conditions develop from consistent spells of precipitation-free weather.

373. Where do most dust storms occur in the United States? Dust storms often plague the southwest deserts and plains regions of the United States when the topsoil or sand has become loose from parched earth following long dry spells. Sometimes air moves in from the northern Pacific and crosses the Rockies to arrive in the plains region quite dried out by its passage over the mountains. As it rolls south and east behind a front, great clouds of dust are carried with it and forced aloft. The sweeping black dust clouds move over thousands of square miles, turning day into night and causing enormous hardships. These *black blizzards* were a common and fearful sight in the 1930s in the Midwestern dust bowl. The fierce dry winds stripped topsoil and growing grain, forcing many farmers to abandon their farm lands.

374. Can dust-bowl conditions strike the United States again? Unless positive large-scale plans are put into effect, there is no guarantee that severe dust storm conditions may not repeat themselves. As long as drought cycles exist, dust will be blown. From 1949 to

1956, the American southwest farm and grazing land was barely wet by rainfall.

375. What will the United States government do about drought and dust? One administration official said, "We hope to see developed a long-range remedial program rather than a crash program for temporary relief. Many of us think the dollars should go into canals, reservoirs, and irrigation projects which will give farmers, cattlemen and communities sources of water on which they can depend to carry them through the dry months and years." The Department of Interior is working on three major water conservation projects: (1) the huge Gulf basin development in Texas which would bring water from east Texas rivers to the dry plains of central and southwest Texas; (2) a project to divert water by tunnel from the western Continental Divide into the Arkansas River in the upper plains of eastern Colorado; and (3) the Washita River basin project in Oklahoma where two multipurpose dams would provide water storage, produce electric power and irrigation.

376. How far and high can dust be carried? Dust raised in storms over Texas, Oklahoma and Kansas has traveled to Vermont and New Hampshire, borne aloft by the westerlies as high as two to three miles above the earth.

377. What basic factors result in floods in the United States? Disastrous floods may occur in the eastern United States when an extensive cyclone (low) stalls in its eastward course with the air in the warm sector of the low extremely moist and unstable and with excessive rain falling in advance of the warm front. The excessive rain may combine with mountain and valley country which is either bare of waterholding forest and brush or which is oversaturated and cannot hold additional precipitation.

378. When do most of the worst American floods occur? In winter or early spring. From December to March, heavy continuous-type rains develop, aided by melting snows. Although summer rains may be heavy, they are more scattered and showery. Also, in the summer, the ground is in a better condition to absorb and hold water. Important floods in the summer usually are preceded by monthly rains well over 10 inches. Half as much rain will suffice in the winter. The exception for summer floods may develop with Atlantic hurricanes which may carry torrential rains inland as far as the northeastern United States and overload the land, rivers, lakes and streams with water.

379. Where do greatest flood losses occur in the United States?
Statistics show that over a period of many seasons, the greatest losses because of floods are confined to the Mississippi Valley, with the Atlantic and Gulf coasts also suffering appreciable damage. The cost of a flood to a community includes these principal items: (1) loss of human life; (2) danger to the public health; (3) damage to immovable property; (4) damage to movable property; and (5) intangible losses, such as disruption of business and transportation. By far the greatest of the losses subject to classification are those of property—largely buildings, highways and railroads.

380. What is a flash flood?   It is a local flooding condition, especially of narrow valleys and arroyos (gullies). Flash floods are typical of the arid Southwest in mountainous sections where the valleys are narrow canyons and the high slopes are relatively bare of forest or brush (as when after a forest or brush fire). Sometimes a severe thunderstorm will occur over the mountains and let loose a sudden deluge of water in the form of a cloudburst. The heavy rain tumbles into the innocent-looking dry arroyos or narrow valleys and can transform a dry gully into a maelstrom of rushing waters. These flash floods are not, of course, as severe as large-scale floods. But, for their size, they can be intense and cause drowning of an unwary traveler. They can also wash out bridges and roads and sometimes inundate small communities.

381. What is a monsoon?   The term describes winds which blow with great steadiness and regularity at specific times of the year. It usually describes a seasonal wind which blows from a large land mass to the ocean in winter and in the opposite direction in the summer. The monsoons of India and southern Asia are most typical although they occur in lesser extent in north Australia, parts of western, eastern and southern Africa and North and South America. The word monsoon comes from an Arabic word for season.

382. What causes the Indian monsoons? In summer the broad land areas become heated more quickly and to higher temperatures than the Indian Ocean to the southwest. Rising warm currents, generated over India, result in a low pressure area created over the land. Relatively, then, the air over the cooler Indian Ocean represents a high pressure area. The winds start to blow from high pressure to low, from southwest to northeast. As the moisture-laden winds pass

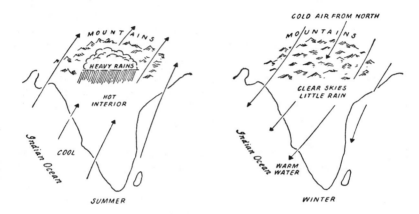

over the steadily rising land to the mountains of Tibet, the rising air begins to condense and to precipitate its water vapor. Thus, with the southwest summer monsoon, torrential rains occur during the period from June to October. The winter, or northeast monsoon, is the reverse when cold dry air pours seaward from the interior of Asia and very little rain occurs.

383. What causes land and sea breezes? The familiar breezes of day and night associated with the seashore during the summer are miniature examples of the monsoon effect described in the preceding question. Instead of seasonal, they are diurnal (daily) and instead of affecting an entire continent or large land mass, they occur in limited boundaries. The basic reason is the same for a sea breeze as for a monsoon—differences in heating between the land and sea, leading to pressure differences. The air from the sea, being cooler and hence heavier in pressure, flows in to replace rising warmer air of the coast line.

384. What are the characteristics of a sea breeze? A flow of air starts from sea to land when a pronounced pressure gradient exists between air over the sea and air over the land. It is shallow to start with, but may circulate to possibly 2,000 feet. Sea breezes may blow

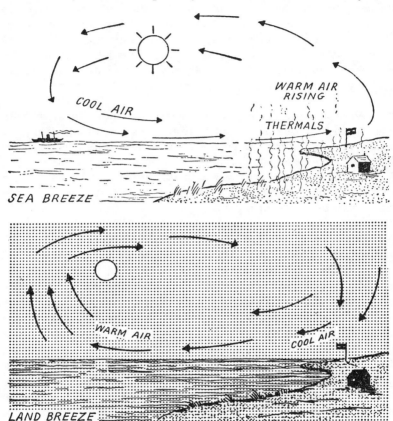

from 15 to 20 miles from the coast, both inland and seaward. They usually start from about two miles out to sea and extend inland about 10 miles. The breeze is rarely more than about 15 miles per hour. It can reach 25 or more miles an hour on some occasions. It blows hardest when the ocean is still quite cold from the winter and when an early hot spell may occur, such as in late May or June, the sea breeze becomes noticeable about two to three o'clock in the afternoon.

It generally decreases toward evening when the land starts to cool, thus easing the pressure gradient.

385. What are the characteristics of a land breeze? The land breeze blows from the shore to the sea when the land area has radiated its heat away at night. In this case, the adjacent water area may be warmer than the land and a flow of air starts from the land to sea to equalize the pressure differences set up. It is purely a reverse of the sea breeze. Land breezes are generally not as intense as the sea breezes or as highly developed in distance and depth of the current.

386. What causes lake or forest breezes? Winds similar in principle to land and sea breezes blow from lakes and forests. They are usually much weaker. The lake acts like a miniature ocean so that shoreward breezes prevail during the day and on-lake breezes at night. Forests may be relatively cooler than surrounding open country and a light breeze may, therefore, blow from the forest during the day and toward the forest at night.

387. What wind currents are typical in mountains and valleys? During warm clear days, valley sides and slopes become heated and the air in contact is warmed and expands up the mountainsides as an updraft. This drift of air upslope is usually gentle but can vary greatly, depending upon the height of the mountain and the topography. The updraft winds slide up the sides of the valley and when cooled by expansion return to the valley center as a downdraft. At night, the opposite occurs, valley sides cool quickly with downdrafts of air flowing along the sides and helped by gravity. In this situation, updrafts then form in the center of the valley. Ascending winds blowing up an incline are called *anabatic*. Winds blowing down an incline are called *katabatic*. When anabatic winds are made stronger by the funnel effect of a valley they are called *valley winds*.

388. What is a wind eddy? It is a separate current of air moving against the main current usually with a circular motion. Eddies can be large and significant such as hurricanes, but the term eddy is mostly used to describe small currents of air. Typical small eddies are currents of air which rise and are carried along with the wind when wind blows over a rough surface at moderate or high speeds.

Wind Currents

These eddy currents may range in size from a few feet to a hundred feet in diameter.

389. What is gustiness?   Gusts of wind are local irregular variations in a main wind flow. They are a sudden brief increase in the speed of the wind, followed by a lull or slackening of the wind. Gustiness can be caused by wind blowing over rough country with irregular terrain which produce eddy currents carried along by the wind. Such effects are minimized over open, level country or sea. Rising warm currents of air (thermals) can also bring a gusty factor to the wind by superimposing their own eddy currents on the main wind flow.

390. What is the difference between a gust and a squall? Gusts are transient changes in the mean wind usually lasting for only a few seconds. A squall usually is a wind of considerable intensity, an increase in the mean wind lasting for some minutes and then dying away. Squalls are generally not due to mechanical factors such as terrain effects. The most dangerous and sudden type of squall is usually associated with severe thunderstorm activity along a cold front. This is called a *line squall* (See Questions 528, 529).

391. What is air turbulence? Turbulence is a state of agitated or disturbed air, usually referring to irregular vertical air currents or *bumpiness* of air. Turbulence may be caused by wind flowing over an uneven surface or in thermal currents set up over areas of different heating. Turbulent air may also be caused when it is forced aloft mechanically as on the windward side of a hill or mountain. Severe turbulence also occurs along a cold front where cold air may force warm air aloft with a violent burrowing motion.

392. What is a jet stream? Jet streams are belts of high-speed winds occurring at upper levels of the troposphere with narrow cross sections relative to their length. Pilots flying high-altitude missions during World War II reported that their planes were buffeted by fantastically strong upper air winds. These winds blew from west to east at tremendous speeds at levels about eight miles above the earth, near the top of the troposphere, and seemingly a part of the prevailing westerlies wind belt. It was not until 1946 that the jet stream was fully recognized as a meteorological entity.

393. What are the characteristics of a jet stream? These narrow filaments or rivers of air exist between 10,000 and 40,000 feet above the Earth's surface. They move in a general west to east direction at high velocities with individual jets joining each other to form a great river of rushing air which sweeps around the earth in a wavy course between the Arctic Circle and the Tropic of Cancer. Jet stream winds blow at speeds of 150 to 300 miles per hour. Speeds as high as 400 miles an hour have been reported. These streams seem to spurt sharply from a central core which has a sort of serpentine movement and shape. They stream from side to side in ropy waves around the earth, wandering north or south above the westerlies with the season.

In winter in the Northern Hemisphere, they move south, blowing hardest at about 30 to 35 degrees north, or over the southern United States area. In the summer, the core of the strongest winds lies about 40 to 45 degrees north. Corresponding jet stream movements appear to exist in the Southern Hemisphere.

Jet stream

394. What causes the jet stream?   Although much is being learned about jet streams, an exact explanation for their origin is lacking. One theory is that they are the results of large-scale horizontal mixing processes brought about by the lows and highs of the middle latitudes (cyclones and anticyclones). Another theory contends that jet streams may form by the coming together of vast streams of equatorial and polar air at great heights. When these warm and cold air flows are side by side, resulting pressure differences may cause a strong circulation which becomes a jet.

395. Why is the jet stream important to aviation?   Military and commercial aircraft are flying ever faster and higher. If a pilot could "ride" a jet stream going his way, the crusing range could be extended enormously, especially with jet planes, because of the limited fuel supply in these planes.

396. Can jet streams be forecasted with accuracy?   In some areas, especially important areas in North America, the North Atlantic and Europe, upper air observations have been sufficient to detect jets and to forecast their behavior in a general way for about a day in advance. But jet streams in many other parts of the world, such as in the Pacific, are undetected because they are so narrow and, therefore, a dense network of upper air soundings is needed. The usefulness of the jet stream to aviation is limited by severe buffeting, turbulence regions about 50 to 100 miles wide and about 3,000 feet thick. This turbulence may be caused by the jet disturbing its surrounding air much as a ship does when it plows through water.

397. How does the jet stream affect the weather? The jet streams affect on weather in the middle latitudes is a subject of much thought and controversy. Its movements are closely related to the Polar Front which separates tropical and polar air masses at about 60 degrees north. The Polar Front is a continuous "breeder" of weather for the Temperate Zones. This may indicate that the jet stream plays an important part in the possible formation, steering, or intensifying of such weather phenomena as cyclones, anticyclones (lows and highs), hurricane tracks and abnormal weather spells.

398. Can insurance be purchased as protection against life and property loss or damage due to winds, storms, floods, drought, etc.? It is axiomatic that anything can be insured—if premiums are met. This applies to damages or losses sustained because of weather phenomena. The type of insurance available which covers such loss or damage to crops, homes and business properties varies greatly throughout the United States. In Florida, for example, insurance against *wind* damage caused by a hurricane may be purchased readily, but insurance against *tidal* or *flood* damage caused by a hurricane will rarely be written by an underwriter, except, perhaps, under sufficiently high premium rate.

399. Is there an all-weather insurance plan against crop damage? Plans are being developed in a few test areas throughout the United States by some insurance companies. It is hoped that present standard insurance coverage against rain and hail damage to crops can be expanded to include consideration of other damaging elements such as drought, extreme heat, excessive moisture, insects, wind, sleet, tornado, frost, etc.

400. What insurance against weather losses or damages is available to homeowners? In general, there are three types of insurance for homeowners which include insurance against loss or damage by weather perils. The first type, called *fire and extended coverage,* has the most amount of limitations. The second type, called fire and extended coverage plus *additional perils,* is less limiting on insurance coverage. And the third type, described as all-physical *loss,* offers the most comprehensive kind of insurance available to home owners against weather perils. The insurance underwriter must be guided by consideration of a wide range of weather abnormality statistics for specific areas. He may not, for example, be able to offer the all-physical loss type of insurance where the chances of severe damage because of weather extremes are increased.

401. What should the property owner do about weather insurance?
Any property owner, farm, industry or home, should study the range of weather extremes in his particular area. Past flood conditions, wind storms, drought, dust, frost, excessive rain, hail, etc. should be carefully analyzed so that he can estimate the possibilities of these damaging elements occurring again. These statistics may be obtained from a local Weather Bureau office or from the Office of Climatology of the U.S. Weather Bureau in Washington, D. C. The services of a private meteorological consultant are also available. Once having established the pattern of weather extremes peculiar to his location, the property owner should then discuss with an insurance representative the best possible type of insurance available to cover damages or losses which may be sustained by weather extremes.

402. How were local winds of different lands named?   Through many hundreds of generations, people of many places in different areas of the Earth were deeply impressed by different winds. These winds seemed purely local in effect. It was not understood that many of the winds were Earth-circling or parts of huge circulations of air which covered half a continent. Nor were people aware of effects of compression, gravity, topography and a host of factors which affected smaller movements of air. But their practical knowledge of winds was excellent and the names they gave winds are colorful and descriptive, showing the deep imprint winds have made on groups of people in some cases from time lost in antiquity. Some names were based on the direction of a wind and some on the wind's warmth or cold, dryness or moisture, gentleness or fury. Other wind names describe the sound of the wind or its effect on land or people. Some winds, too, were named for ancient gods.

403. Is there a simplified classification for these winds? In essence, all winds of the world can fit into most of the categories of the primary, secondary or local wind effects as described in preceding questions of this section. A more basic classification would be to call them either hot or cold—governed by just a few basic meteorological principles. *Hot winds* may occur when hot air is simply carried from

CHINOOK OR FÖHN WIND

a warm source region into another area at more or less frequent times. Hot winds may also form when air heats by compression as it descends from a high region to a low one. *Cold winds* may develop by an outbreak of cold air from a cold source region or by cold air flowing down an elevated area to a lowland area under the influence of gravity.

404. What are the characteristics and local areas of the following winds? Abroholos   A squally wind which blows on the southeast coast of Brazil, more frequently from May through August.

405. Austru   A west wind blowing over the lower Danube lands in the winter frequently bringing dry, clear and cold weather.

406. Bad-i-sad-o-bistroz   A violent downslope wind which affects the region around Afghanistan, blowing from a northwest direction from May to September. Sometimes called the *wind of 120 days*.

407. Bali A strong east wind which blows over the eastern end of Java across the Java Sea.

408. Barat A squally, occasionally violent northwest wind which blows across the Celebes Sea to the northeast coast of the island of Celebes. This wind occurs most frequently from December to February and often causes severe damage.

409. Barber A term used in sections of the United States and Canada to describe a strong wind which carries precipitation that freezes upon contact with objects, especially the beard and hair.

410. Barine Unusual winds from the west which blow over eastern Venezuela.

411. Bayamo A violent gust-type of wind usually associated with thunderstorms forming on the windward side of the Sierra Maestra range in the southern part of Cuba. The winds travel with the storm southward from the land over the south coast of Cuba.

412. Belot A strong land wind from the north and northwest which blows along the southeast coast of Arabia during the period from December to March. Sand picked up from the interior by this wind often imparts a hazy look to the atmosphere.

413. Bentu An east wind blowing in the Mediterranean along the coast of Sardinia.

414. Berg Hot and occasionally very dusty winds usually coming from the eastern portions of southwest Africa, from the north on the south coast of Africa and from the northwest in Natal. They are most frequent in the winter season and bring high temperatures for a period of several days. They carry dry land air to the coastal sections.

415. Bhoot In India, a term describing a relatively small-scale counterclockwise whirling of air filled with loose dust.

416. Bise A strong outbreak of cold dry air from the north which blows over the mountainous regions of southern France and Switzerland.

417. Blizzard A term usually applied in Canada and northern United States to describe a howling, cold, piercing wind, usually of gale force, out of the north or northwest. These winds are intensely cold and punishing. They sweep in strong surges behind cold fronts moving from territories usually between Hudson Bay and Alaska. They are usually laden with blinding, powdery snow, mostly picked up from the ground.

418. Bohorok A dry and warm downslope type of wind which blows frequently during the period from May to September in Sumatra from the leeward side of the Barison Mountains in the southwest to the northeast coast of Sumatra. This wind comes from the Indian Ocean but loses much of its moisture on the windward side of the backbone mountains in Sumatra.

419. Bora Excessively cold winds which sweep southward from the valleys of the Karst and Dinaric Alps across the Adriatic Sea and affect the entire Dalmatian coast from Trieste to Albania. Sometimes the bora's icy winds exceed 100 miles an hour as they sweep down the abruptly changing land height from the Alps to the coastlines of the Adriatic.

420. Borasco Winds associated with violent thunderstorms, especially in the Mediterranean.

421. Bornan One of many valley winds in the Swiss Alps. The Bornan blows over the central part of Lake Geneva from the valley of the Drance.

422. Brave West The strong, often stormy winds from the west and northwest—the prevailing belt of westerly winds of the Temperate Zone of the Southern Hemisphere.

423. Breva A valley wind which blows on Lake Como in northern Italy.

424. Brick Fielder   A hot, dry and dust-laden wind blowing across the south of Australia from the deserts of the interior during the summer. The name originated in Sydney because of the dust the hot winds raised from the brick fields to the south of the city.

425. Brisa   Trade winds which blow from the northeast on the coast of South America or from the east on Puerto Rico.

426. Brisote   A northeast wind, part of the trades, blowing at a stronger than normal rate over Cuba.

427. Broboe   A dry wind blowing from the east over the southwestern part of the island of Celebes. It is caused by the warming and drying of air by compression as the prevailing wind rises over the hilly peninsula in the southwest of the island from June to October from the east.

428. Brubu   A squall in Indonesia.

429. Brüscha   A northwest wind in the Besgell Valley, Switzerland.

430. Bull's-eye squall   A sudden squall forming in apparently fair weather. This type of squall is characteristic over the ocean off the Cape of Good Hope, South Africa. Its name is derived from the appearance of a small and isolated cloud seen at the beginning of the squall marking the top of an otherwise invisible vortex of the storm.

431. Buran   A dreaded, wildly violent and intensely cold wind which breaks out throughout Siberia and into south Russia from the northeast. Similar to the blizzard of the United States and Canada, the buran is dangerous because of its snow-filled character.

432. Burga   A strong windstorm in Alaska, usually attended by snow or sleet.

433. Cat's paw   A light puff of wind in America, just barely noticeable—enough to cause a patch of ripples on water. The cat's paw is a light breeze that affects a small area.

434. Cacimbo   A cooling sea breeze which blows from the south-west to the port of Lobito on the coast of Angola in western Africa. The breeze is very frequent in July and August, starting from about 10:00 A. M. and lasting through most of the day. It is cooled by its passage over the Buengala water current lying to the west of Lobito in the South Atlantic.

435. Challiho   A strong southerly wind experienced in parts of India during the spring months. It is the forerunner of the southwest monsoon wind which prevails during the summer months.

436. Chergui   An intrusion of hot air into Morocco in northwest Africa from the Sahara Desert areas to the east.

437. Chili   A hot wind from the deserts of North Africa and Arabia which blows over middle and south of the Mediterranean. Chili is the name applied when this wind passes over Tunisia. It is a dry wind and carries much dust and stand. It occurs more frequently in the spring when the Mediterranean is normally much cooler than the desert areas far to the south.

438. Chinook   A dry, warm wind which blows from a westerly direction down the east slopes of the Rockies in North America. This Föhn-type wind loses its moisture on the windward side of the Rockies and heats by compression as it slides down the lee side. The chinook sometimes raises the temperature as much as 40° F. in a quarter of an hour, causing a rapid melting of snow. On eastern slopes of the Rockies during the winter in United States and Canada, the chinook is sometimes called the *snow eater*.

439. Chubasco   A violent squall-type of wind associated with severe thunderstorms which frequently occur on the western coastal sections of Central America and Mexico between Costa Rica and Point Eugenio (Lower California). These local thunderstorms are most frequent during the rainy season from May to November.

440. Churada   A fierce rainy squall in the Mariana Islands occurring most frequently during January, February or March.

441. Cockeyed bob   A squall wind associated with thunderstorms on the northwest coast of Australia, occurring most frequently from December through March.

442. Collada   Strong northerly winds which blow over the Gulf of California.

443. Contrastes   Winds which blow from opposite directions even though a short distance apart. The western Mediterranean area is frequently subject to these contrasting winds in the spring and fall seasons where air masses may move toward each other from the European Continent and the desert regions of Africa.

444. Coromell   A nighttime offshore breeze occurring with great regularity in the La Paz area at the southern part of Lower California from November to May. At these times the land area cools rapidly at night and the overlying air flows toward the relatively much warmer water areas around La Paz.

445. Coronazo   Strong south winds blowing along the west coast of Mexico. These winds are usually the eastern peripheries of tropical storms which are located well offshore to the west.

446. Criador   A west wind in northern Spain which is usually associated with traveling disturbances and which brings rain.

447. Crivetz   A wind which blows from the northeast over the lower Danube lands from the Russian interior.

448. Datoo   A westerly sea breeze which blows over Gibraltar from the adjacent waters of the Atlantic Ocean.

449. Doctor   A term originating in England to describe the cooling sea breezes which occur in the tropics. The term *Cape doctor* refers to a strong southeast wind which blows on the South African coast.

450. Elephanta   A strong wind which blows from the south or southeast along the Malabar coast at the extreme southwest end of India during September and October. It heralds the beginning of the dry season and marks the ending of the southwest monsoon.

451. Etesian   Outbreaks of pleasantly cool air which blow over the eastern Mediterranean Sea, particularly the Aegean Sea. They blow rather frequently in July and August, moderate to strong, and occasionally of gale force, although the stronger velocities are usually limited to the northerly portions.

452. Föhn   A wind which is characteristic of many mountainous regions of the world and is called by many different names in different countries (Chinook, Santa Ana, etc.). The term Föhn is more specifically applied in many Alpine valleys, notably in the upper Rhine, the Reuss and upper Aar in central Europe. It is a warm dry wind which blows down the lee side of a hill, mountain or mountain range after it has mechanically risen over the windward side where it has lost its moisture. Such Föhn winds have a marked effect on the climate of a region, sometimes creating an oasis of warm pleasant weather in what would normally be a cold region. In the Reuss Valley in Switzerland, Föhn winds blow on an average of 48 days a year and mostly in March, April and May.

453. Gallego   In Spain, a cold wind from the north.

454. Gharbi   Sometimes winds from the Sahara Desert blows northward over the northern and eastern Mediterranean. As these hot winds blow, they occasionally pick up moisture en route and arrive on the north coasts of the Mediterranean as strong winds, warm and damp. These winds are called Gharbi in the Adriatic and Aegean Sea regions. They bring heavy rain, especially on mountainous coasts. Some of the dust of the Sahara is often mixed with these rains causing "red rain."

455. Ghibli   A dry hot wind in Tripoli, originating from the deserts of North Africa and Arabia.

456. Gregales   A strong polar outbreak of wind from the northeast which is pumped into the northeastern coasts of the Mediterranean, usually in spring and autumn and associated with extremely variable weather.

457. Haboob   Extremely severe dust storms occurring mostly in the summer in the north of the Anglo-Egyptian Sudan. When strong squall winds blow in this area, walls of loose dust advance with the wind rising to heights of several thousand feet, about 15 miles along a front and moving at speeds of 35 miles per hour. It is mostly a dry squall caused by the interplay of southwest monsoon winds at the surface and the dry hot winds from the northeast above.

458. Harmattan   The Harmattan is a continental part of the globe-encircling trade winds. It dominates the Sahara Desert and impresses its extremely dry and warm characteristics upon a huge area of North and Northwest Africa. In the summer, the Harmattan's parching winds blow moderate to strong from the Mediterranean Sea southward and eastward to about latitude 17 degrees north. In the winter, the warm and dry air blows from latitude 30 degrees north to the Guinea coast in West Africa and sometimes penetrating to the African equator.

459. Helm   A strong wind which blows in the Pennine Chain in north central England. One portion of the mountain chain, east of Westmorland County, runs northwest-southeast for about 10 miles, with an average steep-descending southwest slope of about 2,000 feet. Extremely strong wind currents blow from the northeast over the Pennines to the Westmorland-Cumberland region on the central west coast of England. These helm winds are sometimes associated with a roll or series of rolls of clouds which overhang the crest of a wind wave to leeward of the hills. These mountaintop clouds that form in the windstorm are called *helm clouds*.

460. Howling fifties   A term probably originating with sons of the whalers of the nineteenth century who sailed the oceans of the Southern Hemisphere. They found the winds to be punishing over their routes southward, unhampered by earth or mountains. The different latitudes of the general area over which these strong prevailing westerly winds blow (40 to 50 degrees south of the equator) were given names which reflected their impression of these battering winds— thus, the howling fifties and the roaring forties.

461. Imbat   A sea breeze which tempers the heat of the North African coasts.

462. Karaburan From early spring till the end of summer, these gale-force winds form each day in the Gobi Desert and surrounding regions of the heart of Asia. They blow with violent strength from the east-northeast, carrying clouds of dust up from the desert. This blowing sand often darkens the air and is the reason for the Karaburan being sometimes called the "black storm." The lighter dusts from these stirred sands carry far beyond the desert areas and provide a characteristic summer haze. Throughout the course of time, deep deposits of loess have been built up from this dust. The Karaburan rages by day only. At night the desert air calms and the skies clear rapidly. In this respect, the Karaburan of the Gobi and the Harmattan of the Sahara are somewhat similar.

463. Kapalilua A prevailing type of sea breeze in Hawaii.

464. Kaus A wind blowing from the southeast over the Persian Gulf during the winter. This wind is usually temporary in character and associated with low-pressure areas moving over the Gulf region from across the Mediterranean Sea. The Kaus wind brings cloudy skies, some rain and above normal temperatures.

465. Khamsin In the winter half of the year, Egypt is invaded by irregular outbreaks of cool and hot winds. The Khamsin is a hot wind which is sometimes pulled into Egypt from Arabia, the Gulf of Aden and possibly the Arabian Sea far to the south and east. The Khamsin wind blows into lower Egypt as an east or southeast wind, very hot, extremely dry and so hazy with fine dust that lights are sometimes required at midday. It usually continues for two or three days and then is swept away by an invasion of cold air moving in from the northwest behind a cold front. As the cold air strikes, dust and sand is raised, the sky is clouded and sometimes showers fall. The Khamsin winds are usually moderate in force but may reach gale force. The mean frequency is about three a month in February, March and April.

466. Kharif A strong, often gale-force wind which blows from the southwest in the Gulf of Aden. It is called Kharif on the Somaliland coast on the south shores of the gulf where the wind descends sand-laden and uncomfortably hot from the African interior.

467. Knik   A strong southeast wind in the vicinity of Palmer, Alaska, less than 50 miles northeast of Anchorage.

468. Leste   A hot dry wind blowing from the south and east sections of the North Central African desert regions to the Madeira and Canary Islands.

469. Levanter   A strong east wind which frequently blows through the Strait of Gibraltar from the Mediterranean. When the winds are particularly strong and stormy, the wind is sometimes called *Llevantades*.

470. Levanto   A hot wind from the southeast which blows over the Canary Islands—similar to the Leste wind.

471. Leveche   One of the many hot and dry winds which originate in the hot deserts of North Africa and Arabia and affect the general regions of the middle and south Mediterranean. All of them may be grouped under the general term *Sirocco,* but they are so important in different areas that many of these Sirocco-type winds have been given different names. The Leveche is such a wind which blows from the south and southeast over Spain.

472. Leung   A cold wind from the north which blows over the China coast.

473. Marin   The Marin is in many respects an opposite kind of wind to the Mistral. It is a Sirocco-type of wind, blowing with strong intensity from the southeast in the Gulf of Lions and the neighboring shore lines of southeast France. It is warm and brings unpleasantly cloudy weather and heavy rain.

474. Mistral   A particularly well-known wind is the Mistral, the "masterful" north wind of the Gulf of Lions, which surges southward in outbreaks from polar regions above north and central Europe to affect a wide area over the northwest coast of the Mediterranean. This is a cold and usually dry wind of damaging violence in certain areas. It often rushes southward in the winter over the lower Rhone Valley in such force as to threaten the stability of railway trains in the Rhone delta. It blows throughout the year, although it is most prominent in winter and spring. Marseilles is exposed to the Mistral for about 100 days a year. Sometimes the Mistral is wide enough in extent to affect the coast between Barcelona and Genoa, the Gulf of Lions (with squalls up to 100 miles an hour) and the Balearic Sea. It has even crossed the Mediterranean to the African coast.

475. Narai   A cold wind in Japan blowing from the northeast and polar regions of the Asiatic land mass.

476. El Norte.   The Norte wind is an outbreak of cold northerly air over the Gulf of Mexico and Central America which flows over huge distances in the winter from pile-ups of polar air in north central United States and the Canadian basin. It also refers to cold winds from the north which sweep over eastern Spain in the winter.

477. Northeaster   A wind which blows from moderate to strong force from the northeast over the New England coastal regions. This wind is Polar Maritime in character, generally moist and often chilly or cold. It is frequently accompanied by cloudiness and precipitation caused by warmer, semitropical air which rises over the northeasterly surface winds from a south or southwest direction.

478. Norther   A strong cold wind from the north which sweeps over the Gulf States of the United States, the Gulf of Mexico and eastern coasts of Mexico. Like the Norte of Central America, it is a winter outbreak of air which reaches far south from the polar regions of Canada.

479. Northwester   A moderate to strong wind from the northwest used more specifically to describe a cool or cold wind which blows from the northwest over North America east of the Rockies. The name is also applied to frequent gale winds which batter the Cape region of South Africa from the northwest, attended by overcast skies and heavy rain in the winter.

480. Oe   A localized type of whirlwind which occurs off the coast of the Faeroe Islands in the northeast Atlantic.

481. Pampero   A violent squall which attends cold fronts as they sweep from the southwest to the northeast in the pampas of Argentina and Uruguay. It is something like the norther of the United States plains section in the sense that it is an outbreak of air from polar latitudes. Behind the violent frontal squall, the wind eases, blowing moderate and steady. The sky clears and the air is cool or cold. Buenos Aires has about 12 pamperos a year, mostly in spring and summer; Montevideo about 16; the River Plate area has about 20 a year.

482. Papagayo   A violent wind from the north which invades the Gulf of Papagayo on the northwestern coast of Costa Rica. It is a kin to the Norther of the United States and the Norte of Mexico and results from large southward surges of polar air from the North American continent.

483. Ponente   A westerly wind over the Mediterranean, particularly as a refreshing sea breeze on the western Italian coast line.

484. Purga   Another name for the dreaded Buran of the tundra regions in northern Siberia in the winter. The Purga wind sweeps down from the north with extraordinary violence throughout Siberia and sometimes to south Russia, particularly violent over the open plains sections. The air is filled with swirling snow picked up from the snow-covered tundra sections of Siberia and cuts visibility to zero. The Purga is very similar to the North American blizzard.

485. Reshabar   A strong wind which blows from the northwest over the Caucasus Mountain range between the Black and Caspian Seas.

486. Roaring forties   The area of the oceans between 40 and 50 degrees south latitude where day after day winds exceeding 40 to 50 miles an hour blow from the west over oceanic areas. They are the prevailing westerly belt of winds which circle the earth in the Southern Hemisphere as part of the Earth's primary atmospheric circulation.

487. Santa Ana   A Föhn-type wind named from a community southeast of Los Angeles in the coastal area of Southern California. During the winter, when north and east winds blow from the deserts and plateaus of lower eastern California, they cross the Coast Ranges and descend through such passes as the Cajon and Santa Ana to reach the coast as hot and dry winds, often laden with piercing particles of dust.

488. Seistan   Another name for the wind of 120 days or the Bad-i-sad-o-Bistroz which blows strongly from the north in summer over the Seistan Basin in eastern Iran. It sometimes blows in almost continuous fashion for about four months.

489. Shamal   A northwesterly wind which blows down from the Mesopotamian Corridor over Iraq and the Persian Gulf in the summer. It often blows strongly during the day, carrying clouds of dust and sand, but decreases its intensity at night.

490. Sharki   A wind from the southeast which occasionally blows over the Persian Gulf.

491. Simmoom   One of the Sirocco-type winds, hot, dry and dust-laden, which blow over the middle and south portions of the Mediterranean. When the wind is abnormally strong in the southeast Mediterranean, it is called a Simmoom. It blows from the south, originating in the hot desert sections of North Central Africa. Also sometimes called Simoon.

492. Sirocco   A warm wind of the Mediterranean area usually sweeping northward from the hot and dry Sahara or Arabian Deserts. The Sirocco wind usually invades the Mediterranean shores in the spring when many low pressure areas move over the Mediterranean. As the low-pressure area moves, its forward part pulls air northward which originated hundreds of miles to the south. Frequently, therefore, the Sirocco is dusty. It often picks up moisture as it crosses the Mediterranean and often arrives on the north Mediterranean shores as a warm and damp wind.

493. Sno   Cold, swift-moving air currents which fill the Scandinavian valleys in the winter from the highlands, attaining considerable velocities in the fiords.

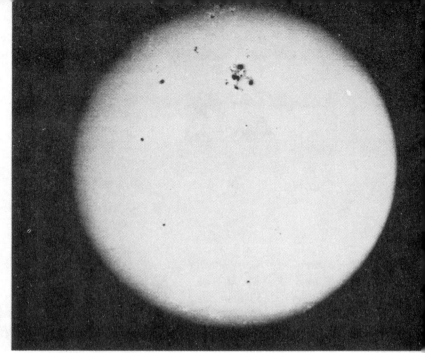

Official U.S. Navy photo

Sunspots

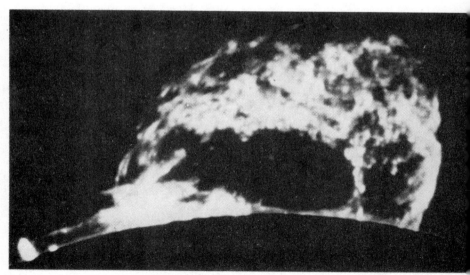

High Altitude Observatory, Univ. of Calif.

Solar prominence

STORMS ON THE SUN (Questions 14–20)

U.S. Air Force photo

The radiosonde is a tiny radio transmitter and weather observatory. As the instrument is carried aloft by a balloon, it radios back signals indicating temperature, pressure and humidity of the level through which it is passing.

U.S. Army Signal Corps photo

Weather balloon is released and will carry radiosonde equipment in little white box to about 100,000 feet above the Earth's surface.

U.S. Air Force photo

The balloon-borne radiosonde is tracked with this dish-shaped radar antenna as it rises through the atmosphere. In this manner the speed and direction of the wind at various levels above the earth can be computed.

SOUNDING THE UPPER AIR
(Questions 217, 218, 225–228)

THE WORLD OF CLOUDS
(Questions 113–133)

The Munitalp Foundation

Cirrus

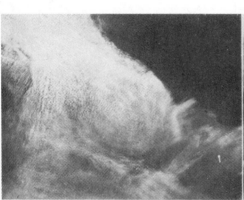

The Munitalp Foundation

Cirrocumulus

U.S. Weather Bureau, H. T. Floreen

Cirrostratus

U.S. Weather Bureau

Altocumulus

U.S. Weather Bureau

Altostratus

Stratocumulus

Nimbostratus

Stratus

Cumulus (fair weather)

Heavy cumulus

Cumulonimbus

THE WORLD OF CLOUDS (Questions 113–133)

Official U.S. Navy photo

Cumulus cloud forms from rising warm air currents over forest fire

Official U.S. Navy photo

Anvil top of thunderhead (cumulonimbus) cloud poking
above towering cumulus clouds (Questions 113–133)

Circular pattern of hurricane clouds seen from U.S. Air Force weather reconnaissance plane. In this medium altitude view, the storm center or "eye" is at upper left.

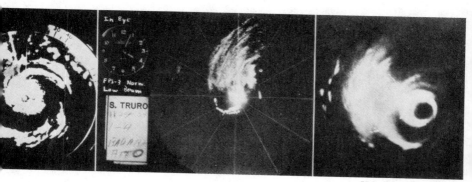

This is the way hurricanes look on radar scopes to navigators of U.S. Air Force Air Weather Service reconnaissance aircraft. (Questions 320–346, 1146–1149)

Threatening pendulous-bottomed cumulus cloud
warns of severe thunderstorm or tornado activity.

Tornado rips through Dallas, Texas, on April 7, 1957. The toll:
10 dead, more than 100 injured and 500 homeless, in addition
to millions of dollars in property damage. (Questions 348–365)

Official U.S. Navy photo
Heavy sea fog blankets USS Missouri

Official U.S. Navy photo
Seven U.S. destroyers that ran aground in fog off Honda Point, California

FOG—SCOURGE OF TRANSPORTATION (Questions 134–142)

Official U.S. Navy photo

A test atom bomb mushroom cloud at Bikini. Are atomic explosions changing the world's weather? (Questions 964, 965)

General Electric Laboratories

A historic moment in weather research. Current artificial rain-making experiments began at General Electric Research Laboratory in 1946. Dr. Vincent J. Schaefer blows his breath into a freezer, causing a supercooled cloud. A few small particles of dry ice dropped into the freezer turn the supercooled cloud into snow. Watching are researchers Dr. Irving Langmuir (left) and Dr. Bernard Vonnegut (right). (Questions 1164–1167)

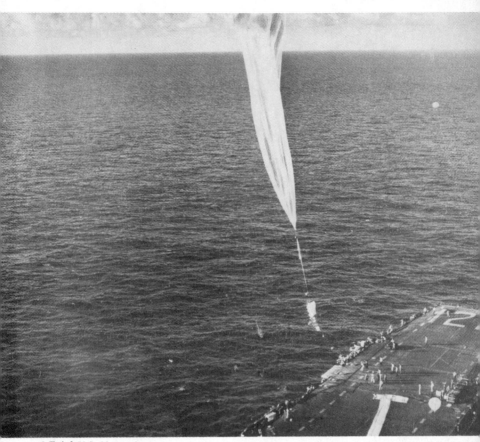

Official U.S. Navy photos

High-soaring balloons are playing a most important role in sounding our air ocean, the atmosphere. Here are photographs of the U.S. Navy's "Skyhook" balloon. (Questions 1150–1154)

Scientific assault on the upper air. Test rocket firing at Air Force Missile Test Center, Cape Canaveral, Florida. Preparation for launching of artificial Earth satellite as part of the United States participation in the International Geophysical Year. (Questions 1168–1180)

Man-made moon. Plastic model of scientific Earth satellite. Launched from 3-stage rocket, the instrument-crammed device will orbit 300–800 miles above the Earth to provide information about the frontier of space. (Questions 1181–1189)

Official U.S. Navy photo

In research gondolas like this, carried by huge plastic balloons, man is being carried higher into the stratosphere. (Question 208)

Official U.S. Navy photo

Aviation medical studies try to reveal man's limitations in today's rocket age. (Questions 910–924)

494. Solano An oppressively warm and dusty east wind which blows over Gibraltar and southeast Spain.

495. Southeaster Strong winds from the southeast, sometimes of gale force, which blow near the extreme southwest end of the Cape of Good Hope, South Africa. They usually occur in the winter and often in advance of a cold spell. Clear skies and bright sun attend the southeaster at the Cape but the surface layer of air often carries a whitish haze of salt particles and sea spray.

496. Southerly burster Cold winds which move in from the polar zones northward over Australia. The air moves behind cold fronts which are attended by strong to gale-force winds accelerating and intensifying over the highlands of New South Wales on the southeast coast. About 30 bursters a year occur over southern and southeast Australia, most of them in spring and summer.

497. Steppenwind A cold northeast wind which sometimes sweeps over Germany from the steppe regions of Russia.

498. Stikine A strong and gusty wind of the extreme southern coastal areas of Alaska near Wrangell. The wind is named for the Stikine River or Stikine Mountains to the northeast of Wrangell in Canada.

499. Sudestades (Suestado) Strong to gale-force winds from the southeast which affect the coastal area of Uruguay, Argentina and Brazil. They are part of frequent traveling cyclonic circulations in that area and are accompanied by considerable cloudiness and rain.

500. Suhaili A strong wind from the southwest which blows over the Persian Gulf, bringing thick clouds and rain.

501. Sumatra Strong thunderstorm squalls which move over the Malacca Straits from the southwest during the southwest monsoon season. The squall winds attended by severe lightning and thunder move along a front with a northwest-southeast axis sometimes 100 miles long. They last for a few hours, blowing in gale tendency gusts. They occur mostly at night and appear to be surges in the southwest monsoon given extra push by the mountain ranges of Sumatra which lie parallel to the Malacca Straits about 200 miles to the southwest.

502. Surazos   Cold polar winds of the Andes Plateau in Peru. The wind sometimes blows very strong, sweeping through the mountain passes with violence. Temperatures at such times are often below freezing and the sky clear.

503. Taku   A strong wind from the east or northeast which blows in the vicinity of Juneau, Alaska. The name is taken from the Taku River, the mouth of which is in Alaska but which flows over the Alaskan line into Canada. At the mouth of the Taku near Juneau, winds sometimes reach 75 miles an hour.

504. Tehuantepecer   A violent north wind in the region around the Gulf of Tehuantepec on the extreme south coast of Mexico. In the winter, polar winds sometimes pour southward in great surges from the North American continent (Norther or Norte). The winds roar southward across the Gulf of Mexico and reach the Gulf of Campeche in the region just southeast of Vera Cruz. The wind then enters a low-level pass existing on the Isthmus of Tehuantepec in the mountain chain of the Central American Cordillera. It is intensified by this funnel effect and pours southward out over the Gulf of Tehuantepec at gale strength. This wind is a scourge to boatmen in the gulf because there are few, if any, precursory weather signs.

505. Thalwind   A pleasant valley breeze in Germany.

506. Tramontana   A pleasantly cool wind from the north or northeast, blowing with fresh velocities in winter over the Mediterranean. The name is applied more specifically to this wind when it blows off the western coast of Italy.

507. Vardarac   Cold dry polar winds which blow in the winter over the north Aegean Sea. They blow down the Vardar River Valley in southeastern Yugoslavia from which the name Vardarac comes.

508. Vendevales   Strong to gale winds which often sweep the coast of Spain from the southwest. They are often attended by heavy rain and high seas.

509. Virazon   A regular and prominent sea breeze which blows from the Pacific Ocean to the coast of Chile. The Virazon is particularly strong on summer afternoons at Valparaiso where occasionally harbor work must be stopped. The opposite-blowing land breezes are called "Terral."

510. Waff   In Scotland, a slight puff or air or gentle breeze, similar to the Cat's paw of the United States.

511. Warm braw   A Föhn-type wind which moves from a southerly direction over Schouten Island just northeast of New Guinea in the South Pacific. The wind crosses the Nassau and Orange Mountain ranges of New Guinea which lie on a long curving west-southeast axis. Some of the mountain tops reach 16,000 feet. The air loses its moisture on the southern side of the ranges during the southwest monsoon season and sweeps northward as a warm and dry wind.

512. Whirly   A small but violent storm in the antarctic. The whirling winds may measure up to a 100 yards in diameter or more. They occur most frequently near the time of the equinoxes.

513. Williwaw   A sudden, violent squally wind of the Aleutian Mountains. The Williwaw also describes such winds in the Strait of Magellan.

514. Wisper   A well-defined valley wind of the Rhine.

515. Zephyr   A term originating in the Mediterranean regions to describe a soft, gentle breeze, especially from the west.

516. Zonda   A strong west wind which blows over the western region of Argentina. The wind is hot, dry and dusty and most frequent in the spring. It acquires its dry characteristics from the Föhn action as it descends to Argentina from the Andes Mountains lying to the west.

IV. THUNDERSTORMS

Introduction. Thunderstorms are so interesting in themselves, so stupendous in their development and their destructive effects and so illustrative of many meteorological principles, that they deserve the closest study.

Perhaps of all weather phenomena, lightning and thunder have been the most impressive and fear-inspiring from the earliest times. The jagged streaks from the heavens seemed like the fiery fingers of some angry god whose voice was the sound of thunder. The ancient Greek god Zeus was worshiped as a god who wielded thunder and lightning as well as rain. So, too, were Jupiter and Thor deities who hurled bolts from the sky. History records that spots which had been struck by lightning were regularly fenced in by the ancient Greeks and consecrated to Zeus the Descender. They believed that he came to the Earth in lightning visitations.

People still fear thunderstorms, varying greatly in their reactions. Perhaps some insight into the causes and mechanism of a thunderstorm will dispel a degree of time-ingrained fear and reveal the terrible beauty of a thunderstorm as a gigantic experiment in aerial physics.

517. Which clouds produce thunderstorms? Cumulonimbus clouds, popularly called thunderheads, produce the various weather phenomena which are characteristic of a thunderstorm; lightning, thunder, hail, squalls, etc. To be classified as a thunderstorm, the storm must have audible thunder, visible lightning, or both.

518. What are the general characteristics of cumulonimbus clouds? They are heavy masses of cloud with great vertical development, whose cumuliform summits rise in towering, mountainous form. The upper parts have a fibrous icy texture and often spread out in the shape of an anvil which frequently points in the direction of the cloud's movement. This frozen cap on a cumulonimbus is called *incus* or *cirrus nothus*. Masses of cumulus, however heavy they may be or however great their vertical development, should never be classified as cumulonimbus unless the whole or part of their tops is transformed or is in the process of transformation into a cirrus mass.

519. What weather conditions generally exist in thunderstorms? Several types of weather phenomena reach great intensity in a thunderstorm. Heavy showery rain may fall at a rate of several inches within a few minutes. Gusty winds may blow out of a thunderstorm at speeds in excess of 75 miles an hour. Violent up-and-down currents blow within a thunderstorm. Lightning flashes may occur several times per second. In violent thunderstorms, large hailstones may fall. Shifting squall winds are typical in advance of a thunderstorm and clouds of nearly every type may develop.

520. What meteorological conditions favor thunderstorm formation? Reduced to a bare minimum, two basic requirements favor the breeding of a thunderstorm. One is the presence of a warm and moist air mass which is unstable (warm in the lower levels and cool aloft). Tropical Maritime air in the summer usually provides this requirement (See Questions 287, 288). The second requirement is the existence of one or more processes which will add to the instability of the air or steepen the lapse rate. The lapse rate refers to vertical temperature change aloft. The colder the air is with gain in altitude than the air at the surface, the sharper or steeper will be the lapse rate (See Question 276). This will allow a parcel of warm air at the surface to rise vertically to great heights and to condense into a thunderstorm cloud. Instability for an air mass with high temperatures and high humidities may be brought about by strong surface heating, lifting of air over a hill or mountain (orographic lifting), or by overrunning of warm air by cold currents aloft or lifting along a frontal surface.

521. How are thunderstorms classified? Thunderstorms are of two main classifications; *air mass* or *frontal*.

522. What is an air-mass thunderstorm? Air-mass thunderstorms occur well within a given air mass, unaffected by frontal activity. They form, generally, in isolated local conditions as a result of strong convectional columns of rising heated parcels of air.

523. How does an air-mass thunderstorm form?   The first essential
for a local thunderstorm is strong heating of the land or water surface
with consequent heating of the air overlying the surface. Expansion
of the heated air and its forced ascent by the surrounding cooler air
result. The swiftly rising column of air cools until the dew point
is reached and a cumulus cloud starts to form, capping the updraft
and marking the first stage of the thunderstorm *cell*. The cumulus

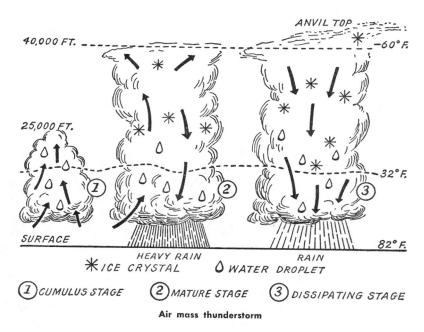

Air mass thunderstorm

cloud continues to develop into a more congested and complex form,
rising vertically just as long as the outside or *free* air is cooler than
the drafts of air within the cloud. With continued ascent of the rising
parcel and with the building of the cumulus cloud, raindrops start to
coalesce and fall. The growth of surging vertical currents is intensified
by latent heat which is released as continued condensation takes
place. Under favorable conditions, therefore, the thunderstorm grows
fast and high, reaching into the below-freezing levels where ice
crystals, supercooled water droplets and water droplets begin a

mixture. The full-grown thunderstorm, capable of producing the various phenomena associated with clouds of such structure and depth, can tower six to eight miles above the earth with a diameter at its ragged base about four or five miles. In this mature stage, the original single thunderstorm cell has developed and budded into several cells, each merging and fusing into cloud cauldrons of violent vertical winds and containing the ingredients for the formation of electrical phenomena.

524. When do most air-mass thunderstorms form? In the Temperate Zone, they are common only during the warm season. They occur over land when the surface is most heated and maximum convection occurs—usually about from two to four P. M. Over the ocean, the favorable time for convection occurs between midnight and 4:00 A. M. so that thunderstorms tend to form over oceans in the early morning hours. Land thunderstorms tend to start dissipating in late evening or after sundown when surface heating lessens.

525. Where do air-mass thunderstorms form most frequently? They occur most frequently during the rainy season of the tropics where warm, moist and unstable air masses prevail. They are often a daily and regular occurrence in the tropics. They seldom occur above 45 degrees latitude where intense heating and unstable air are lacking. In the Temperate Zones, air-mass thunderstorms occur most frequently in the late spring and summer when considerable invasions of tropical maritime air take place.

526. How does a frontal thunderstorm start? When a wedge of cold air moves into a region of warm, moist air behind a cold front, the cold air acts as an inclined plane, with the result that the warm air is *mechanically* forced aloft and then undergoes the process of cooling, growth and structure that finally produces a thunderstorm. Such storms may extend as a line of thunderstorms hundreds of miles in length, traveling along the cold-front line. In general characteristics, air-mass and frontal thunderstorms resemble each other. A big difference is that frontal thunderstorms need not necessarily occur at the time of maximum thermal convection although surface heating may serve as an additional trigger.

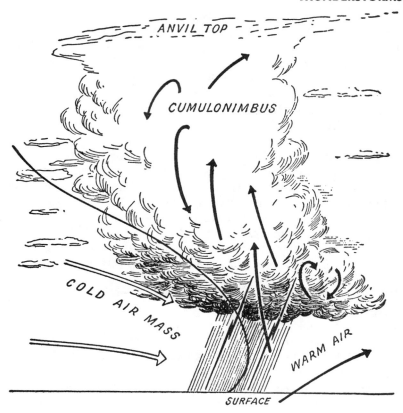

ANVIL TOP

CUMULONIMBUS

COLD AIR MASS

WARM AIR

SURFACE

Frontal thunderstorm

527. What is the nature of winds within a thunderstorm?   Violent up-and-down drafts of air and gusts are characteristic of thunderstorms. Strong initial buoyant currents of rising air which feed into a thunderstorm cell reach velocities which sometime exceed 80 feet per second. These columns of air reach their greatest speeds at altitude levels of about from 10,000 to 30,000 feet with velocities observed of about 30 feet per second at heights of 50,000 to 60,000 feet. The horizontal extent of the draft columns may range from about 1,000 to about 25,000 feet. Correspondingly swift air columns sweep downwards in cold drafts caused by frictional effects of falling water drops or ice. Evaporation adds to the cooling so that they reach the surface of the ground and spread forward and outward as cold gusty winds. Temperatures are low in these gusts and are the reason for the sharp relief from summer heat as a thunderstorm approaches. The turbulence within the thunderstorm is severe and dangerous as the result of the simultaneous strong up-and-down air movement.

528. What is the thunderstorm squall head? Strong updrafts in the forward part of the thunderhead and cold downdrafts produce a roll cloud which is attended by strong shifting winds just in advance of a storm. As the thunderstorm approaches, the winds blow *toward* the storm. As the roll cloud passes, squall winds, gusty and cold, sweep forward and *out* of the storm—sometimes in gusts of about 60 miles an hour.

529. What is the pre-cold-front squall? It is a line of gusty shifting winds and heavy showers or thunderstorms which frequently move well in advance of cold front thunderstorms. They might be caused by a cold air tongue pushing out from the main cold front line and causing a convergence with warm air far in advance—forming a "pseudo cold front." A typical pre-cold-frontal squall may last from about 12 to 24 hours, moving about 150 miles ahead of the cold front.

530. What is the rainfall pattern in a thunderstorm? The average duration of thunderstorm rain at a given location is about 25 minutes, although highly variable from case to case. The most intense rain falls under the core of the main thunderstorm cell within two or three minutes after the first rain from that cell reaches the ground. The rain usually remains heavy for about 5 to 15 minutes and then decreases in rate, but much more slowly than it first increased. Around the edges of the thunderstorm cell, lesser rainfall occurs. The mass of water released in a thunderstorm is tremendous. In a thunderstorm which drops about ¾ of an inch of rain over an area of about 3 x 3 miles, it is estimated that slightly more than a half million tons of water are precipitated.

531. What is a cloudburst? A cloudburst is a sudden and excessive downpour of rain. It is caused by a mass of rain held up in a portion of a thunderstorm by strong updrafts of air and then released downward in a huge concentration when the supporting air column gives way. Local flash floods are frequently caused when cloud bursts occur over dried valleys and arroyos (See Question 380).

532. How does lightning form? The mechanisms which produce lightning are extremely complicated and numerous and are still imperfectly understood. Essentially, many processes occur simultaneously within a thunderstorm cloud which lead to separation of positive and negative electrical charges. From time to time the great electrical stresses which are set up by the interaction of wind and enormous amounts of moving water drops and ice crystals are relieved by discharges of lightning. Some of the processes which lead to charge separation are friction, splitting of raindrops and the freezing of water or the melting of ice.

533. What is the rate of electric current flow in a lightning bolt? The amperage is enormous. Peak values as high as 200,000 amperes have been recorded. Scientists at the University of Pittsburgh trapped a superbolt consisting of five separate charges, at least one of which contained an estimated 345,000 amperes of electricity—enough to service about 200,000 homes! One measured thunderbolt had a current which totaled more than 160,000 amperes of electricity, delivered at a pressure exceeding 15 million volts. The electric energy dissipated in a thunderstorm is tremendous, frequently being released at the equivalent of a continuous expenditure of a million kilowatts. Scientists have produced artificial lightning in laboratories comparative in voltage to some of nature's strongest efforts.

534. How are lightning strokes studied? The study of the mechanism of lightning strokes to the ground has been accomplished mostly by high-speed cameras which are specially constructed to photograph and examine natural lightning in the wink of an eye as it strikes. A particularly ingenious optical device is a rotating camera called the *Boys camera* named after its inventor, Sir C. V. Boys. Engineers use an instrument called a *fulchronograph* to measure the power of lightning strokes. It applies the principle that hard steel will be permanently magnetized by current passing through adjacent coils.

535. How do lightning strokes travel?   The discharges of lightning which relieve great electrical stresses set up by a cloud usually consist of a number of separate strokes which are preceded by a *faintly luminous streamer* or *leader*. From one of several regions or cells located in the negatively charged base of a cloud, the pilot leader forces a channel a few inches in diameter earthward. This leader of the first stroke does not strike earthward in a single shaft but burrows downward in a series of steps, each about 150 feet long. Between each step the streamer pauses for about 50 microseconds. Additional

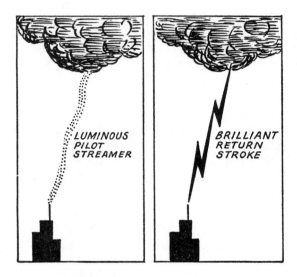

LUMINOUS
PILOT
STREAMER

BRILLIANT
RETURN
STROKE

streamer surges help the leader along called *step-leaders* which pulse down to the advancing tip in rapid succession. The combined effect ionizes, or makes electrically conductive, a pencil-thin path through the air.

When the pilot streamer reaches the earth, an intensely brilliant flash we call lightning surges *back along the path* created by the streamer. This is called the *return* stroke, from ground to cloud, from positive to negative. So swiftly does it move that the human eye is confused and sees the stroke as coming downward from the cloud. Discharges of lightning from cloud to cloud or within a cloud do not appear to have return strokes.

536. What is the speed of lightning? The faintly luminous leaders or streamers burrowing earthward as described in the preceding question travel about 2,000 feet in about $\frac{1}{100}$ second. The brilliant discharge of lightning which answers back through the ionized path created by the leader travels at the rate of about 100 million feet a second or about 100,000 times as fast as sound.

537. How far do lightning strokes carry? They flash through perhaps a few thousand feet and ordinarily last a few millionths of a second, made up of several pulses following each other in rapid succession. They come so close together that the eye cannot always distinguish them. These separate pulses make lightning flashes seem to flicker at times. It has been estimated that an average of 10 lightning strikes occur per square mile in the United States each year. Most lightning flashes are from one cloud to another. Only a small part of them reach the earth.

538. Can lightning strike twice in the same place? The old adage about lightning never striking twice in the same place is untrue. Lightning is not limited to a one-bolt action. Many lightning flashes are of the multiple variety and may strike repeatedly in the space of a few seconds. As many as 22 successive strokes have been observed in a multiple flash. The top of the Empire State Building tower in New York often intercepts several bolts during a severe thunderstorm.

539. How are lightning discharges popularly described? Names like *streak, bead, ribbon, fork, heat, sheet* and *ball* lightning have been used to describe various observed forms of lightning discharges.

540. What is streak lightning? This is the most common type observed—the ordinary bolt as described in preceding questions. It appears as a rather sinuous shape although commonly drawn as zigzag. Its shape is very much like the outline of a river through uneven broken country. It may be a single streak but most of the times it is split off into smaller jagged branches which seem to terminate at times between cloud and ground.

541. What is bead lightning? Bead lightning is a form of streak lightning. The appearance of beads may be caused by variations in the luminosity along the channel of the stroke which takes on a brushlike discharge quality.

542. What is ribbon lightning? Ribbon lightning is streak lightning with multiple discharges where the channel is blown sideways by the wind.

543. What is fork lightning? The term forked lightning is used to describe strokes which seem to have several apparently simultaneous paths to the ground.

544. What is heat lightning? Heat lightning is a form of streak lightning far enough away in distance so that thunder is not heard.

545. What is sheet lightning? It is a diffused, glowlike lightning, whitish in color, extending in clouds over a considerable area between the upper and lower atmosphere. The wide distribution of sheet lightning and its relatively persistent glow mark the principal differences between it and heat lightning.

546. What is ball lightning? Ball lightning has been described as a luminous ball, somewhat reddish in color with an average diameter of about four or five inches which some observers have seen emerging from clouds at several hundred feet per second, floating horizontally through the air or moving along the ground in stalled motion. Only visible briefly, it either disappears quietly or with a loud explosion. There is considerable controversy about ball lightning. Some authorities believe it to be an optical illusion created in the retina of the eye by retention of vision of a heavy lightning discharge. Some apparently well-qualified observers, however, have reported its occurrence. Photographic evidence has not been revealing.

547. What color is lightning? Lightning is whitish in color, having the combined spectrum of oxygen and nitrogen. It sometimes appears differently colored against different backgrounds and surroundings. In contrast to yellowish artificial lights, lightning may appear bluish or the reverse if lights are blue-toned. When very moist air is ionized so as to produce the spectrum of hydrogen, lightning flashes may seem somewhat reddish.

548. How many fatalities occur in the United States because of lightning? Approximately 500 people are killed each year in the United States. The number of injured by lightning effects is about 2,000 persons a year. The probability of death or injury from lightning is far greater in rural areas than urban. Nine out of ten deaths occur in places with inhabitants of less than 2,500 because, primarily, city dwellers have steel structures which act as conductors and people do not have far to go for protection. Men and boys are the chief victims of lightning, owing to their greater participation in outdoor activities. Deaths from lightning in the United States are relatively five times as frequent among males as among females. The highest death rates occur in the mountain states of Idaho, Montana, the Dakotas, Wyoming, Colorado and New Mexico, ranging between a high of 14.6 per million in New Mexico and 6.5 in North Dakota and Colorado. Arkansas, Mississippi, Alabama, South Carolina and Florida also have high rates compared to the national average.

549. What first-aid measures should be applied to a lightning victim? If someone is indirectly struck, he should be given artificial respiration because the diaphragm or lung muscles will probably be paralyzed by the bolt. Burns can be treated after respiration resumes.

550. Why is it dangerous to seek shelter under a tree? Trees, and particularly isolated trees, are dangerous havens in a thunderstorm. A tree cannot carry large volumes of current. The lightning bolt may jump away from it at any point toward a more desirable conductor. Because of their height, trees initially attract the bolt and then the bolt may flash sideways or, after reaching the base of a tree, may run along the ground and strike anyone in its path. Also, flying branches and splinters cracked by lightning are an additional hazard. About one third of all lightning victims lose their lives when seeking shelter under a tree.

551. What are the lightning hazards to swimmers or persons in small boats? Being in or on the water is a dangerous hazard when lightning is nearby. Persons in small wooden boats are conspicuous targets. While the chances of a swimmer being hit direct are slim, the flow of current carried by the water from a bolt striking at some distance can cause electrocution.

552. What are some other hazardous conditions? The practice of seeking refuge in small isolated sheds in exposed areas is dangerous. Small barns, wooden beach bathhouses, ticket booths, telephone booths, and similar structures should be avoided when they stand quite apart from other structures or tall trees. Carrying golf clubs on the course during a storm is dangerous or, for that matter, any polelike object with metal in any open field area. Overhead wires and telephone poles should be avoided. A noted lightning expert, Dr. K. B. McEachron, has offered this philosophy about lightning: "If you heard the thunder, the lightning did not strike you; if you saw the lightning, it missed you; and if it did strike you, you would not have known it."

553. What general safety precautions should be taken in the following cases? Inside a house? Stay clear of stoves, fireplaces, attics, doors and windows. Keep out of bathtub or shower during a storm. Make sure radio or TV antenna are properly grounded. In large buildings and modern homes, very little chance exists of being hurt or killed by lightning.

554. If caught outdoors? Hurry to a sizeable building for shelter—the bigger the better. Roofs and walls of buildings usually make a better path than the human body for lightning to reach the ground. But if available shelter is small in size or if no shelter exists, lie flat on the ground. Avoid isolated trees, water towers and exposed ridges and peaks. Keep away from metal fences, wires of all kinds and metal pipes. Keep off golf courses, open beaches. Stop outdoor games. Do not ride a bicycle or a horse and do not operate an exposed machine such as a tractor.

555. In a car, bus or train? Persons in such vehicles should remain there unless they must get out, in which case touching the ground and car at once should be avoided. The reason for the relative safety of such enclosures is that the metallic frames offer the protection, in principle, of the *Faraday cage* which is the best possible form of lightning protection. The Faraday cage is a grounded shell of metal not less than $\frac{1}{20}$ inch thick completely surrounding an object and several feet away from it on all sides.

556. How does lightning affect various substances? Insecure or irregular rods or wires may be displaced, bent, burned or melted. Conductors, such as radio antenna which are insufficiently large, are usually vaporized or melted by direct strokes. Nonconductors, such as wood and stone, are subject to splintering and explosion. Lightning is terribly destructive when it strikes large conducting masses separated by nonconducting ones, such as an ungrounded metal roof on top of a wooden house. The conducting mass seems to intensify the bolt's effect with resultant great damage to the nonconducting mass underneath. Lightning causes about 12 million dollars' worth of damage to farm buildings alone each year, causes a great part of all oil-tank fires and does untold damage to valuable forests.

557. What is the principle and function of the lightning rod? The fundamental principle of lightning protection for structures is to provide means by which a discharge may enter or leave the earth without passing through a nonconducting part of the structure, as, for example, parts which are made up of wood, brick, tile or concrete. Damage is caused by the heat and explosive forces generated in such nonconductive portions by the discharge. Lightning discharges tend to travel on those metal parts which extend in the general direction of the discharge. Hence, if metal parts are provided, of proper proportion and distribution, and are well grounded, damage can be largely prevented. The lightning-rod system simply receives the lightning stroke and discharges it harmlessly to the earth.

For almost 200 years, since the famous experiments with lightning by Benjamin Franklin, the traditional type of lightning rod has consisted essentially of a metal object placed above the structure to be protected. This object, whether knobbed or pointed, is connected to the ground by wire. Another form of lightning protection is the metal frame of the modern house which conducts lightning bolts to the ground.

Power lines are protected by grounded shielding wires which are erected parallel to the lines. These lightning interceptors are called *arresters*. They range from cigarette-lighter size which protect railway signal lines and fire-alarm systems to huge 30 foot high arresters which guard generators in hydroelectric plants. The latter are earthquakeproofed by being suspended from cables above the earth.

558. How high should lightning rods be placed? In general, lightning rods should extend at least two feet above all the higher parts of a structure and should be well grounded. They should be bound to each other as well as to other large metallic bodies in the structure. For those who would like detailed specifications for the proper installation and grounding of rods, a pamphlet of nominal cost is available from the Superintendent of Documents, Washington, D.C., entitled *Farmer's Bulletin No. 1512.*

559. Does lightning produce any beneficial effects? Thunderbolts actually may do more good than harm. Nearly 16 million thunderstorms rage annually over the earth. The lightning discharges create about 100 million tons of fixed nitrogen compounds annually by the breaking down of air which is composed of, roughly, four parts of nitrogen to one part of oxygen. The nitrogen is deposited on soil and plants with the rain and acts as a valuable fertilizing agent. When men manufacture fixed nitrogen, electric sparks 15 or 20 feet long are used. Nature's spark, lightning, may be thousands of feet long. If it were not for the extremely short duration of the lightning current, it would produce immensely greater quantities of this essential agricultural chemical.

560. What causes thunder? Lightning strokes measure in length from a few hundred feet to a few miles and in thickness from less than an inch to about a foot. As the bolt streaks through the air, its intense heat causes a sudden and violent disassociation of the air molecules in its path. This sudden disruption and ionization results in an increase in air pressure along the path and causes the vibration which we hear as thunder. It is literally an abrupt explosion from end to end of the lightning path, so sudden that the surrounding air cannot ease gently away but is forced, pushed, and crowded in waves.

561. What causes rolling thunder? If lightning strikes quite closely, the sound of thunder is a single sharp crack. The ear is too deafened temporarily by this near blast to notice the different gradations of reverberating thunder from more distant portions of the bolt. Most thunder does not reach the ear as a single crack because of the speed at which sound travels. One portion of a lightning stroke in the clouds may be six seconds away, while the portion of the stroke near the earth may be only one second away. The sound, then, usually reaches the ear as a continuous rumble. More complicated and drawn out rumbles of thunder may develop when certain differences of air stratification or the presence of steep mountains or cliffs may reflect, intensify and, in general, reinforce the original sounds.

562. How far can thunder be heard? Thunder is seldom heard over horizontal distances farther than about 15 to 18 miles. If the flash occurs at high altitudes, it may not produce audible thunder at the ground. The distance to which the sound of thunder can be heard varies greatly with many factors; whether the stroke is cloud to earth, cloud to cloud, or open space; the discharge volume; height of the stroke above the surface (different air density); wind force and direction; and any other factor which bears on sound wave alteration.

563. How can the distance to a lightning flash be estimated? Light from the flash reaches your eye almost instantly. The sound of thunder travels at only about 1,100 feet per second. Start counting seconds as soon as the lightning flash is seen. Stop when the thunder is heard. Multiply the number of seconds by 1,100. The answer is the approximate distance in feet to the lightning stroke. By similar subsequent timings of the main flashes from the core of a storm, the direction and movement of the storm may be determined.

V. COLOR IN THE SKY

Introduction. We generally think of weather in terms of its more obvious effects—the blowing wind, the falling rain or snow or the sudden changes of temperature and humidity. This is understandable because these elements often directly affect our bodily comfort. There is, however, a more subtle and more gentle area of meteorology, one not usually thought of in terms of ordinary weather, yet one without which no discussion of meteorology would be complete. It includes the many colorful, sometimes startling and bizarre, phenomena in the form of glows, halos, arcs, flashes and streamers which seem to make of the sky a kind of aerial tapestry into which Nature has woven an ever-changing embroidery of color.

This part of meteorology might be called *atmospheric optics* because, for the most part, these phenomena result from an optical interplay between the light rays and the atmosphere. Dust rides the air, including pollen, smoke, salt, ice, water, swirling air currents of different densities—a mixing, twisting vat into which plunge the Sun's rays. The light of the Sun is bent, reflected, broken, absorbed and scattered by the teeming particles of air. From the effect of the atmosphere on light comes a host of strange and wonderful sights— bits of optical magic in the air which sometimes causes us to see things where they are not and even things which are not there to see!

564. What is visible light? Visible light is a form of radiant energy of a specific range of wave length in the electromagnetic scale of energy (See Question 71). Only a small amount of the radiation from the sun is visible to the human eye. It is this narrow band of energy which stimulates the organs of vision and produces sight.

565. How do light waves travel? Light travels in waves which are called *transverse* or perpendicular to the forward motion of the wave itself. The motion is similar to that in a waving flag. A ripple runs from one end of the flag to the other, but each particle of cloth moves only from side to side—crosswise to the motion of the wave itself. Transverse waves are easily set up in a rope attached at one end to a post and shaken at the other. The rope will move only up and down, but the wave impulse travels from the hand to the post.

566. How much do visible light waves measure? Light waves are the result of many millions of vibrations of electrons, parts of atoms which are in rapid motion about the center of the atom. The number of waves per second, or frequency of vibration, is determined by dividing the velocity of light (186,000 miles per second) by the length of the wave. The longest wave length of visible light is about eight ten-thousandths of a millimeter (0.0008 mm) or about three 100-thousandths of an inch (0.00003). The shortest wave length of visible light is about 0.0004 mm or 0.0001 inch. This means that vibrations which produce light waves must range between about 400 trillions and 750 trillions of vibrations per second! The eye is extremely limited in its ability to respond to waves that enter it, hence the greater portion by far of different waves coming from the sun in forms of energy cannot be seen (the short wave lengths of gamma and x rays and the long wave lengths of heat and radio waves).

567. What is color? Since all light travels at the same speed, the shorter waves will vibrate more often in the same period of time than do the longer waves. A different color sensation is the result. In physics, the color of a light is measured by its wave-length. White light can be resolved into a *spectrum* or series of wave lengths, the longest of these producing the effect of red upon the eye, and the shortest that of violet. The spectrum is continuous from end to end, but the eye divides it roughly into seven distinct colors—red, orange, yellow, green, blue, indigo and violet—which blend into one another. White light, which is the total or essence of all colors, may be separated into the colors that compose it by sending the light through a triangular prism, or any transparent substance having nonparallel sides. This separation of light into its component colors is called *dispersion*. It is a frequent occurrence in weather phenomena because of the presence of many particles in the atmosphere which have dispersing effects upon light from the sun.

568. Does color exist in objects? Color is not something which exists in an object, but is rather a *sensation* produced in the brain by light sent out from an object to the eye. Colored opaque bodies are *selective* in their reflection of light and their color is determined by the wave length or lengths that they reflect to the eye. Colored objects owe their color to the fact that they absorb from white light certain wave lengths, and reflect or rediffuse others, the latter producing a color sensation in our brain. Thus, a red object is one which absorbs all light falling on it except red, and if we examine it by light which does not contain red, as, for instance, pure green light, it will appear *black*. Transparent bodies absorb some waves and transmit others. The color of the transparent substance is judged by the wave length of the light it transmits.

569. What is light reflection?   A ray of light that strikes the surface of a dense but transparent substance like glass, water or clear ice will in part be reflected by the surface and in part be transmitted through it. Reflection is simply a process whereby light bounces off the surface at the same angle that it struck—like the carom of a billiard ball from a table.

570. What is light refraction?   Light travels in a straight line as long as the medium through which it goes is of the same substance and constant density. But when light passes obliquely (at a slope or

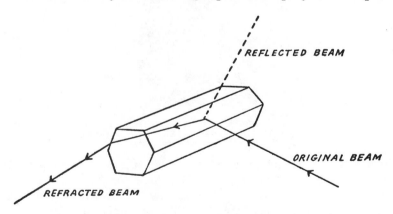

angle) from one medium to another, the rays are *refracted* or *bent* in its path. Light in the atmosphere can be refracted when it passes through layers of air of different temperatures and densities. It can be bent by tiny ice crystals in cirrus clouds which act as prisms to the white light, dispersing it into different colors of the spectrum and changing its path. Light passing through a water droplet is refracted. When white light is refracted, the blue light is bent more than the red with the colors in between bent by intermediate amounts. It is this unequal bending of light that causes the separation of colors in the rainbow and in solar or lunar halos. Refraction lies at the root of many optical phenomena in the skies (discussed in following questions). The atmosphere is usually in motion and frequently lies in layers of air of different density. This creates large natural air lenses which bend light waves in different degrees and angles.

571. What is diffraction of light?   It is the breaking up of a ray of light into dark and light bands or into the colors of the spectrum when the ray is deflected or bent at the edge of an opaque object. This curious effect may result in a portion of light illuminating a normal shadow region and causing a colored or whitish radiance to outline

the object. An example of a diffraction effect is seen when a person who is standing in front of a dazzling light seems momentarily to take on a silhouette which is outlined by a radiance. The light so deflected may be intense if the angle of deviation is small, but lessens rapidly with larger angles. Diffraction of light by water droplets in certain clouds causes a radiance around the sun or moon called a *corona* (See Question 589).

572. Why is the sky blue? The visible light rays of the sun passing through the turbid atmosphere lose a considerable portion of the blue waves, which are the shortest of light waves. The minute particles of matter and molecules of air within the atmosphere intercept and *scatter* a larger portion of the blue than any other color. By this *selective* scattering, the sky takes on its characteristic blue color.

573. At what altitude does the sky cease appearing blue when the Sun is shining? Air molecules thin out drastically with ascent. The intensity of blues and indigo tones deepens with altitude until a general region is reached at which not enough air molecules exist to scatter even the shortest blue waves of the Sun's light. At approximately 13 miles, the sky color during the day or the horizon is a whitish haze followed then by a normal sky-blue for about 25 degrees above the horizon and emerging gradually to dark blues, virtually black. Major Stephens, veteran balloonist who rose to 72,000 feet in 1935, reported the following concerning the darkening of the sky through the tropospheric depths ". . . at the highest angle that we could see it, the sky became very dark. I would not say that it was completely black; it was rather a black with the merest suspicion of very dark blue." Recently, two U.S. Navy balloonists who ascended 76,000 feet reported that "It was dark as night."

574. What causes twilight? Twilight is a period of incomplete darkness following sunset (evening twilight) or preceding sunrise (morning twilight). Shortly after the Sun sinks below the horizon or just before it rises, its rays of light are reflected from the gas molecules and impurities of the atmosphere to impart to the earth a faint luminescence. Twilight is designated as *civil* when the Sun is 6 degrees below the horizon, as *nautical* when it is 12 degrees below, and as *astronomical* when it is 18 degrees below the horizon. The broad high arch of radiance or glow seen occasionally in the western sky above the highest clouds in deepening twilight is called the *afterglow*. It is caused by the scattering effect of very fine particles of dust suspended in the upper atmosphere.

575. What is the twilight limit? The limit of twilight is the greatest height at which the density of the air still is sufficient to cause a perceptible amount of scattered sunlight. This height is about 40 miles above the Earth.

576. What causes the sky colors associated with twilight, sunrise and sunset?   When the Sun is low, its rays of light must travel a long distance through the turbid atmosphere, and it loses its shorter wave length colors of blue, indigo and violet. The longer wave-length colors, therefore, predominate during these times of day. The clear upper twilight sky then may be streaked with pale green, paralleled at lower air depths by yellowish and pink tints along the horizon. The light that nearly touches the surface is so top heavy in red tones by the time it gets halfway across that a distinct ruddy glow may mark the place where the sun just set or where it soon will rise. Reddish or pink tones of this selective scattering of light are often reflected back to the earth by clouds, particularly high ice clouds which catch the slanting rays of the setting or rising Sun and impart a great variety of crimson hues to the Earth's landscape.

577. What causes a red Sun?   The orange or red colors of the rising or setting Sun are caused by the increased length of the path traversed by its rays before the light reaches our eyes. The shorter, more refrangible rays (blue tones) become almost completely scattered by the lower atmosphere's thickened impurity-filled depths and only the red tones remain. As the Sun rises higher in the sky, its light passes through a lesser distance of dense atmosphere so that it loses its burnished redder tones.

578. Why is a low Moon often an orange color?   The conditions described in the preceding question, that cause a red Sun, affect the color of the Moon. While the Moon is low in the sky, its light must penetrate through the lower layers of the atmosphere where smoke, dust and other particles of matter cut off the shorter wave-length colors. As the Moon ascends higher in the amphitheater of the sky, the saffron colors give way and fade to a silver brightness, tesselated by the craters and plains on its surface.

579. Why does the Sun or Moon appear so much larger on the horizon?   One of the strongest of optical illusions is that the rising Moon and Sun are larger than when they are higher in the sky. The illusion is so powerful that it is difficult to accept the fact that every kind of accurate measurement shows them to be just as large high in the sky as they are when low. It apparently is a psychological phenomenon, based on a diameter we ascribe to the Sun or Moon by its apparent nearness or distance with respect to horizons, clouds, and general shape of the sky vault.

580. What is the green-flash phenomenon?   The green flash or ray is a sudden and brilliant sparkling green color from the last tip of the setting Sun. It is seen when the air is cloudless, exceptionally clear and when a distant, well-defined horizon exists as on an ocean. As the last tip of the Sun sets, its light rays are chromatically separated. As the western horizon comes up, it cuts out in swift succession (in a few seconds) the separated colors of white light—first the red and then the yellow, green and blue. The total effect being an abrupt change from white to brilliant green and sometimes to blue-green when the air is exceedingly clear. The green ray is seldom seen because dust and pollution of different kinds are usually in the lower atmosphere, especially over land. This makes the setting Sun only appear red, the color that can penetrate a dusty atmosphere far better than any other.

581. Does the image of the Sun appear in the sky before it actually rises?   It is quite true that just before the Sun actually rises, when it is still physically below the horizon, its complete image appears in the sky above the horizon. Similarly, after the setting Sun has dipped below the horizon, its image lingers on. This is another phenomenon which is caused by refraction. The Sun's light is gradually bent into a more perpendicular direction as it enters the outer air and passes through the denser air layers below. This literally means that our day is given a few minutes of more light each day from the appearance of the sky-mirrored sun.

582. What causes a halo (ring around the Sun or Moon)?   The ever-beautiful and striking rings around the Sun (solar halo) or Moon (lunar halo) indicate the presence of some form of cirriform clouds. These clouds, existing high in the troposphere (about 6 or 7 miles above the earth) where below-freezing temperatures prevail, are composed of myriads of microscopic ice crystals. These crystals act as miniature prisms to the light of the Sun or Moon, bending or refracting the light and dispersing the white light into different colors.

583. What is the most commonly observed halo? Of all the many kinds of phenomena which appear in the sky when covered with thin high clouds of ice crystals, the "22-degree halo" is the most commonly observed. It appears as a circle of 22 degrees radius around the Sun or Moon, red on the inside, followed by yellow and green, to blue on the outside. It is formed in the following fashion. Many of the tiny ice crystals are of the six-sided column variety. When the light of the Sun or Moon enters one of the side faces, it may be bent so as to come out at the second face beyond, after passing through the crystal in the shortest direction parallel to the face skipped. When this happens, the direction of the light is changed by about 22 degrees as it passes through the crystal, the red least bent and the blue the most. Frequently, the color separation is not too sharp, especially in the case of the lunar halo, and there is a diffused blending of color so that a wan ghostly whitish ring is formed. The ice crystals themselves, because of their multiplicity and vertiginious movement, may reflect a whitish scattering sheen to the sky.

584. Why do halos form with different sizes? The patterns and sizes of halos are often complicated because of the great variety of ice crystal forms which may be present in the cirriform cloud. Some crystals are six-sided columns, pyramids, plates and so on and, therefore, may bend the light passing through them in different angles. Also, they may be twisting and turning, now horizontal, now perpendicular, or they may be relatively quiescent.

585. What causes a 46-degree halo? Some ice particles may have ends which are perpendicular to the sides causing a bend of light of the Sun or Moon to a 46-degree radius circle, thus producing a proportionately larger ring effect than the 22-degree halo.

586. What is a parhelic circle? It is a halo consisting of a faint white circle through the Sun and parallel to the horizon. This is called a *parhelic* circle (from the Greek *para helios,* beside the Sun). It is caused by reflection, not refraction, on occasions when a sufficient number of brightly reflecting faces of ice crystals are present.

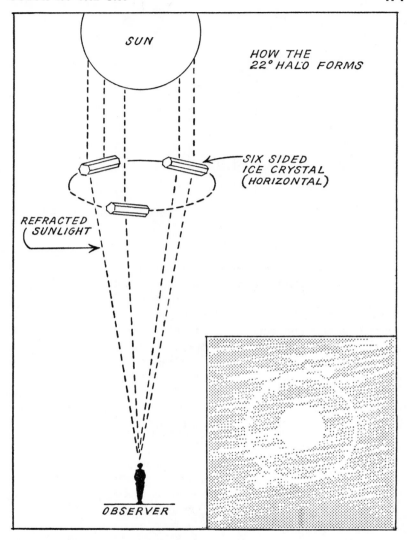

SUN

HOW THE
22° HALO FORMS

SIX SIDED
ICE CRYSTAL
(HORIZONTAL)

REFRACTED
SUNLIGHT

OBSERVER

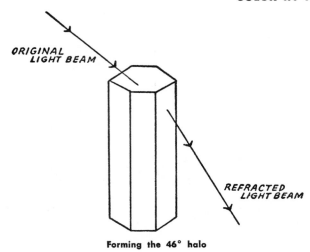

ORIGINAL LIGHT BEAM

REFRACTED LIGHT BEAM

Forming the 46° halo

587. What are mock suns and moon dogs? At times along the parhelic circle described in the preceding question two bright spots may appear as echo images of the Sun or Moon. One appears to the right of the Sun or Moon and one to the left, occurring near the intersection of a 22-degree halo with the parhelic circle. These ghostly Sun images are called *parhelia* (side suns), *mock suns* or *sun dogs*. Similar phenomena in relation to the Moon are called *paraselanae, mock moons* or *moon dogs*. Coloration is sometimes observed in a sun dog —red on the side nearest the Sun and blue on the side farthest away. They possess color because they are caused by the bending of the Sun's light by the same ice crystal faces which formed the parhelic circle.

588. What is a sun pillar? It is a glittering vertical shaft of light, white or reddish, extending below and above the Sun, most frequently observed at sunrise or sunset. It is probably caused by the Sun's reflection from the upper or lower surfaces of rapidly rotating small flat snow crystals. If the Sun is very low, the pillar may generally possess the sunset colors and show a reddish light. If a sun pillar occurs at the same time as a parhelic circle, the effect is called a *sun cross,* a rare and beautiful phenomenon which appears as a radiant cross with the Sun as a point of intersection of the arms of

Mock suns

the cross. A phenomenon similar to a sun pillar but in connection with the Moon is called a moon pillar.

589. What is a corona?   Not all clouds are composed of ice crystals. The greater majority are composed of minute water droplets. Drops of water within a cloud can cause an interference with light known as *diffraction* (See Question 571). When light is diffracted within a cloud, it is as though the light pours around each water droplet, highlighting its outlines and illuminating the region behind the droplet where shadow might normally be. This effect, plus refraction of light, produces a series of encircling colored rings around the sun or moon called a *corona*. It surrounds the luminary body closely with a diameter of perhaps a few degrees inversely determined by the size of the water droplets within the cloud—the smaller the droplets, the larger the corona. Colorwise, there is a reversal of the halo. The corona exhibits a reddish color on the outside and blue on the inside. Occasionally, two sets of rings may develop when droplets are exceptionally small. The corona is often confused with the halo. It is easy to distinguish the difference if it can be remembered that in the case of a halo there is a distinct space separation between the sun or moon and the ring. The corona *hugs* the sun or moon. Of the two types of coronas, solar or lunar, the solar is usually the more vivid and best observed with strongly filtered glasses. When the lunar corona is poorly developed, only an *aureole* is visible—a blue-white disk with a dull brown rim.

Corona

590. What circumstances must exist for the formation of a rainbow?   Perhaps the most familiar and one of the most beautiful examples of sky color is the rainbow. The ingredients requisite to producing this multicolored, arched ribbon are: (1) the presence of water droplets, preferably heavy rainshower clouds; (2) the Sun must be shining; and (3) the observer must be between the Sun and the water.

591. What are the color characteristics of a rainbow?   When circumstances as described in the preceding question occur, two rainbows may be seen in the glistening light-refracting shower cloud—an inner (primary), and an outer (secondary) bow. Each appears as a group of tightly adjacent arcs containing the primary colors and a common center to which a straight line drawn from the sun would pass through the observer's eye.

When two such bows appear in the opposite side of the sky from the Sun, the inner or primary one is the much brighter of the two, having an outside rim of reddish tones blending through parallel bands of yellow, green and blue to the inside in that color order. The secondary rainbow contains a reversed color scheme with blue on the outside rim and red on the inside.

592. What are supernumerary bows? While two rainbows are ordinarily seen, there are additional ones called supernumerary bows —arcs that are inside and parallel to the primary bow. They seldom show more than two colors; red and green. As the distance from the main bow increases, the supernumerary arcs become closer together and fainter in intensity. There are usually from one to four of these attending inner bows. They may also be found by the keen observer as companions to the secondary bow, but on the convex or outerside.

593. What is the size or length of the rainbow arc? The altitude of Sun determines the amount of rainbow arc which will be visible. If the Sun is low on the horizon, either just risen or about to set behind the observer, the rainbows will appear as half circles. The radius of the primary arch will be 42 degrees, with a 51-degree radius for the secondary bow. It follows that if the Sun is higher, a smaller arch of the bows will be seen until, when the Sun reaches an altitude of 42 degrees, only the very top rim of the primary bow can be seen on the horizon. If the Sun climbs to an altitude of 51 degrees, only the top of the secondary rainbow will be visible. This is why the observer in temperate or tropical areas will look in vain for a rainbow in the sky when the Sun is high at mid-day—higher than 51 degrees above the horizon.

594. Can circular rainbows be seen? On a small scale, completely circular rainbows can be seen horizontally as, for example, on an expanse of grass lawn containing heavy dew. A rainbow thus formed is called a *dew bow*. It is entirely possible, also, for an observer in an airplane to see brilliant circular rainbows riding the tops of clouds below.

595. What causes the rainbow? Rainbows result from the refraction and reflection of sunlight by water droplets. The primary bow is caused by *two* refractions and *one* reflection as light passes through the miniature prism of each drop. The ray is refracted as it enters the drop, is reflected from the drop's opposite side and is again refracted as it leaves the drop and passes to the observer's eye with a 42-degree radius direction change. The secondary bow is caused by an *extra* internal reflection of the drop and, therefore, is paler than the primary bow. This extra reflection also causes a different change of the direction of the emerging ray to a 51-degree radius.

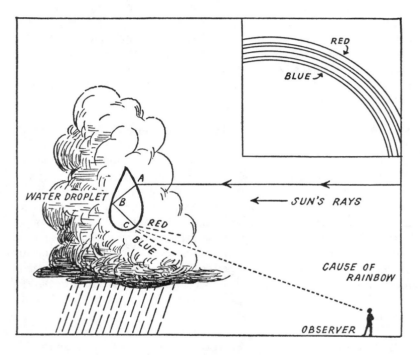

Larger droplets, measuring perhaps $\frac{1}{25}$ inch or larger, form an excellent stage and backdrop for well-defined rainbows. In condensed air where droplets are much smaller, about $\frac{1}{100}$ inch in diameter, the coloration and general definition of the rainbow become poor until, finally, in the case of fog, where droplets may measure about $\frac{1}{500}$ inch or smaller, the rainbow deteriorates to a pale, colorless *fogbow* —a broad, faint white arc.

596. Can a rainbow form at night? Rainbows may form at night with the Moon replacing the Sun as the luminary. When this occurs, it is called a *moonbow*. They have a paler and more subtle blend of colors than the daytime rainbow and are not so easily observed. This is understandable when the dimmer light of the Earth's satellite is considered. On fine, clear, full-moon nights at Niagara Falls, visitors are sometimes treated to arcs of moonbows forming delicately on the misty spumes carried up by the cataract.

597. How far away are rainbows? It is like asking how far away is the constellation of Orion, the Hunter. The stars forming the pattern of Orion are each separated by vast distances, nearer or farther in relation to the observer. So, too, with the rainbow in its entirety. It is as far and as near as the water droplets themselves, spaced dimensionally from drops nearby to those further away. A rainbow may exist in a small spray of fountain a few yards away or arch grandly against a thunderhead some two miles distant.

598. Is it possible to go under the span of a rainbow? The bridge shape of a rainbow seems to lead to this question. The answer must be in the negative, since the very formation of a rainbow implies that the bow is always in front of the observer who has the Sun at his back. No other position is possible. If one reaches the actual water droplet area in which the rainbow has been seen, it will vanish.

599. Can two people ever see the same rainbow? Two people may both see rainbows together, but what they see will not be the same rainbow. The bow is born of reflected and refracted light from different water drops. It is not a material thing fixed in position and visible three-dimensionally. As the eye changes a focal or line-of-sight position, the rainbow will undergo an alteration. Since the eyes of two individuals never coincide, they cannot observe the same rainbow.

600. What is the Brockenspecter? This weird and impressive sight is named for the highest peak of the Hartz Mountains, from the summit of which the phenomenon can be observed several times a year. It has also been seen in many other mountainous regions including the Adirondacks.

When one is standing at the top of a hill or peak in the early morning or late afternoon with the setting or rising Sun in a low position behind him, his shadow may be cast in huge proportions against

a bank of clouds, mist or fog before him. The distorted shadow reflected on the clouds may assume grotesque shapes and motions and
be surrounded by concentric rings of color called *glories* or *brocken-
bows.*

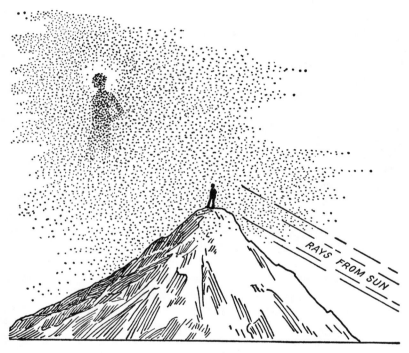

Brockenspecter

In western China, devout Buddhists by the thousands visit a mountain called Gin Din, the Golden Summit, to see what they call Bud-
dha's Glory, a particularly colorful type of the Brockenspecter. The
modern counterpart of this phenomenon is observed by aircraft passengers who may occasionally see their planes reflected on a cloud
deck below as a swiftly moving rainbow-circled shadow.

601. What is a mirage?   It is an optical phenomenon in which objects appear distorted, displaced (raised or lowered), magnified,
multiplied, reflected or inverted, due to varying atmospheric refrac-

tion of light when a layer of air near the Earth's surface differs greatly in temperature and density from surrounding air.

602. How are mirages classified? In general, there are two types of mirages: *superior* and *inferior*. The superior mirage relates to the appearance of an object as if seen in a horizontal mirror overhead and the inferior mirage pertains to the reflection of an object as if seen in a horizontal mirror beneath the observer.

603. What atmospheric conditions favor the formation of mirage appearances? When intense surface heating occurs during a hot, clear day, as over a desert, the heated ground surface will impart very high temperatures to a narrow air layer immediately in contact with the Earth. A marked difference, therefore, will exist in temperature (and density) between the bottom layer of air and the air layer above. At times this profile is reversed. On a clear night over a desert region, for example, the Earth radiates its daytime absorbed heat quickly and cools rapidly. The air layers just overlying the land are then cooled by contact. At these times, it is the higher levels of air which are warmer. This reversal, warm air over cool, is called an *inversion*. In both instances, the differences of densities of the different air layers produce an *aerial lens* capable of refracting and reflecting light waves in a manner similar to glass lenses and mirrors. An inferior mirage can be caused by the first-mentioned circumstances, intense heating in a lower air level. A superior mirage may form when a sharp inversion exists, that is, when a warm layer of air rests atop a cool layer.

604. What causes the lake mirage on a desert? One of the most common examples of an inferior mirage is the lake mirage over a desert area on an extremely hot day. Light rays from the distant horizon never reach the eye. They are refracted upwards by the surface area of extremely hot air just above the sand. The horizon appears amazingly foreshortened or contracted. The effect on an observer is an isolated desert area with a shimmering expanse stretching out where arid sand should be. The eye is not accustomed to seeing the sky below eye level, except when it is reflected in water. Hence the eye will interpret the shimmering expanse as a lake, an effect which is heightened by the quivering of the superheated air over the desert.

Inferior mirage

605. What other appearances may be created by the inferior mirage?   One strange deception, caused by the contraction of the horizon because of light bending upwards from heated surface air, is a *sinking* appearance or blind spot area. For example, if the observer watching a lake mirage has a companion who walks toward the mirage, the companion will appear to be gradually disappearing or sinking into the lake as if he were being swallowed up. As he walks along, his feet disappear first, followed by more of his body until his image cannot be seen at all beyond the restricted area caused by light refraction.

Under some conditions of calm and steady air, an inverted image of the man's head and shoulders may join on to the visible portion sticking above the horizon. The lower image may then seem to writhe or be distorted in various ways.

Similarly, very distant objects in deserts, like hills or mountains, may show this sinking reflection and will appear at times to be reflected in water below or to be floating with an inverted attached image in the air. The distortion often elongates the image of a mountain into a sausage-shaped floating island in the sky.

606. What are the effects produced in a superior mirage?   The inferior mirage causes the apparent contraction of the horizon and sinking effect due to a narrow heated layer of air at the surface. The superior mirage, on the other hand, causes a *looming* of objects. It has the effect of increasing the horizon. When a sharp inversion exists (a cold air layer at the surface with warm air aloft) the light from images which may actually be *beyond* the horizon is bent back toward the observer by the warmer air layers aloft. At such times he sees an object some distance off as it ordinarily appears and, at the same time, also sees just above it its duplicate turned upside down. It is as if the warm air acts like a giant curved celestial mirror reflecting a double image from objects on the far horizon or below the horizon. This looming effect frequently has been noticed over sea surfaces on warm days in the spring when water temperatures are cold. Inverted ships, masts down, appear to be floating in an aerial sea.

Superior mirage

607. What is the Fata Morgana?   It is a form of superior mirage, usually referring to the appearance of inverted images of ships in the neighborhood of the Strait of Messina. The name is derived from Morgan le Fay who was fabled by the Norman settlers in England to dwell in Calabria. One might wonder if the expression "castles in the air" originated from the observations of superior mirages.

608. What are crepuscular rays?   They are beams of light from the Sun passing through openings (interstices) in clouds and made visible by the illumination of dust in the atmosphere along their paths. While these rays are actually parallel, they seem often to diverge. The effect is an optical illusion due to perspective which makes parallel lines seem to radiate from a single point. The name crepuscular (from the Latin *crepusculum,* twilight; dusk) has been applied because it is at times of sunset and sunrise when the atmosphere is often tranquil and dusty that these beams are most frequently observed.

Sailors once called crepuscular rays *backstays of the Sun.* Polynesian natives call them *the ropes of Maui.* In parts of England they are known as *Jacob's ladder.* Perhaps the Homeric phrase *rosy-fingered morn* refers to these rays. They most frequently occur in assotion with dissipating banks of cumulus clouds on calm summer evenings. The cloud types are called *vesperalis.*

609. What are nacreous clouds?   They are luminous and iridescent clouds occasionally seen in the stratosphere about 14 to 19 miles above the Earth's surface during twilight while the observer is in the Earth's shadow. Reflected sunlight makes the clouds luminous and its colors, which are pure and striking, are caused by diffraction of light. The beautiful colors in nacreous clouds, including rose, blue, green and blue-violet, with red tints predominating, have prompted the name *mother-of-pearl* clouds. Although strong evidence exists in favor of these clouds being composed of water droplets, their composition still remains a mystery. The problem to be solved is how the water vapor can reach such great heights and retain its water form.

610. What are noctilucent clouds?   Noctilucent clouds are the highest of all observed clouds. They are faintly luminous when the Sun is a short distance below the horizon and are illuminated by the Sun. They appear at about 50 miles above the Earth's surface and are seldom seen higher than about 10 degrees above the horizon. The clouds apparently consist of very fine dust particles, at one time thought to originate solely from volcanic dust thrown by violent eruptions very high into the atmosphere. Another cause now generally accepted is that they are collections of extremely fine meteoric dust.

611. What is the St. Elmo's fire?   It is a luminous brush discharge of lightning from objects charged to such high potential that the resistance of air breaks down. This spearlike or tufted flame often occurs during thunderstorms from high-pointed objects such as masts of ships, lightning rods, steeples, mountain peaks, tree branches, chimney tops, etc. St. Elmo's fire is also known as corposant or ghost from the Latin *corpus sanctum,* body of a saint. The name originated with sailors who were among the first to witness the displays on the tips of masts and yards of their ships. St. Elmo is the patron saint of sailors, so when they saw the flames, they named the fire after him.

612. What is the light of the night sky? In the absence of the Moon and artificial light, the sky is capable of emitting a measurable luminosity which is called the light of the night sky. There are several sources of such light: radiations from atmospheric gases at high altitudes, the auroral glows, cosmic dust particles, the zodiacal light, and stars and nebulae. Studies of this light of the night sky are important to meteorologists because it yields information about the composition and nature of the high atmosphere.

Modern studies have included the release of chemicals such as sodium vapor from rockets at high altitudes at night. The nighttime glow caused as a result can be studied so as to yield information about the movement, temperature and density of the upper air.

613. What is the zodiacal light? The zodiacal light is a cone of faint light in the sky which is seen stretching along the zodiac (the band in the sky along which path the Sun, Moon and planets move in their courses). It is seen from the western horizon after the twilight of sunset has faded, and from the eastern horizon before the twilight of sunrise has begun. In Temperate Zones it is best seen from January to March after sunset, and in the autumn before sunrise. In the tropics it is seen at all seasons in the absence of interfering moonlight. The light is generally fainter than the light of the Milky Way.

The zodiacal light still remains a mystery. One older theory states that the light may be attributed to sunlight scattered from a lens-shaped disc of dust particles lying in a plane well beyond the orbit of the Earth. A more recent theory attributes it to sunlight absorbed and re-emitted by the rare ionized layers of the upper atmosphere.

614. What is a shooting star? Shooting stars neither shoot nor are they stars! Our upper atmosphere is continuously bombarded by meteoric fragments of matter, particles generally no larger than grains of sand. They enter the Earth's atmosphere at speeds of from 10 to 20 miles a second. As they plunge through the air layers, their surfaces are heated by friction to incandescence and so they leave luminous thin streaks before they are completely disintegrated. Most

cannot be seen. Faint shooting stars are common. On a clear moon-
less night an observer may see five or ten in an hour. When the particle is a bit larger, it can cause a brilliant flare and is then termed a
fireball. An exploding shooting star is called a *bolide*. Some large
ones reach the Earth with explosive force and are called *meteorites*.

615. What causes the aurora? Giant explosions on the Sun and
other forms of solar activity cause the auroral glows termed *aurora
borealis* in the Northern Hemisphere and *aurora australis* in the
Southern Hemisphere. These solar storms emit streams of electrically
charged particles which bombard the ionized, highly rarefied gases in

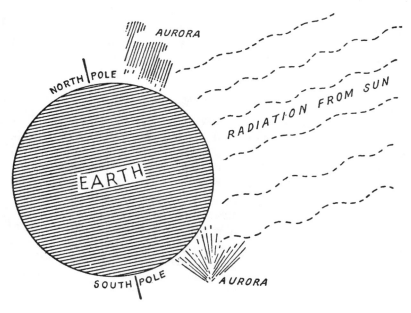

the upper air and which, in turn, are excited to glowing radiance.
The effect is something like that which occurs in commercial fluores-
cent lighting. Red neon lights shine, for example, because electrons
are driven through the gas in the tube, smashing into the neon atoms
with sufficient force to cause a glow of radiation. In auroras, it is the
oxygen and nitrogen molecules of the upper air which are excited to
radiation by the Sun's discharges.

616. When are auroras observed most frequently? They seem to be at their brightest and are most abundant when an unusually large sunspot directly faces the Earth. There are frequent strong magnetic storms on the Earth at such times as well as severe disruption or fade-out of radio and telegraphic transmission emphasizing the electro-magnetic character of the aurora.

617. At what heights do auroras appear? Dr. Carl Stormer of Norway has spent many years in auroral study, and with the aid of a crew of observers has determined by means of simultaneous photo-graphs from various points the heights at which auroras occur. They glow at two general atmospheric levels: the lower zone of activity between 50 and 200 miles; and the upper, between 350 and 630 miles above the Earth's surface. The higher ones are rarely seen in fairly low latitudes. It is interesting to note that the more common, lower-zone auroras appear at about the same heights as the radio-reflecting layers in the ionosphere (See Question 57).

618. Why does the aurora borealis occur most frequently in northerly latitudes? The luminous streamers of the northern lights (another name for aurora borealis) are caught in the great magnetic field which surrounds the Earth. They follow the lines of the Earth's magnetic forces toward the north magnetic pole which, incidentally, is about 20 degrees from the geographic pole. The auroral activity is concentrated, therefore, in a great elliptical ring around the area of the north magnetic pole. The aurora area extends from about 70 to 60 degrees latitude around the earth. Observers in this belt—Iceland, Greenland, Hudson Bay, Alaska, the Bering Strait and across northern Siberia to the Scandinavian countries—are favored by their location in the magnetic belt to see auroras almost two nights out of every three. Moving southward, this frequency breaks down so that an aurora over the southeastern United States or over southern European countries would be unusual, and as far as tropical regions are concerned, very rare.

619. How is the appearance of the aurora classified? The colored streamers, glows, arcs, bands, rays and curtains of the aurora stagger the imagination by their multiplex array. Generally, the aurora forms are divided into two main classes: auroras without ray structure and those with ray structure. Another classification has been suggested, flaming auroras.

620. Which are the most commonly observed types of auroras?
The most common, quiescent and stable ones are contained in the forms without ray structure. These are the diffuse arcs, bands and glows, usually exhibiting a considerable homogeneous character. Wavelike, luminous pulsations unfurl in irregular motion across the sky. Sometimes a feeble glow, like dawn, appears usually in the northern sky as the upper portion of an arc.

621. What is the characteristic appearance of the aurora with a ray structure? The ray structure type of aurora presents a more startling form and a more vivid range of color. Rays and streamers may seem to fan out toward the zenith from arched bands above the horizon. Sometimes dominating the sky in bursts of long brilliance, they then may appear to shrink and pulsate with varying degrees of rhythmic oscillations.

622. What are the colors of the aurora? There is hardly a color in the spectrum that does not add its beauty to the auroral phenomenon. In cases of a faint aurora, it is usually white, and as the brightness intensifies, golden yellowish tones predominate. Delicate pastel tints—cerise, coral and turquoise—sometimes burnish the drape-type aurora. In the brightest auroras, crimson tones prevail.

623. Does the aurora make any noise? There are some observers in the arctic and antarctic regions who, while watching the aurora, claim to have heard a crackling or swishing noise. It is extremely doubtful that these radiations produce any sound waves, occurring as they do at altitudes where the air is so thin as to hardly be able to propagate sound waves. Some have suggested that such noises are due to disturbances such as brush charges occurring on certain portions of the Earth's surface in dry cold air not too far from the observer. Others have suggested that imagination helps the observer to hear a rustling sound, especially when the aurora has a curtain or drape appearance. It remains an unsolved controversy.

624. What causes the stars to twinkle? Air segments do not always lie in even, large-scale layers. Small, shifting air parcels, acting as many concave and convex air lenses, add to the dispersion, reflection, magnification, and bending of light. It is not by mere coincidence that the world's great telescopes are placed on mountain summits which shaft through much of the denser, distorting air elements and even so, must contend with a warping of light when trained on celestial objects. Starlight, after having moved in straight lines through the deep of space, is affected so by the variations of atmospheric densities, that the phenomenon known as *twinkling* or *scintillation* occurs. It is based upon the simple fact that the arriving light of a star may alternately miss or strike the pupil of the eye. The changes in position are caused by curvature in the light rays through different layers of hot and cold air, both of which are always present in the atmosphere.

Scintillation is least near the zenith. The closer the stars are to the horizon, the more they scintillate because they are then seen through a thicker layer of air and, therefore, through more strata of air. Brighter stars in the lower portions of the heavens may seem to change color because the lower depths are densest and hence result in greater refraction and separation of color. Tennyson, admiring this colorful scintillation of the star Sirius, always low in the latitude of England, wrote:

> ". . . fiery Sirius alters hue
> And bickers into red and emerald . . ."

VI. CLIMATE AROUND THE WORLD

Introduction. Of all the branches of meteorology, climate probably has met with the greatest amount of indifference from the layman or the beginning student. Too often it is presented as a phalanx of grim statistics. Yet, with the application of a little understanding, patience and imagination, climate becomes a clue to the history of man.

The influence of climate on the Earth—human, animal and plant —can hardly be overemphasized. Historians, anthropologists and archaeologists have consistently seen evidences of the linkage of climate to man's mental and spiritual development and to the rise and decay of civilizations. Climate has even been thought to have molded in some ways changes of sound in certain languages during an historical period. Some phonologists think that a new cold climate encourages greater energy of articulation or causes one to keep his mouth as closed as possible whereas a warm climate leads to laziness or relaxation of the speech organs! While this is certainly debatable, it hints at some of the fascinating implications of an understanding of climate. The drama of climate does not lie with a recitation of statistics but rather with what these statistics represent in terms of man's adaptability and development in the face of an astonishing range of prevailing weather conditions on the Earth.

625. What is the difference between weather and climate? The difference is mainly one of time. Weather essentially is the existing or instantaneous occurrence of the various weather components such as temperature, precipitation, humidity, wind, pressure, cloudiness, etc. Climate (from the Greek *klima* meaning *region* or *zone*) might be considered as the history of these components reflecting their prevailing characteristics over a long period of time. Climate is thus a synthesis of the weather elements which results in a characteristic weather pattern for a given area over a long period of time.

626. How is climate presented? Standard, accurate and regular measurements and recordings of weather data at a given locality over a long period of time are made. From this collection of data, statistics are prepared in the form of summaries, tables, graphs, maps or charts which represent the different values of each weather element such as extremes, averages, totals and, in general, the behavior or range characteristics of the elements.

627. How is climate information used? Climate statistics are rarely of much use in themselves until they are *interpreted* or *digested* for a particular application or as a guidance in some specific endeavor. For example, construction engineers are very much concerned with *extremes* of weather elements which will have a bearing on the safety or effectiveness of a project. In building a dam for a water-storage reservoir, the engineer is not as interested in average rainfall statistics as much as the largest amounts of rainfall which might conceivably fall within a short period of time based on previous records. In designing a bridge, he is concerned with those weather extremes which will place the greatest stress on the structure. Climate records, for instance, will provide him with a clue to the strongest winds which may be expected.

Farmers have other special interests in climate—knowing the periods from the last killing frost in spring to the first killing frost in autumn—or knowing the ranges of temperatures which are characteristic of the region so as to provide helpful information regarding growth of crops.

The applications of climate information to various branches of industry, agriculture and other pursuits are almost endless. But, essentially, the raw climate data must be specially interpreted or tailored to the needs of the user.

628. What are the elements of climate? Elements which combine to make up the climate of a region are: temperature, precipitation, humidity, winds, pressure and cloudiness. In the case of weather, we are concerned with their instantaneous interaction. In the case of climate, we are concerned with their average occurrences over a period of time and how they create an environment.

629. Which are the most important climatic elements? Temperature and rainfall are the most important of the elements which form to produce climate. These two elements provide a climate framework of reference. They have been more carefully and frequently observed than other elements such as sunshine, cloudiness, fog or humidity. While many other elements are by no means negligible in the creation of a climate, they are not as fundamental as temperature and rainfall.

630. What determines the character of a climate region? The climate of a region is shaped by the average values of the different weather elements over the year and their distribution throughout the year.

631. How are climate values for a locality expressed? Weather elements vary constantly at any place. To choose at random the weather conditions of any particular day as being representative of the climate of a particular region for the whole year would be completely meaningless. In climatic descriptions, therefore, it is usual to state the *mean* value or average figures of temperature and precipitations. A mean value can be determined annually, monthly or daily and is used according to the desired purpose.

632. What is a mean? It is the average of a number of quantities, obtained by adding the values and dividing the sum by the number of quantities involved.

633. How is the daily mean temperature obtained? The mean temperature for a day usually represents the mean of the highest and lowest recorded temperatures within a 24-hour period, taken in the shade under certain standard conditions in order to insure uniformity. Sometimes it is stated as being the mean of 24 hourly readings. Still others are obtained by a combination of readings at various hours, according to the mount of available observations. Each method may not produce exactly comparable results.

634. How is the monthly mean temperature obtained? The mean for any month is determined by taking the separate daily means, totaling them, and dividing that figure by the number of days in the month.

635. How is the annual mean temperature obtained? The mean for a year is determined by totaling the monthly means and dividing the total by 12. To get a true picture of any mean temperature, the

data for the days, months or years should be collected over a long
period—35 years, if possible.

636. What is the annual range of temperature? The annual range
of temperature is an important clue to understanding the climate of
the region. It is the difference between the means of the warmest
and coldest months. The mean diurnal range is the mean difference
between the highest and lowest temperatures for each day in a period,
usually a month, for a series of years.

637. How are extreme temperatures expressed? In climate tables
of temperature, it is useful to indicate the highest and lowest temperatures that normally occur in each month. These values are termed
the *mean maximum* and *mean minimum* for each month. They represent the means of the daily maximum or minimum temperatures
occurring during the month. It is also useful to indicate the actual
highest or lowest temperatures which have occurred at a locality during a month. These extremes are called the *absolute maximum* and
absolute minimum temperatures.

638. How is precipitation expressed in climate totals? The average annual amount of rainfall for a given area is usually shown. This
can be misleading and inadequate. Two cities, for example, may each
have about 35 inches of rain during a year, but one may have most
of the rainfall occurring during a two-month period while the other
may have the rainfall spread evenly throughout the year. The annual
values are insufficient for an adequate appraisel of moisture conditions,
especially, as they affect crops, produce floods, etc. The *seasonal distribution,* that is, whether or not the rain falls at a time of years when
it will be most beneficial to growing crops, must be known, as well as
the *intensity* of the rainfall which determines whether the amounts
falling are absorbed largely by the soil or mainly lost into streams and
rivers, possibly producing floods; also the frequency of droughts; and
many other aspects of precipitation that cannot be determined from
annual or even monthly totals.

Useful precipitation tables and maps usually reflect these requirements and show the average annual, average monthly, and average
seasonal rainfall. A number of auxiliary maps also show many types
of variations in the amount and intensity of rainfall from average

conditions from one region to another. It should be borne in mind that the inadequacy of showing only an annual average amount of rainfall for an area extends to tables of temperature.

639. How are other weather elements expressed in climate tables? Similar to daily, monthly, and annual averages and extremes of temperature and rainfall, various corresponding values are expressed for atmospheric pressure, humidity, amounts of sunshine and cloudiness, and wind speeds and direction for different localities. When the averages of these elements are considered over long periods of time, a resulting type of climate emerges.

640. What factors control the climate?   In a large scale, the climate of a region is controlled or decided by the type of air mass which flows over the region. Locally, topography and geography are important factors in stamping the climate. The proximity of large water bodies, mountains and valleys creates many local variations of climate and weather conditions.

641. How are the different climatic regions of the world classified? Many classifications have been suggested by climatologists, based on considerations of the average distribution of different weather elements around the Earth. Some classifications include vegetation types related to different climate zones. Most classifications, however, essentially consider only temperature and precipitation. Perhaps the most widely used and best known of the many systems of climate region classification is the method devised by Wladimer Köppen based on temperature and precipitation.

642. What is Köppen's climatic classification?   The following are the different types of climate according to Köppen, with some modifications:

   A. *Tropical Rainy Climates* (Hot-rainy)
      1. Tropical rain forest
      2. Tropical savanna (dry season in winter)
   B. *Dry Climates*
      1. Steppe
      2. Desert

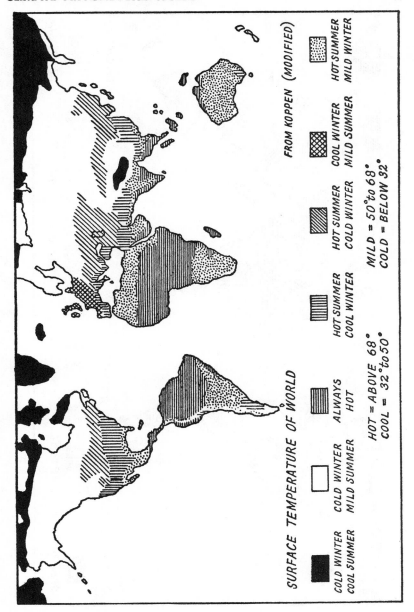

SURFACE TEMPERATURE OF WORLD

FROM KOPPEN (MODIFIED)

COLD WINTER
COOL SUMMER

COLD WINTER
MILD SUMMER

ALWAYS
HOT

HOT SUMMER
COOL WINTER

HOT SUMMER
COLD WINTER

COOL WINTER
MILD SUMMER

HOT SUMMER
MILD WINTER

HOT = ABOVE 68°
COOL = 32° to 50°

MILD = 50° to 68°
COLD = BELOW 32°

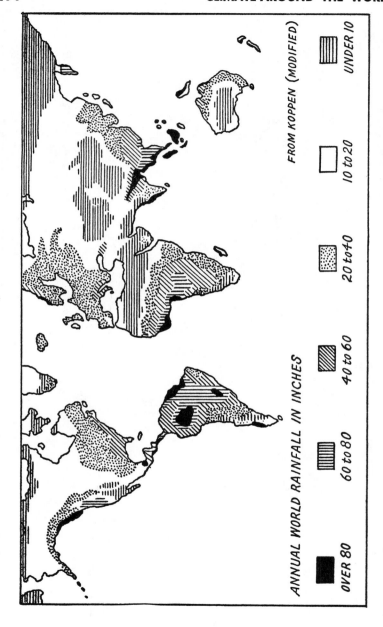

FROM KOPPEN (MODIFIED)

ANNUAL WORLD RAINFALL IN INCHES

OVER 80   60 to 80   40 to 60   20 to 40   10 to 20   UNDER 10

C. *Humid Mesothermal (medium temperature) Climates*
  1. Mediterranean climate (warm) with dry summer
  2. Humid subtropical (temperate)
  3. Marine west coast.
D. *Humid Microthermal (cold) Climates*
  1. Cold moist
  2. Cold with dry winter
E. *Polar Climates*
  1. Tundra
  2. Icecap

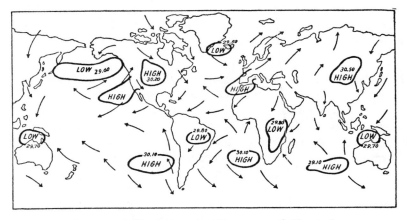

Average wind and pressure centers over earth (January)

643. What are the characteristics of the following climate types: Tropical Rainforest Climate (A1)   This climate is marked by high temperatures, oppressive humidity and rain throughout the year, typical of such doldrum regions straddling the equator as the Amazon and Congo Valleys, portions of Central America, and the East Indies. The mean temperature of the *coolest* month is 64.4° F. Rainfall, usually in the form of daily showers or thunderstorms, is heavy, averaging over 60 inches a year with the driest month having not less than 2.4 inches. Giant trees, many of great commercial value, tower above heavy undergrowth. Vegetation is generally luxuriant.

644. Tropical Savanna Climate (A2)   The word *savanna* has a Spanish or Carib Indian origin and refers to a treeless plain or relatively flat grassland area. It is typical of lands of seasonal regions near the tropics. This climate prevails in regions between the doldrums and the trade-wind belts north and south of the equator. While copious rainfall marks the wet season, there is a dry period in which at least one month has less than 2.4 inches of rain. The annual temperature range is less than 22° F. The hot rainy season encourages the growth of very tall grass, such as elephant grass, which sometimes towers to a height of 10 feet or more. Typical savanna lands form a halfway transitional stage between hot deserts and equatorial forests. Scattered umbrella-shaped trees, well-adapted to survive the dry season, give these savannas a parklike appearance. The *veldts* of south central Africa, northern Australia, the plains of central and eastern India, southern Florida and the *campos* of Brazil are typical of tropical savanna climates.

645. Steppe Climate (B1)   This is a semiarid type of climate prevailing between the savanna lands and the low-latitude deserts. Total rainfall is usually less than 20 inches a year and concentrated during a wet season which is quite pronounced from a dry season. Steppe climate vegetation is sparse, the grass being stunted and wiry and thorny bushes rather than trees being characteristic. There are great variations of steppe climate temperatures. In tropical and subtropical steppe country, the mean temperature for all months is above 32° F., typical of portions of eastern India, northwest Africa and central Mexico. In middle-latitude steppe regions such as the vast treeless plains of southeastern Europe, west central Asia and a narrow belt in the United States and Canada east of the Rockies, temperatures are frequently quite low. A fair amount of grass may exist in the spring season which dries out in the summer.

646. Desert Climate (B2)   Desert climate zones are found in subtropical and middle latitudes, generally coinciding with the subsiding and cloud-dissipating airs of the horse latitudes, about 25 to 30 degrees latitude, north and south of the equator. They usually are adjacent to the steppe regions and are characterized by high summer temperatures, large daily temperature ranges and scanty rainfall. Vegetation is sparse, usually of a cactus or sagebrush type which can exist in the hot dry climate. Representative areas are the southwest deserts of the United States, the vast Sahara Desert in northern Africa, the Arabian Desert and the interior of Australia.

647. Mediterranean Climate (warm) with dry summer (C1)   This is a fine, medium-temperature climate with the mean temperature of the coldest month remaining above 26.6° F. and below 64.4° F. The mean temperature of the warmest month is above 50° F. There is a definite dry season during the summer. At least three times as much rain falls in the wettest month as in the driest. The driest month has less than 1.2 inches of rain. The Mediterranean coastal lands, the west coast of California, the Chilean coast and portions of southeast and southwest coastal areas of Australia enjoy this equable climate which favors citrus fruit growth.

648. Humid Subtropical (temperate) Climate (C2)   Temperatures of this climate are moderate and similar in range to the climate described in the preceding question. There is no pronounced dry season and rainfall occurs throughout the year. Large amounts of cone-bearing trees and shrubs, some hardwoods and thick grasses are typical of the vegetation in this climate zone. Agricultural development is aided greatly by ample rain and the favorable temperature range. This climate type prevails over most of the eastern half of the United States, including the rich corn, wheat and cotton-producing states. Much of Uruguay, Paraguay and southern Brazil have this climate.

649. Marine West Coast (C3)   This climate has a medium range of temperature similar to the preceding two types (mean temperature of the coldest month between 26.6° F. and 64.4° F. and mean temperature of the warmest month about 50° F.). Most of the regions with this climate are coastal areas in more northerly portions of the Temperate Zone. Their positions near the oceans result in relatively cool moist summers and mild winters. Representative areas of this cool marine type of climate are the British Isles, west coast of France, coastal areas of British Columbia in Canada, south coastal sections of Chile, southeast coast of South Africa and Australia, and New Zealand.

650. Cold moist (D1)   In this climate zone, the mean temperature of the coldest month is below 26.6° F. and the mean temperature of the warmest month is above 50° F. Precipitation occurs throughout the year with no pronounced dry season. Many types of coniferous (cone-bearing) and deciduous (seasonal leave-shedding) trees and tall grasses comprise the vegetation of this climate. Fertile agricultural land lies within this zone which generally stretches over large central portions of North America, Europe and Asia. Representative areas are the north central states in the United States, central and southeast Canada, southeast portions of the Scandinavian countries, large areas of eastern Europe and central Russia.

651. Cold with dry winter (D2)   The temperatures in this climate zone are similar in range to those of the preceding question. Precipitation, however, is generally more concentrated in the summer period with severely cold and dry winters prevailing. North central Canada and the vast expanses of northeastern Asia are representative of this type of climate.

652. Tundra Climate (E1)   This polar climate of far northern latitudes has a mean temperature of the warmest month below 50° F. Vegetation is limited to stunted mosses, lichens, some berry-bearing bushes and dwarf willows. Not even the cold-resistant coniferous trees can live in this polar climate since the temperature of the warmest month fails to reach the 50° F. summer limit for tree growth. The treeless hummocky lowlands are called tundra, a word of Russian origin referring to wasteland conditions of the arctic regions. Tundra climate conditions dominate the arctic fringes of northern Alaska, northern Canada, coastal Greenland and along the entire arctic fringe of Siberia.

653. Icecap Climate(E2)   This climate dominates the regions of the polar caps of permanent ice and snow. The mean temperature of the warmest month is below 32° F.

654. How far back can the Earth's past climate types be estimated? It is not until the beginning of the Cambrian period of geological time, about 500 million years ago, that a picture of the world's climate begins to emerge. All that can be said about the pre-Cambrian period is that at intervals of a few hundred million years, glaciers or ice sheets covered various parts of the world. Of those pre-Cambrian intervening times, hardly any inferences can be drawn.

655. Has the Earth's climate always been similar to the present?
At the present time, the Earth's over-all climate is far more rigorous and marked by greater extremes of heat and cold than it has been for the last 500 million years.

656. What kind of climate has characterized the history of the Earth in the last 500 million years? A general sequence of climates during past geologic ages as developed from many lines of evidence shows a striking preponderance of warm and moderate climates. Uniform and genial temperatures prevailed. The polar regions were ice-free and even fertile. Animals ranged more widely and plants spread over wide areas which are barren today. This benign climate had intervening interludes of severe but brief climate changes in the form of glacial periods, but only representing about one tenth of the total period of time. Two typical warm periods in the geologic age are the Eocene, about 50 million years ago, and the Jurassic period, about 150 million years ago. During the Eocene, land vegetation was abundant as far north as northern Greenland. The flora included yew, pine, spruce, poplar, birch, hazel and grass—remarkable examples of flora in consideration of the barrenness of this region at the present time. The Jurassic period was marked by the existence of corals into fairly high latitudes, indicating that in 50°–60° north the temperature of shallow sea water was about 60° F., or 10° F. higher than the highest present-day ocean temperatures in those latitudes.

657. What are the ice ages? The great ice ages stand out in sharp contrast to the long mild periods of geologic time. The latest and best known is the Quaternary Ice Age which began, roughly, about a million years ago. In this last period of a million years, four major advances and retreats of great sheets of ice occurred, spreading from high latitudes southward and causing periods of great upheaval of the earth's multifarious animal and plant life.

658. How far south did the ice sheets advance? Ice sheets thousands of feet thick forged southward from Scandinavia to spread over eastern Britain. Ice from Scotland occupied Northern Ireland. In

North America the southern margin of the ice at its greatest extension was everywhere south of the Canadian border, and in the Mississippi Valley it reached almost to the latitudes of central Missouri. In Europe the Scandinavian ice extended as far south as 50° north, the approximate latitudes of northern France, central Germany and Poland. Mountain glaciers reached far below their present limits in

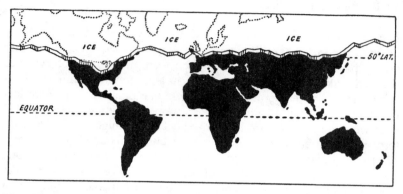

The Himalayas, the Russian mountain ranges, in the Rockies and costal ranges of North America, the Andes, New Zealand and even on the equator in East Africa. Glaciers formed on many mountains now ice-free, including those in New Guinea and southeastern Australia.

659. How many miles of area did the ice sheets cover at their maximum advance?   It has been estimated that nearly one tenth of the Earth's surface must have been covered by ice sheets. This included 5 million in the antarctic, 4½ million in North America, 1¼ million in Europe, at least as much in Asia and over 800,000 square miles in Greenland. The present ice-covered area is about 6 million square miles, almost entirely in the antarctic and Greenland. It is unlikely that the ice everywhere reached its maximum extension and thickness simultaneously.

660. When did the last ice age end?   It is difficult to state an exact period of time. Scandinavian geologists consider that the ice age ended about 6500 B. C., marking a time when the last remnants of the ice sheet split into two parts. By this time, however, most of

Europe was enjoying a temperate climate. As far back as about 20,000 B. C., the periphery of the glaciated areas of the ice ages had ended. Greenland and the antarctic, on the other hand, are still in the ice age. Climatologists disagree as to whether we are existing in the remnant stages of an ice age or in a temporary cycle of recession within the greater glacial epoch.

661. What kind of climate existed in Europe after the ice ages to the time of Christ? Several fluctuations or broad cyclical changes of climate occurred after the last maximum ice recession. From about 8300–7800 B. C., arctic climate prevailed over most of Europe, with a turn to dry and cool weather from about 7800–6800 B. C. European climate then was characterized by dry cold winters and warm summers, from about 6800–5600 B.C. Warm humid conditions marked the period from about 5600–2500 B.C. with a return to dry cool weather from about 2500–500 B. C. Cool and wet conditions prevailed in Europe from about 500 B. C. to the beginning of the Christian Era.

662. What climate changes occurred during the Christian Era? Variations in climate after the beginning of the Christian Era have been documented in considerable detail. Many fluctuations from periods of warm to cool or rainy to dry have occurred in varying degrees and through many cycles of different lengths of time. The level of the Caspian Sea, for example, reflects the many different cycles. In the year 900 it was 29 feet over its present level. In the year 1100 it was 14 feet below the present level. One outstanding climate factor was the advance of the glaciers in the Alps, Scandinavia and Iceland which began near the middle of the sixteenth century and went through six periods of advance and retreat, ending in the late nineteenth century. In Norway and Iceland the glaciers have not yet retreated to the positions they occupied in the fourteenth century. So striking was this recent glacier advance that it has come to be known as the "Little Ice Age."

663. What major affect have climate variations had upon civilization? Recurring intervals of wet and dry climates have shifted the areas of optimum production of basic food plants. These changes, from fertile areas to drought-devastated lands, have forced large migrations of people from one region to another, in some cases marking a tragic end to a flowering civilization. Some climatologists believe, for example, that the abandoned pueblos of the American Southwest represent the remains of a civilization which prospered 1,000 years ago. Increasing dry climate reduced crop fertility by the twelfth century to a point where cultivation was impossible. Drying and renewal of lakes in the Basin of Mexico have caused similarly important cultural shifts. Another example of a civilization altered by climate change is Yucatan. At present that country is covered by dense forests. It is hot and the humid climate is enervating and certainly not conducive to the development of a high culture. Yet, buried in the forests, are the ruins of magnificent Mayan cities. The Mayan culture developed to a high degree during successive centuries of relatively dry weather from about 400 B. C. to 300 A. D. Thereafter, increasing periods of rainfall occurred and gradually the Mayan culture was literally engulfed by the rapid growth of tropical forests until, finally, it succumbed to oppressive conditions of climate by the fifteenth and sixteenth centuries. Droughts of even a few years duration in recent times in the United States have caused internal migrations.

One of the most abrupt and greatest of climate changes since the end of the Ice Age occurred about 500 B. C. in Europe when a colder and wetter climate developed. Forests over a large area were killed by a rapid growth of peat. Alpine lake levels rose suddenly, flooding entire villages, and most of the mountain area became uninhabitable. Traffic across Alpine passes which had gone on since 1800 B. C. came to an end.

664. Is our present climate changing? There are regions, particularly Scandinavia, which show a definite warming trend in temperature in the last 50 years. Glaciers have retreated rapidly. Sea lanes in northern oceans are open much longer than they were. It is difficult to say whether this trend is due to a general warming of the earth or a relatively local effect such as a change in an oceanic drift. The study of all evidences involving past climate variations indicates that our present warming trend is merely an insignificant ripple of change in the many large-scale variations which have been occurring since the Earth was created some four billion years ago.

665. How can climate changes be dated over so many years?
Every large-scale climate regime leaves its identifying imprints on the
face of the Earth in a variety of ways. Studies of deep ocean-floor
cores and glacial sediments make it possible to obtain some knowledge of climate back for about a million years. The evidence of fossil
vertebrates is very important because it covers the widest possible
range of ecological environments. Vertebrates have lived at all levels
in the oceans and at almost all levels on land; in continental waters,
in dry upland environments and in the air. Periods of rain or drought
are dated also by a record of trees. Each year adds either a thicker
or thinner ring to a tree, depending on the heavy or limited amounts
of rain.

The detailed knowledge of climatic changes during the Christian
Era has come from evidence which includes fluctuations of lakes and
rivers, growth of peat bogs, succession of floras, rate of growth of
trees as shown by annual rings, advances and retreats of glaciers, locations and migrations of people (motivated by climatic changes in
many cases), literary records and old weather journals and, finally,
instrumental records.

**666. What are some theories of terrestrial causes for climatic
changes?** One theory which has run into many snags of opposition
is that continents have drifted long distances during geologic time,
causing extensive temperature changes. Another theory is based on
the occurrences of major volcanic explosions which have blown dust
into the high atmosphere, scattering sunlight and thus reducing the
Sun's radiation reaching the Earth. Periods of extensive mountain-
building which may have caused important changes of wind flows,
and differences of temperature is cited as another reason. Changes
of ocean currents which might possibly have affected polar icecap
temperatures is still another terrestrial cause offered for climatic
change.

**667. What are some theories of solar radiation causes for climatic
variations?** Since it is the energy of the Sun that fires the atmosphere
into motion and is the very basis of weather formation, it seems
logical and obvious to many meteorologists that changes of solar
radiation lead to changes of climate. Glacial periods, it has been
suggested, were caused by decreases in solar radiation and an increase

in solar radiation should lead to a warm or mild climate. This rather simplified cause-and-effect postulation is challenged by some who show strong reasons why the *opposite* will occur. They point out that if solar radiation increases, the effect will be larger at the equator than the poles, thus leading to increased horizontal temperature differences. This leads to increased circulation, increased evaporation and finally to a build-up of precipitation developing to an eventual preponderance of snow and ice.

668. Would sunspots cause climate changes? Some meteorologists strongly believe that sunspot cycles are linked with large-scale climatic variations. This seems logical because of the pronounced effects which solar storms have on the upper atmosphere. But the problem of correlation of sunspots with climate cycles is extremely complicated and no clear-cut result has been achieved.

669. Can the climate be affected by a change in the Earth's orbit? The Earth bulges slightly at the equator. This bulge is acted upon by the Sun, Moon and planets so that a slight wobble motion, like a top spin, is imparted to the Earth. It takes about 26,000 years for the axis of the Earth to complete one wobble circle. This change in the orbit results in three different effects which, in turn, might conceivably result in climate changes over long periods of time. These effects are: (1) a change in angle occurs between the Earth's orbit and the Earth's equator amounting to about $2\frac{1}{2}°$ over a period of 45,000 years; (2) the eccentricity of the Earth's orbit has a period of about 90,000 years and can become as large as .05 as compared to its present value of .016; and (3) at present, the closest approach of the Earth to the Sun occurs in January; in about 10,000 years it will occur in July; and in January again 21,000 years from now.

These changes which mark an alteration in some manner of how the Earth receives solar energy undoubtedly have some effect on the climate. Whether they are large enough in scale to result in the Ice Ages is extremely questionable.

670. What are the average temperature and precipitation conditions in the United States for each of the following months? January January is the coldest month of the year. In the extreme north-central portion, the monthly average is near 0° F., while a mean in excess of 43° F., the critical vegetative temperature, obtains over about one tenth of the country, mainly Florida and the Gulf sections, as well as lower elevations of Arizona and California. In the extreme upper Great Plains cold waves occasionally bring temperatures of −50° F. or lower. The extreme subzero line extends to the northern portion of the Gulf States.

In Tennessee, Mississippi and parts of adjoining states, average precipitation for January is more than 4 inches. Over considerable western areas, especially in the Plains States, the average is less than one inch, mostly in snow form. In Upper Michigan and central New York, snowfall for this month averages about 30 inches but decreases gradually southward to about one inch in northern portions of the Gulf States.

671. February The mean February temperature ranges from less than 10° F. in North Dakota to about 45° F. in the Gulf States. In most of the Central and Eastern States, February precipitation averages over four inches, exceeding six inches in parts of Alabama and Georgia. From these areas the amounts decrease to about one inch in central and western Oklahoma and the cotton districts of western Texas. Precipitation is light in the Great Plains and the rainy season continues in the Pacific Coast states.

672. March In March the spring rise in temperature becomes rather pronounced, especially in the Central and Northern States. In the most northern portions of the north-central districts, the monthly average is about 20° F., or about 15° higher than in February. Along the Gulf Coast, temperatures average about 60° F. Occasionally temperatures reaching 90° F. or over may occur in the interior of the country. Above 80° F. has been recorded in all sections except the higher elevations of the Western States.

Precipitation averages from four to six inches in the central and eastern cotton states and about two inches in much of the western cotton belt. A noticeable increase in precipitation sets in with the rising temperatures in the northern Great Plains. Rainfall continues heavy along the Pacific Coast. Thunderstorms become more frequent, especially in the lower Ohio Valley and the central and southern part of the Mississippi Valley.

673. April Normally April brings a marked rise in temperature, especially in the Central and Northern States. The monthly averages are about 40° F. in the extreme north, some 20° F. higher than for March, and in most of the Gulf area they range from 65° F. to 70° F. Temperatures as high as 90° F. or above have been recorded in April in nearly all sections of the country, while extremes of 100° or over have occurred in the southwest. South of the Ohio and central Mississippi Valleys temperatures seldom fall below 20° F. in April.

674. May The average temperature for the month ranges from about 50° F. in the extreme upper Lake region and the interior of the northeast to 75° F. along the Gulf Coast.

An outstanding characteristic of the normal rainfall of May is the marked increase over the Great Plains with monthly averages ranging from about three inches over the eastern part of the area to about two inches in the west. May rainfall averages more than 3 inches practically everywhere east of the 100th meridian. East of the Rockies the heaviest falls, more than four inches, occur in the Ozark region, including eastern Oklahoma. The dry season is approaching in Pacific Coast sections with pronounced decrease of rainfall in California.

675. June The average temperature for June ranges from 60° to 70° F. in the Northern States, is lower at high western elevations, and rises to about 80° F. in the extreme south.

In the far west the dry season becomes well-established, but in the Great Plains States June is one of the wettest months of the year. It is noteworthy that a large percentage of the total annual rainfall is the Great Plains region, particularly in the northern part, comes during the spring and early summer months when it is of great value to growing crops.

676. July July, as a rule, is the warmest month of the year except along the Pacific Coast, where highest temperatures usually come later. The average temperature for the month ranges from 65° F. (lower in the higher western elevations and locally east of the Rockies in the extreme north) to more than 80° F. over the lower Great Plains and most of the Gulf area. Despite the general warmth in

July, parts of the Northern States sometimes experience frost and freezing weather. Subfreezing temperatures have occurred in July locally as far south as southwestern Virginia. Parts of Wyoming and Colorado have reported temperatures as low as 10° F. while 134° F. was notched in Death Valley, California.

In much of the Great Plains there is, on the average, more rainfall in July than in June, the amounts ranging from two to three inches. Scanty rainfall occurs in Pacific Coast sections in contrast to normally heavy amounts in the southeastern states where the maximum, eight to ten inches, occurs in Florida. July rainfall results largely from local thunderstorms, this being the month of their greatest activity. The greatest number occurs in the southern Rocky Mountain area and the east Gulf sections.

677. August The normal August temperatures do not differ greatly from those of July, being perhaps a degree or two cooler, except in the Pacific Coast States, where the reverse is true. Mean monthly temperatures range from about 65° F. in the most northern part of the country (somewhat lower in elevated western sections) to more than 80° F. in most of the South.

Rainfall still results largely from local thunderstorms, with the heaviest falls, about eight to ten inches, along the eastern Gulf Coast. Precipitation noticeably falls off in the Great Plains and the dry season continues in the Pacific Coast States. In the far southwest, July and August largely comprise the rainy season; New Mexico and Arizona usually have much more rainfall than in any of the other months.

678. September In September the recession of temperature becomes quite noticeable, amounting to 10° F. in the northern portion of the country. In the extreme south, however, September is only about 3° F. cooler than August.

In the Great Plains States the dry season is approaching with a monthly average of less than two inches of rainfall over much of the area. The heaviest average September rainfall is in the extreme southeast where monthly amounts range from six to eight inches. The rainy season also begins in the North Pacific area, some localities having an average of four to six inches.

679. October   The midfall recession of temperatures in October is rapid, corresponding to the April, or midspring, rise. October averages range from less than 45° F. in northern sections of the country to about 70° F. along the Gulf Coast, or an average of about 13° F. cooler than September in the North and about 8° F. cooler in the South.

Average October rainfall is greatest in southern Florida and along the North Pacific Coast, where amounts range locally up to 10 inches. Over the Great Plains region there is a marked falling off in precipitation, much of the area having monthly averages of less than one inch.

680. November   East of the Rockies the average temperature for November ranges from about 25° F. in the north to 60° F. along the Gulf Coast. Severe cold waves sometimes penetrate the north-central states, but usually lose energy rapidly when moving south and east and are seldom of long duration. Zero temperatures have never been recorded in November at a first-order Weather Bureau station south of the Ohio River, but freezing has occurred as far south as Tampa, Florida.

East of the Rockies the heaviest rainfall occurs in the central and lower Mississippi Valley, where normals in some areas are slightly more than four inches. The Great Plains States usually have less than one inch of precipitation, but in the Pacific Coast area the rainy season becomes fully established, with some localities having a monthly average of 20 inches. This is the month of maximum rainfall in some North Pacific sections. Rainfall is light, however, in the southern end of the Great Valley in California.

681. December   December mean temperatures range from about 10° F. in the most northern parts of the north-central districts to about 55° F. along the Gulf Coast and 70° F. in the most southern part of Florida. More frequent cold waves appear in the interior. Readings of −50° F. have been recorded in the northwest. Subzero weather has occurred as far south as North Carolina.

East of the Rockies, heavy precipitation occurs in the lower Mississippi Valley and in southern Appalachian Mountain districts. The dry season is on in Florida and the monthly average in the Great Plains States is only about one-half inch. Rainfall is heavy in the

North Pacific and in northern Rocky Mountain sections. Heavy snows are frequent in the elevated regions of the Western States, with occasional heavy falls in Michigan and in the Northeast. December has the fewest thunderstorms of any month.

682. What are the basic climate features of the following nations or territories? Afghanistan This mountainous country, about the size of Texas, is split east to west by the Hindu Kush range of The Himalayas, rising in the east to heights of about 24,000 feet. Except in the southwest, most of the country is covered by high snow-capped mountains and deep valleys. In the north, temperatures vary greatly with extremes from 0° to 100° F. It is not so extreme in the south, although snowfall is heavy in all portions in the winter. Rainfall occurs chiefly in the spring and is relatively light.

683. Alaska Alaska is a land of great contrasts of temperature and precipitation. Relatively mild and equable temperatures prevail over the extreme southern portions, ranging from a mean of 32° F. in January to about 54° F. in July. These high temperatures are caused by the warmth of the Japanese Current of the Pacific Ocean. To the north, however, winter temperatures become progressively severe with extremely cold weather frequent in the interior basin and the Arctic area. In the Yukon and Tanana Valleys, average temperatures range from about −20° F. in January to about 60° F. in July. In these valleys, an average of more than 100 days a year occur with minimum temperatures of zero or lower. On the arctic coast the average annual range is from −18° F. in January to about 40° F. in July. In the Bering Sea area to the west, average temperatures range from about 3° F. in January to about 50° F. in July.

Rainfall is very heavy along the southern coast, averaging from 100 to 150 inches a year. Inland rainfall decreases rapidly, ranging in the interior basin from about 7 to 14 inches a year. Precipitation on the Arctic Coast is less than 5 inches a year. Snowfall has a similar distribution.

684. Albania The climate is typically Mediterranean with dry hot summers and moderate winters. Temperatures in the interior mountain plateaus and basins are generally lower than temperatures on the coast.

685. Algeria Most of Algeria consists of plateau land between 2,500 and 5,250 feet above sea level except for some low plains near the Mediterranean coast. The climate is generally shaped by the Mediterranean Sea to the north and the Sahara to the south. On the coast, winter temperatures average about 55° F. Summer temperatures on the coast average about 80° F. Farther inland, the winter average is about 40° F., and summer about 81° F. Summer temperatures in the Algerian Sahara average from 95° F. to 100° F.

About 20 to 30 inches of rain a year falls on the coast and decreases to virtually none inland to the south.

686. Argentina Except for the northern Gran Chaco, which has mild winters and very hot summers, Argentina lies in the South Temperate Zone. The pampas region has an average temperature of about 60° F. and freezing there is rare. Temperature extremes increase farther southward. January is the warmest month for Argentina with June and July the coolest months. The mean temperature for January and February at Buenos Aires is about 73° F. and in June and July, about 50° F.

Heaviest rainfall, over 60 inches a year, occurs at the Gran Chaco, while on the pampas it ranges from about 20 inches in the west to 40 inches in the northeast. Annual rainfall at Buenos Aires averages about 38 inches.

687. Australia The northern third of Australia lies within the tropics and the other two thirds within the Temperate Zone, but because of its position and island form, it has a more temperate climate than other regions in the same latitude. The coolest portion in Australia is Victoria, in the southeast. The climate there is similar to the Mediterranean coasts of Italy and Spain. The average temperature for Australia as a whole is about 70° F. and the northern coastal areas average 82° F. Only in the continental interior does the annual range of temperature exceed 30° F. The eastern highlands and Victoria experience the most rainfall, about 48 inches a year, while large sections of the continent, especially in the deserts of the western interior, receive less than 10 inches of rain a year.

688. Austria Austria's mountainous topography (more than ⁹⁄₁₀ of Austria is made up of large portions of the eastern Alps) includes many snow fields, glaciers and snow-capped peaks. In the north, the mean annual temperature is about 47° F., and in no month does the average exceed 68° F. In the Tyrol, mild winters and warm summers prevail. About 40 inches of rain falls over much of Austria during the year, and most of it occurs during the summer.

689. Bahamas The Bahamas, an archipelago of about 3,000 islands, islets (cays) and rocks east of Florida and north of Cuba, enjoy an agreeable climate. At Nassau, mean temperatures range from 60° F. (January to March) to 88° F. (June to September). The rainy season is May through October. Nassau receives about only 18 inches of rain a year.

At Barbados, temperatures range between 70° F. and 86° F., rarely below 65° F. The cool season (December through May) is also the dry season. Average annual rainfall is 60 inches with September the wettest month. The islands are breeze-swept by the trade winds, although hurricanes threaten from July to October.

690. Belgium The climate of Belgium is temperate. At coastal Ostend the average annual temperature is about 49° F. In the Ardenne heights in southeast Belgium where a plateau rises to about 2,300 feet, the average annual temperature is somewhat lower, about 43° F. Rainfall on the coastal areas averages slightly more than 27 inches, but in many of the plateau localities in the southeast, nearly 60 inches of rain falls annually.

691. Bermudas This archipelago of about 360 small islands 580 miles east of North Carolina possesses one of the most enjoyable climates in the world. At Bermuda or Main Island, the mean annual temperature is 71° F. with extremes of 49° F. and 94° F. Rainfall averages about 58 inches annually.

692. Bolivia The climate varies in land-locked Bolivia from the humid heat of the equatorial lowlands in the east to the arctic cold of the Andean peaks in the west. In the lowlands temperatures average about 77° F. with no great seasonal departure. Rainfall in this warm section ranges from about 30 to 50 inches a year. At higher elevations in the west (to 11,000 feet), the climate is temperate with occasional winter frost. In the great plateau of central Bolivia, moderately cool temperatures average about 50° F. with about 23 inches of rain a year.

693. Borneo   The climate of North Borneo is tropical, with a mean annual temperature range of only 3° F., although extremes of 64° F. and 91° F. have been recorded. The total annual rainfall is heavy, varying between 60 and 180 inches, heaviest in the last three months. Brunei, in the northwest, has a climate similar to North Borneo, except that the wet season lasts longer, sometimes into March. Sarawak Colony in the west has temperatures which seldom rise above 90° F. and fall to about 70° F. at night. Average annual rainfall at Kuching in the extreme west amounts to about 160 inches a year.

694. Brazil   Although much of Brazil straddles the tropical latitudes, its large coastline and wide variety of topography which includes the heavily wooded basin of the Amazon, the tall mountains (almost 10,000 feet) on the Venezuela-Guiana border, and the large central plateau areas combine to vary the climate from tropical to temperate.

Manaos on the Amazon has an average temperature of about 81° F. and annual rainfall of about 72 inches a year. Rio de Janeiro's average annual temperature is about 72° F. with 44 inches of rain. Rio's warmest month is February. In much of the Amazon Basin, rainfall averages about 80 inches a year.

695. Bulgaria   Bulgaria's climate is characterized by cold winters and warm summers. Temperatures at Sofia in the west where much of the land is about 3,000–5,000 feet above sea level, average about 28° F. in January and 69° F. in July. In the lower lands of the southeast a warmer climate prevails. Rain and snowfall average from 20 to 40 inches a year in Bulgaria.

696. Burma   Burma is land-surrounded except for its western coast which faces the Bay of Bengal. This results in monsoonal weather which can be divided into three seasons: (1) cool and dry (November through February); (2) hot and fairly dry (March through May); and (3) rainy (June through October). Burma's varying topography which includes a long narrow mountain range in the east, flatlands in the extreme south and a large plateau to the east modifies the climate. At Rangoon in the lowlands of southern Burma, the annual temperature range is only 10° F.; at Mandalay in central Burma, about 20° F. Annual rainfall at Rangoon is about 100 inches; at Mandalay, about 34 inches.

697. Canada   This country has an extremely varied topography—mountains in the west, then foothills and prairies, the barrens north of Lake Superior, the open lands of Ontario, the rocky Laurentian district in Quebec with the fertile eastern townships to the south of it, and then plains sloping down to sea level in the east. The coastline of Newfoundland is rugged, particularly in the southwest and the mountains of New England extend north into Canada.

The climate reflects these topographical extremes. South of the Gulf of Lawrence, the Maritime Provinces have an average temperature of 40° F. for the summer months. In Quebec and northern Ontario, the winters are cold and the summers average from 60° F. to 65° F. In southern Ontario the average summer temperature is 65° F. with an occasional rise to 90° F. The prairie provinces have a distinctly Continental climate with comparatively short, warm summers and long, cold winters. The west coast has relatively mild winters and moist cool summers. Northwest and northeast of Hudson Bay, the climate is too severe for tree growth.

698. Canary Islands   This group of volcanic islands, about 60 miles off the northwest coast of Africa, is swept by breezes from the North Atlantic. Its climate is extremely agreeable, with temperatures hardly varying all year from 60° F. to 85° F. Rainfall is moderate.

699. Ceylon   Ceylon's climate is monsoonal and modified by mountains in the south and central regions which rise to about 8,000 feet. When the southwest monsoon winds blow in May, heavy rainfall occurs in the southwest, amounting to about 200 inches a year. The northeast part of the island, on the other side, has only about 40 inches of rain. In October and November, the dry northeast monsoon winds blow over Ceylon, bringing pleasant weather in contrast to the generally hot and moist conditions of the summer months.

700. Chile   Chile is a long, narrow and mountainous land. In the extreme north, the days are hot, the nights warm on the coast and cool in the interior. Central Chile which contains a 700-mile-long valley between the Andes and the coastal plain has a pleasant climate similar to that of southern California. Southward in the lake regions, the climate is quite moist and cool. In the extreme south, fogs and storms keep the mean temperature low. Santiago has extreme recorded temperature ranges of 25° F. and 96° F. Rainfall there averages 14 inches a year.

701. China The climate of this vast country varies considerably from north to south and from the coast inland. For its latitude, North China has the coldest winters in the world (23.5° F. average in January at Peiping, about 40 degrees north). The Yangtze Valley is warmer, with winter temperatures like those of Britain, while the south has relatively warm winters. Summer temperatures are uniformly hot throughout China, averaging about 79° F. in July at Peiping and 82° F. at Hong Kong. South China receives regular rainfall averaging from 40 to 60 inches a year, but in the north, rainfall is irregular and not as heavy; droughts and floods are common.

702. Colombia While Colombia lies almost entirely in the doldrums just north of the equator, its climate is tempered somewhat by the high altitudes of the mountains in the west. Bogota's mean temperature stays in the 50's each month with an annual rainfall of 42 inches. But in the lower areas of the east, southeast and along the coast, hot, humid and rainy conditions prevail.

703. Costa Rica In the highlands of central Costa Rica, the weather is pleasant, with average temperatures of about 68° F. From the central elevated tableland about 6,000 feet above sea level, the land slopes sharply to the Pacific and the Caribbean. Along the coasts temperatures are much higher, with a mean annual temperature of about 82° F. About 70 inches of rain falls annually on the Pacific Coast and more than 130 inches of rain on the Caribbean. The driest months are from January to April.

704. Cuba Long, narrow Cuba has mostly flat or rolling areas, except for some mountainous country in the southeast, central and west. The steady trade winds temper the tropical climate. Havana, in the northeast has an average annual temperature of 77° F. The annual range varies only about 10° F. The dry season lasts from November to April and the warmer wet season follows. About 50 inches of rain falls on Havana each year.

705. Czechoslovakia   The climate is moderate and does not show extremes of heat or cold. At Prague, in the west, temperatures average about 30° F. in January to about 66° in July. About 20 inches of rain occurs there. The eastern areas are slightly cooler, with rainfall averaging about 25 inches a year. Snow falls on the mountains which form several of its boundaries.

706. Denmark   The climate is moist and cool, similar to that of eastern England but with somewhat colder winters and slightly warmer summers. Average temperatures range from 32° F. in January to 61° F. in July. About 25 inches of rain falls during the year.

707. Dominican Republic   This area's climate is influenced by a mountain range which crosses from northwest to southeast, with some elevations exceeding 10,000 feet. The lowlands of the northeast receive ample rain and are fertile. The southwest regions are arid because the trade winds drop their moisture on the northeast slopes. Most of the rainfall occurs from May to October. The elevated interior is cooler than the coastlands. Temperatures average about 74° F. in January and about 81° F. in August, with very little departure.

708. Ecuador   Even though it straddles the equator (from which its name is derived), Ecuador presents a contrast of climates unusual in a country so small. Two high and parallel ranges of the Andes traverse the country from north to south, creating three main climate regions. One is the hot, humid and swampy Pacific Coast land where temperatures average 83° F. for the year. The Andean plateau, however, ranges from about 46° to 70° F. with cool and fertile valleys. The eastern section contains the hot and moist jungles of the upper Amazon basin. The rainy season extends from December to April.

709. Egypt   Except for a narrow belt on the Mediterranean, Egypt lies in an almost rainless area in which high daytime temperatures fall quickly at night. The mean temperature at Cairo varies between 55° F. in January and 84° F. in July. At Alexandria, the range is between an average of 58° F. in January and 80° F. South of Cairo, pure desert conditions prevail. At Aswan, the mean maximum temperature for the year is 118° F. Cairo and Alexandria get only about eight inches of rain each year.

710. Ethiopia Ethiopia, lying wholly within the tropics just north of the equator, escapes an extremely hot climate because of its elevation. Several mountains over 10,000 feet high rise from the main plateau land, covering almost the whole of western Ethiopia. The lowlands to the east are hot. The mean annual range of temperature is between 60° and 90° F., except for colder conditions in the higher mountains. Rainfall at Addis Ababa in the plateau land of west central Ethiopia amounts to about 50 inches a year.

711. Finland The flat, lake-filled and heavily forested country of Finland extends from the Gulf of Finland on the south to north of the Arctic Circle. Long, severely cold winters and short, cool summers prevail. Southwest and south coastal area weather is tempered by prevailing southwest winds, although southerly Finnish ports are ice-bound part of the year. Rainfall is generally light with the driest months from May to September.

712. France France's climate is generally temperate but varies from long, cold winters and hot summers in the northeast, to the warm temperatures of the Mediterranean Coast with mild winters. Prevailing Atlantic winds move into the lowlands of western France and blow across the higher mountainous country to the east. Rainfall is adequate throughout France, averaging about 25 to 30 inches a year. At Paris, in north-central France, the average temperature for January is about 36° F. and for July, 66° F.

713. French Cameroons The climate is tropical and oppressive. The temperature rarely falls below 70° F., even in the cooler months. On the coast of the Gulf of Guinea on the Atlantic Ocean, rainfall is very heavy, averaging about 150 inches a year, and is fairly evenly distributed through the year.

714. French West Africa Central and northern parts of the huge area have two seasons, rainy and dry. Two rainy seasons occur in the most southern regions, separated by a short dry season. At Dakar, on the west coast, temperatures average about 70° F. in the winter and in the summer about 82° F., with daily variations of about 20° F.

715. Gambia Temperatures are fairly regular throughout the year, ranging from about averages of 60° F. to 80° F. Maximum rainfall occurs in August and September. About 50 inches of rainfall occurs annually.

716. Germany The climate of Western Germany is intermediate between the oceanic climate of western Europe and the Continental climate farther east. Summer temperatures average about 61° F. The sheltered mountain valleys of the south enjoy a more temperate climate, especially in the Rhine Valley above Mainz. Rainfall is heaviest in the south and west, averaging about 30 inches a year.

Most of the Eastern Germany area is part of a low plain. The climate is mostly temperate but with greater extremes between summer and winter than in Western Germany. Rainfall ranges from 20 to 30 inches a year.

717. Gold Coast (Ghana) Coastal climate is hot and humid, averaging near 80° F. for the year. At Accra, on the coast, about 27 inches of rainfall occurs annually.

718. Great Britain Although Great Britain lies in the same approximate latitude as Labrador, its climate is tempered by the westerly winds blowing off the Gulf Stream. These same winds also prevent excessive summer heat. Rainfall is abundant, especially in the early fall. The gloomy thick fogs of London and other industrial cities occur mostly in November and March. It is estimated that London in December has an average of less than 15 minutes of sunshine each day. Even sunshiny days in the summer are relatively cool and damp. There is generally more sunshine on the coasts than in the interior sections.

The mean annual temperature of England and Wales is about 50° F.; the west coast being somewhat warmer than the east. January is the coldest month (average about 40° F.) and July the warmest (about 62° F.). Highest July temperatures usually occur near London, where the mean temperature is about 64° F. Temperatures at London average about 38° F. to 39° F. in December and January. London has about 25 inches of rain which is distributed rather evenly throughout the year.

719. Greece　Climate is varied, coastal regions having a temperate climate with short winters and little snow or frost. In the interior of the uplands, winters are long and severe. Athens, on the southeast coast, has an average January temperature of about 47° F. and about 81° F. in July. Annual rainfall at Athens is about 15 inches a year. At the island of Corfu, just off the northeast coast, about 50 inches of rain falls each year. Precipitation is generally heaviest in the mountains. Summer heat is moderated by cool sea breezes and northerly winds from the mountains.

720. Greenland　This largest of the world's islands lies chiefly within the Arctic Circle. About ⅚ of Greenland is covered by a glacial icecap from 1,000 to 8,000 feet thick. Most of the island is a lofty plateau, 9,000 to 10,000 feet in altitude. A rugged polar climate prevails with most of the population hugging along the fiord-indented west coast. Many portions in the extreme northeast have yet to be explored. U.S. Weather Bureau studies are being made at Thule in the northwest.

721. Haiti　This narrow, long peninsula is the western end of Hispaniola, southeast of Cuba. A mountain range traverses the center of the long west-east axis of the country. The climate is hot along the coasts but is moderate in the higher mountain areas. Port-au-Prince on the north coast has a mean annual temperature of 81° F. Rainfall varies from about 20 inches to 100 inches a year, with hurricanes frequent in the May to October rainy season.

722. Hawaii　The outstanding features of the climate are the remarkable differences in rainfall over adjacent areas, the persistently equable temperatures throughout the year and the steadiness of the trade winds.

Frequent and copious showers fall almost daily in windward and upland districts, while leeward sides of the mountains are relatively sunny and dry. This is illustrated by an extreme condition in central Kauai, where, near the summit of Mt. Waialeale, at an elevation of about 5,075 feet, the average amount of rain is 450 inches a year while just 15 miles southwest, on the leeward side of where the trade winds blow, it is less than 20 inches. August and September are the warmest months, while January and February are the coolest; but the range is only from about an average of 69° F. to 75° F. In the territory as a whole, more rain falls from November to April than from May to October. About 82 inches of rainfall occurs over the 8 main islands with much of it on the eastern, windward sides of the volcanic islands.

723. Honduras   The climate is oppressively hot and humid in the narrow coastal lowlands but pleasant in the interior highlands. At Tegucigalpa, south of central Honduras in plateau land, highest temperatures occur in May, about 90° F. The lowest occur in December, about 50° F. Rainfall amounts to about 105 inches a year.

724. Hungary   Most of Hungary is a rolling plain lying east of the Danube. It has a moderate climate with a mean annual temperature range of 48° F. in the north to 52° F. in the south. Rainfall varies from about 35 inches in the west to about 15 inches in the east. Budapest, in north-central Hungary, has about 25 inches of rain a year with an average annual temperature of about 50° F.

725. Iceland   Although more than 10% of this bleak volcanic island is covered by snowfields and glaciers, its climate is modified by the Gulf Stream so that the climate is much like that of southern Canada, but with longer winters and shorter summers. Reykjavik, on the southwest coast, has an annual mean temperature of about 39° F. January is the coldest month, averaging 34° F.; and July the warmest, about 52° F. Rainfall varies widely from about 12 to 80 inches a year. Most of the population lives on the relatively fertile coastal areas.

726. Indochina   Jutting out of southeast Asia into the China Sea, this region, which is mostly the flatlands of the Mekong River Delta except for a range of mountains to the east, has a monsoonal climate. Heavy rainfall and winds from the sea occur from May to October. During the rest of the year, dry, cool weather exists, with the winds blowing from the Asiatic interior.

727. Indonesia The climate throughout this group of islands is equatorial and monsoonal, with little variation of temperature which averages very close to 80° F. for the year. Rainfall is plentiful, averaging about 100 inches through the year but falling more on the windward slopes of the mountains in Sumatra and Java during the hot and rainy season from May to October. December and January are relatively cool and dry with hot and dry weather usually from February to April.

728. Iran The central plateau is hot in summer and very cold in winter. The Caspian Sea Coast land in the north has a warm climate throughout the year, sheltered by a range of mountains rising well over 10,000 feet to the immediate south of the narrow Caspian Coast. Teheran, just south of this range, has a mean January temperature of 35° F. and 85° F. in July. The annual average is 62° F.

729. Iraq The climate runs to great extremes and is influenced by the topography. Arid high desert land lies to the west and south, with a broad central valley between the Euphrates and Tigris Rivers and mountains in the northeast. Long, hot summers and short, cold winters prevail, with the valley area fronting the Persian Gulf in the southeast one of the hottest regions in the world. At Baghdad in the fertile central valley, January temperature averages 49° F. and in July, about 95° F. Rainfall there amounts to about only 7 inches a year.

730. Ireland The climate is influenced by the Gulf Stream which keeps winter temperatures mild, with moist and cool summers prevailing. At Dublin, January temperature averages 40° F., and in July, about 58° F. Rain is plentiful, about 30 to 40 inches evenly distributed throughout the year in all regions.

731. Israel Israel is largely a plateau traversed by mountains from north to south. The western lowlands along the Mediterranean are very fertile, but the Negev region in the south which occupies almost half the total area is largely a wide, desert-steppe region where summers are hot and dry. At Jerusalem, somewhat east of central Israel in the plateau, the mean annual temperature is about 63° F. Rainfall averages about 28 inches along the coast with scanty rainfall in the southern inland desert area. The rainfall occurs chiefly in the autumn and spring.

732. Italy Along the Gulf of Genoa, the Italian Riviera enjoys an equable and pleasantly warm climate. Very cold winters prevail in the high Apennines which traverse the length of central Italy. The western coast and slopes of Italy are warmer than the eastern sides. The Po Basin in the north shows extremes of cold winters and very hot summers. Sicily has a warm and equable climate. At Rome, near the central west coast on the Mediterranean, January temperature averages about 45° F. and July, about 76° F. About 35 inches of rainfall occur each year in the coastal lowlands, with much heavier rain in the Alps to the north.

733. Jamaica This largest of the British West Indies has a favorable climate, its tropical weather tempered by the easterly trade winds and by its elevated interior. At Kingston, on the south coast, temperatures are quite high, averaging about 79° F. for the year, but considerably cooler inland. The rainy season occurs in May and October, with total annual rainfall about 65 inches, the north and east usually having more. Rainfall at Kingston amounts to about 35 inches a year.

734. Japan The climate varies widely according to latitude. In the southern portions, warm subtropical climate prevails around Kyushu. In the Hokkaido area to the north, wintry cold and snow reflect the influence of the colder latitudes of the Asiatic mainland. The central islands escape severe winters because of the influence of the warm Japanese Current of the Pacific. At Tokyo, in the central east, the average temperature in January is about 38° F., and 76° F. in July. Rainfall throughout Japan is plentiful. Kanazawa, on the west coast fronting the Sea of Japan, receives about 99 inches of rain a year, with Tokyo's average about 58 inches.

735. Jordan Jordan is mainly a plateau with an average altitude of 3,000 feet sloping gently eastward. The western edge has a steep slope overlooking the valleys of the Dead Sea and Jordan River. Cold winters and dry hot summers are typical of the steppe and desert areas. The mean annual temperature is about 65° F., but with a wide range between high temperatures in August and moderately cool temperatures of January. Rainfall averages from about 26 inches in the north to about only 10 inches a year in the south.

736. Kenya The coastal lowland of Kenya, in the east, is hot and humid, with February and April the hottest months. Mombassa, on the Indian Ocean at the southeast coast, averages 82° F. in these months and is slightly cooler in June and July, averaging 76° F. In the interior, where highlands dominate, the climate is pleasantly temperate. Nairobi, in the interior, has temperatures in the 60's throughout the year. Rainfall averages about 40 inches a year.

737. Korea The climate of Korea is monsoonal. The peninsula has mountain ranges along its eastern portions and lowlands to the west. Warm, moist winds from the Yellow Sea to the west prevail during the summer, with heavy rains falling on the windward sides of the mountains. Cold and dry weather prevails from October through March, with northerly winds from the Asiatic mainland influencing the climate. At Kaesong, near the 38th Parallel in the east, the January temperature averages 23° F. and 77° F. in July, with rainfall amounting to about 47 inches a year.

738. Lebanon Hot, dry summers and cool rainy winters prevail in Lebanon. Except for a long narrow coastal plain area on the Mediterranean to the west, Lebanon is composed of high mountains and plateaus. At Beirut, on the coast, the average temperature is about 80° F. in July and about 55° F. in January. Annual rainfall at Beirut is about 35 inches.

739. Leeward Isles These agricultural islands, southeast of Puerto Rico and curving northeastward from the Windward Isles, have a typical trade-wind tropical climate. The temperature range is not wide over the year, averaging about 76° F. in January and 81° F. in August. Summer rainfall is heavy.

740. Liberia This small country, on the west coast of Africa, has 350 miles of narrow low coast land facing the Atlantic. The land rises to a low plateau which covers most of the country from seven miles inland to the northeast boundary. The climate is tropically hot and humid. Rainfall along the coast amounts to about 150 inches a year.

741. Libya   Libya's northern lowlands on the Mediterranean give way to higher desert regions to the south. The northern coastal regions have pleasantly cool winters and warm summers. Extremely hot climate prevails in the interior southward. At Tripoli, on the northeast coast, the average temperature for January is 54° F. and for July, 79° F. Only about 10 to 16 inches of rain falls on the coast each year; much less in the interior.

742. Malaya   The Malay Peninsula, north of Sumatra, has a tropical climate, consistently hot and humid along the coasts and slightly modified in higher central portions. At Singapore, in the extreme southern coastal tip, hardly any seasonal temperature change occurs. The average temperature in January is about 80° F. and in July about 82° F. Rainfall at Singapore amounts to about 95 inches a year.

743. Mexico   Mexico is a great high plateau, open to the north, with eastern and western mountain chains and with ocean-front lowlands lying outside them. From the coasts inland to the plateau, it is tropical, with temperatures averaging about 80° F. The plateau's temperatures are about 5° F. to 7° F. cooler and in mountains over 6,000 feet, the average annual temperature is 60° F. Rainfall varies greatly, with over 100 inches a year sometimes occurring on the east coast, while hardly any rain falls on Lower California. Plateau rainfall is moderate, about 20 to 40 inches a year, and combined with moderately warm temperatures, creates a pleasant climate for many plateau localities. At Mexico City, temperatures average about 55° F. in January (the coldest month) to about 64° F. in April (the warmest month). About 30 inches of rainfall occurs there each year. At Vera Cruz, on the southeast coast, the corresponding range is 71° F. and 80° F. Rainfall at Vera Cruz amounts to 64 inches a year.

744. Mongolia   Typical Continental climate prevails with very cold winters and dry warm summers. Outer Mongolia, in the north, is a tableland from 3,000 to 5,000 feet in altitude. Temperatures at Ulan Bator range from averages of 15° F. in January to 64° F. in

July. Rainfall is light throughout the country and almost negligible in the Gobi Desert in the southeast.

745. Morocco Morocco, in northwest Africa, has a long, narrow coastal plain which faces the Atlantic on the west, and a short mountainous coast facing the Mediterranean on the north. High mountain ranges traverse broad central sections of the interior. Atlantic coastal temperatures are pleasantly cool, ranging from averages of about 60° F. in January to 72° F. in August. Inland, the climate is more continental with colder winters and hotter summers. Rainy seasons occur in the spring and fall, ranging from about 10 to 40 inches along the coastal regions.

746. Mozambique A warm rainy season occurs from December to March and a cool season from April to August. The coastal lowlands facing the Indian Ocean have average temperatures about 68° F. in July and 79° F. in January. Rainfall amounts along the coast range from about 30 to 60 inches a year.

747. Netherlands This country is exceptionally low and flat except for some hills in the southeast. The climate is marked by excessively moist conditions with frequent fogs and mists. Summers are cool and moist with Amsterdam on the east coast having a July average of 63° F. January average temperature at Amsterdam is about 37° F. In central Netherlands, winter temperatures are somewhat lower. Average annual rainfall is about 28 inches with July to September the wettest period.

748. New Zealand The temperature range of New Zealand's two main islands is very narrow because of the tempering influence of the ocean. At Auckland in the north, temperature in January averages about 66° F. and in July about 51° F. It is a few degrees cooler at Wellington on the south of North Island, and in Dunedin at the extreme southern area of South Island, temperatures average about 10° F. lower than in Auckland. Rainfall is moderate throughout, averaging about 40 inches a year.

749. Nicaragua Nicaragua is mountainous in the west, with fertile valleys. A plateau slopes eastward toward the Caribbean. The Pacific Coast is barren and rocky; the Caribbean swampy. The coasts are hot and humid while the elevated interior is generally cool. The low Mosquito Coast in the east receives up to one hundred inches of rain a year with the wettest season from late spring to late fall.

750. Nigeria Although Nigeria lies in the tropics, its climate varies from tropical in the south to almost temperate on a few portions of the plateau in the north. At Calabar, on the Gulf of Guinea Coast in the south, the temperature varies very little all year, averaging about 80° F. Rainfall is extremely heavy on the coast in the south. At Calabar, over 125 inches of rainfall occurs each year. Rainfall in the central and north ranges from about 25 to 40 inches a year.

751. Norway The Gulf Stream has a moderating effect on the climate. Summer temperatures average about 48° F. in the extreme north to about 62° F. at Oslo in the south. February temperatures in Oslo average about 24° F. Bergen, on the southwest coast, has an average February temperature of 34° F.; but in the extreme arctic latitude portions, February temperatures average about the zero mark. Abundant rain falls on the long coastal area but decreases sharply inland.

752. Panama Panama's narrow length runs east to west for 420 miles with the Canal bisecting the narrowest and lowest point. From coastal lowlands on the Caribbean to the north and the Pacific to the south, the land rises to upland valleys and plateaus. At Colón, on the Caribbean Coast, the temperature range is narrow, averaging about 80° F. for the year. Rainfall there is extremely heavy, more than 125 inches a year. Balboa Heights has similar temperatures, but its annual rainfall averages about 69 inches.

753. Paraguay In the eastern grassy upland country, temperatures average about 81° F. from December to February, and 64° F. in May–August. The marshes, lagoons and dense tropical forests in the Chaco region to the west have higher temperatures. Rainfall at Asunción amounts to about 55 inches for the year.

754. Peru Tropical climate dominates the eastern lowlands, ranging to the much cooler climates in the Andes mountain regions to the west. Western coastal areas have an extremely narrow range of moderate temperatures. At Lima, the average in July is 61° F.; in January, 70° F. Extreme rainfall conditions exist. Lima has less than 2 inches of rain a year and Mollendo, on the west coast, has less than one inch of rain. In the hot and humid Montana of the east, between 75 and 125 inches of rain falls annually.

755. Philippine Islands   Lying in typhoon paths, the archipelago has warm temperatures throughout the year, varying very little from an average of 80° F. Rainfall averages from 75 to 100 inches annually, with the wettest season from June through October. Manila has about 80 inches of rain a year.

756. Puerto Rico   Puerto Rico's north coast is a lowland facing the Atlantic. The land rises to elevations of between 2,000 and 5,000 feet along the southern length of the island. At San Juan, on the north coast, temperatures in January average 75° F and in July, 80° F. At Ponce, in the southeast foothills, a similar range exists while at Cayey, in the elevated interior, temperatures average about 5° F. lower. Rainfall amounts range from about 45 inches a year in the south to 60 inches a year in the north.

757. Poland   The climate is shaped by the Baltic Sea on Poland's north coast and the Carpathian Mountains to the south. Winters are moderately cold and summers fairly cool, with westerly oceanic winds moderating the weather. At Warsaw, in central Poland, the climate is typical of most of the country. In January, temperatures average about 60° F. and in July, about 65° F. Rainfall at Warsaw amounts to about 28 inches a year. Snowfall is not heavy, but temperatures occasionally dip below zero, with rivers generally icebound for about two months a year.

758. Portugal   Portugal's climate is equable and temperate, but in the deep valleys of the mountains to the east and north, the cool Atlantic winds are blocked and very hot summers prevail. Coastal temperatures are pleasantly moderate with Lisbon's average July temperature 71° F. and its January temperature being 51° F. About 29 inches of rainfall occurs at Lisbon each year, with fogs common along the coast. Heavy rainfall occurs on the western mountain slopes in the north and east.

759. Rhodesia   In Northern Rhodesia, average temperature in the south ranges from about 65° F. in July to about 80° F. in October. Most of the rain occurs between November and April, varying widely in different parts of the territory.

Southern Rhodesia's climate is characterized by hot days and cool nights throughout the year. The hottest month is October, and June to August the coolest season. Rainfall, which averages about 28 inches a year, is greatest from October to December.

760. Romania   The most extreme weather in Romania occurs in the southeast plains where summers are hot and winters have severe frosts and blizzards. Variations are less extreme in the Transylvania plateau to the north and west and in the Carpathians farther north. At Bucharest in the southeast, average temperatures range from about 26° F. in January to 73° F. in July. In some winters the Danube, marking Romania's southern boundary, is icebound for three months. Rainfall averages about 20 to 25 inches a year, heaviest in summer.

761. Salvador   The south coastal areas on the Pacific are hot and humid, with a more temperate climate in the interior highlands to the north and east. At San Salvador, in the southwest, monthly temperatures vary little from the annual average of about 75° F. Rainfall there amounts to about 69 inches a year, with May to October the wettest season.

762. Saudi Arabia   The country's desert and steppe land is markedly barren, hot and dry. Agriculture is restricted to the highlands of Asir and scattered oases. The uniformly hot and dry characteristics are reflected by the statistics from Aden on the southern coastal tip where temperatures average 76° F. in January and 88° F. in July. Muscat on the extreme east coast has corresponding temperatures of 70° F. and 91° F. Rainfall in both regions is less than 5 inches a year.

763. Scotland   Although Scotland lies in a high latitude, its climate is kept from being cold by the moderating influence of the Gulf

Stream. Relatively mild winters and cool moist summers prevail. Glasgow's temperatures average about 39° F. in January and 58° F. in July. The area receives about 37 inches of rain a year with lowland mists and fogs common on the north and east coasts.

764. Somaliland (British)   This land extends along the Gulf of Aden for about 400 miles and inland for 80 to 220 miles. The interior is an elevated plateau falling sharply to the coastal plain. The climate is very hot and dry. At Berbera, on the coast, the temperature averages about 77° F. in January and in the 90's in July. Rainfall along the coast averages less than 8 inches a year.

765. Spain   The climate varies considerably. The southeast coast, protected by the Sierra Nevadas just to the north, is subtropical. The southwest coastal area, fronting the Atlantic, has relatively mild winters and hot summers. Valencia, on the east coast facing the Mediterranean, has an average January temperature of 50° F. with July temperatures averaging about 75° F. The plateaus and mountains of central Spain show more extremes, with Madrid's temperatures ranging from 40° F. in January to 74° F. in July. The northeast coast, with climate like that of the British Isles, is the only region with normal rainfall. At Oviedo, near the North Atlantic coast, rainfall averages about 36 inches a year, with cool moist summers and relatively mild winters. Except for this north coastal area, rainfall throughout the rest of Spain is fairly light, ranging between 12 and 20 inches a year.

766. Sudan   The northern region is mostly desert country with some fertile areas in the central zone and most of the south being well-watered by tropical rain. The Nile traverses Sudan in a crooked course, north to south. The climate is hot—more so in the central area and least in the deserts where the temperature range is large. At Khartoum on the Nile in central Sudan, the mean annual temperature is about 80° F.

767. Sweden   Sweden's climate is diversified. The warmest month is usually July with a mean temperature of 62° F. at Stockholm. February is the coldest month with a mean average below the freezing point for all Sweden (about 27° F. at Stockholm). Average annual rainfall in the north is about 16 inches and about 23 inches in the south.

768. Switzerland The climate is temperate and varies with altitude. Basel, with an elevation of 909 feet, averages about 32° F. in January. Santis with an elevation of 8,202 feet averages 16° F. July is the warmest month, with a mean of about 66° F. in Basel and 41° F. in Santis. Precipitation at Santis is almost 96 inches a year. Zurich, in the northeast, has a similar temperature range to that of Basel, with annual precipitation amounting to about 45 inches.

769. Syria Coastal Syria is a narrow plain backed by a range of mountains with a steppe area further inland. The climate is subtropical. Summer temperatures at Aleppo, in the north, average about 88° F., with winter temperatures about 40° F. Wide night and day temperature differences occur throughout the year. Rainfall amounts to about 50 inches a year on the coastal range but diminishes to less than 4 inches in parts of the inland desert.

770. Tanganyika The territory's narrow coastal plain in the east faces the Indian Ocean. Inland, to the west, the land rises rapidly to the eastern side of the Central African Plateau. Mt. Kilimanjaro, the highest point on the African continent, lies on the northeast border of Kenya. The climate is hot and humid on the coast, with the average annual temperature at Dar es Salaam 80° F. Rainfall there averages 60 inches a year. Inland rainfall and temperatures are lower.

771. Thailand The climate is monsoonal, but the full force of the southwest monsoon is broken by a narrow line of hills in the west. The warm rainy season is from March to October, with moisture carried from the Bay of Bengal. Humidity is always high. During the hot season, temperatures frequently rise to 100° F. inland but in the cool season, November to February, they often reach 40° F. Bangkok, on the Gulf of Siam to the south, is hot and rainy. Temperatures in January average 79° F. and 86° F. in April with an annual rainfall average of 52 inches.

772. Tunisia The climate is Mediterranean. At Tunis, near the coast on the north, the average January temperature is 50° F. and in

July, it is about 80° F. Rainfall on the coast averages about 20 inches a year, decreasing to less than 5 inches in the south.

773. Turkey   Along the coast from Antioch to the Dardanelles, the climate is Mediterranean, with mild rainy winters and warm dry summers. Beyond this area to the Bosporus, warm and all-year-round rainy weather occurs. The western plateau has a harsh steppe climate with cold winters, hot summers and scanty rainfall. The eastern plateau is almost Alpine in climate. Istanbul has a mean annual temperature of 57° F. (maximum 99° F., minimum 17° F.). Rainfall averages about 28 inches a year.

774. Union of Soviet Socialist Republics   The climate necessarily is varied, but for the most part is continental. In general, the climate of the northern and central regions is characterized by long, cold winters and by summers which are shorter and cooler than those in the northern part of the United States. Siberia has the coldest climate in the world; the January average at Verkhoyansk is —59° F. In the southern regions, the climate varies between temperate and subtropical. The Uzbek, Turkmen and Kazakh areas are largely desert and semi desert regions.

In the central belt, rainfall is fairly uniform, averaging about 15 inches east of the Urals and 20 inches a year to the west. In the tundra to the north, it decreases to about eight inches and to four inches in the southern regions.

At Moscow, the mean temperature for January, the coldest month, is 14° F. In July, the warmest month, the mean temperature is 66° F.

775. Union of South Africa   Except for the western semiarid regions, the climate is generally subtropical, much like that of northern Florida. Rainfall averages about 40 inches a year on the east coast and diminishes sharply westward.

The mean annual temperatures are remarkably uniform throughout the region, with very little variations occurring during the year. At Johannesburg, in the north central plateau area, it is about 60° F.; at Durban, on the east coast, about 69° F.; at Cape Town in the extreme southeast, about 61° F.; and at Grahamstown, near the southern coast, about 61° F. January is the hottest month, with most of the rainfall occurring from October to March.

776. Union of Southwest Africa   Except for an extremely long narrow coastal plain on the Atlantic, the land is mostly a high plateau with a general elevation of from 3,000 to 4,000 feet. Temperatures are moderate and rainfall very light. At Swakopmund near Walvis Bay on the central east coast, temperatures in July average 57° F. and 64° F. in January. At Windhoek in the Auaz Mountains in the central inland region, corresponding temperatures are 56° F. and 74° F. Less than an inch of rain falls on Swakopmund each year, with about 15 inches at Windhoek.

777. Uruguay   A low rolling plain marks the southern portion and a low plateau the north. Uruguay's southeast coast is on the Atlantic and the southwest coast on the Rio de la Plata. The climate is pleasantly temperate, with frost almost unknown. At Montevideo, on the south central coast, temperatures in July average about 50° F. and in January, about 72° F. Rainfall amounts to about 38 inches a year, heaviest in April and May, the autumn of the Southern Hemisphere.

778. Venezuela   Venezuela's north coast faces the Caribbean. A broad central lowland area divides two mountainous regions. A narrow range of heights lies along the coast. Climate is tropical and oppressive except where modified by altitude. In the higher portions of the Sierra Nevada de Merida, in the west, the climate is almost temperate. Caracas, on the coastal heights, has a narrow range of moderate average temperatures, ranging from about 65° F. to 69° F. throughout the year. Ciudad Bolívar, however, in the east central lowlands, averages about 13° F. higher throughout the year. Maracaibo, on the Gulf of Venezuela in the northwest, is even warmer. Rainfall ranges in amounts from about 20 to 35 inches a year, although some localities on the western sides of the western mountains receive about 75 inches. Most of the rainfall occurs between April and October.

779. Windward Isles   This group of four islands, curving southward from the Leeward Isles in the Caribbean, enjoys a pleasant trade-wind-swept climate, except for extremely heavy summer rainfall occurring in a few localities. The temperature throughout the year varies very little from month to month, averaging about 77° F. in January and about 80° F. in September.

780. Yugoslavia From the long Adriatic Sea Coast lying northwest-southeast, mountains and plateaus rise abruptly from the sea and extend inland and then slope to the plains in the north and northeast. On the Adriatic, the climate is mild but, in the interior, winters are cold and the summers hot. At Belgrade, in the northeast, the average temperature for January is 33° F. and for July, 72° F. Rainfall there averages about 25 inches a year with greatest amounts in the late spring.

781. In tabulating extreme weather records, what is an important consideration? The measurements must be made under standard conditions with accurate instruments, professionally observed and reflecting as long a period of time as possible to insure effective comparisons.

782. What is the highest temperature recorded anywhere on earth? At Azizia, in Tripolitania in Northern Africa, a temperature of 136.4° F. was recorded on September 13, 1922.

783. What is the world's lowest temperature on record? Temperatures of —90° F. were recorded at Verkoyhansk, Siberia, U.S.S.R., on February 5 and 7, 1892, and at Oimekon, Siberia, on February 1, 1933. In 1938, press reports from Russia indicated that a temperature of —108° F. had been recorded at Oimekon. —100.4° F. was recorded near the South Pole, Antarctica, May 11, 1957.

784. Where is the world's highest mean annual temperature? At Lugh, in Italian Somaliland, East Africa, the mean annual temperature is 88° F.

785. Where is the world's lowest mean annual temperature? At Franheim, in the antarctic, the temperature averages —14° F. for the year. The estimated mean temperature at the south polar icecap, elevation 8,000 feet, is considerably below −22° F. An estimation of −60° F. has been made near the South Pole.

786. What are the highest and lowest temperatures on record? (° F)

	Highest		*Lowest*
No. America	134	(*Death Valley, Cal.*)	−81	(*Snag, Canada*)
So. America	115	(*Santiago, Arg.*)	−27	(*Colonia Sarmiento, Arg.*)
Europe	124	(*Seville, Spain*)	−61	(*Ulst Tsilma, USSR*)
Asia	123	(*Baghdad, Iraq*)	−90	(*Verkoyhansk, USSR*)
Africa	136.4	(*Azizia, Tripolitania*)	1	(*Gerryville, Algeria*)
Australia	127	(*Bourke*)	19	(*Mitchell*)
Atlantic	99	(*Las Palmas, Canary Isls.*)	−40	(*So. Orkney Isls.*)
Antarctica	38	(*Little America*)	−89	(*South Pole*)
United States	134	(*Death Valley, Cal.*)	−69.7	(*Rogers Pass, Mont.*)

787. Where does the greatest average annual rainfal occur in the world?   At Mt. Waialeale, Kauai, Hawaii, 471.68 inches of rain falls annually. Perhaps the most famous spot on earth for heavy rainfall is Cherrapunji, India, which held the record for many years (450 inches a year). Cherrapunji still holds many rainfall records, including the following:

Greatest amount for any 12-month period: 1041.78 inches (August, 1860, through July, 1861)

Greatest amount for a calendar year: 905.12 inches in 1861

Greatest amount for a calendar month: 366.14 inches in July, 1861

Greatest amount for 5 consecutive days: 150 inches in August, 1841

788. Where does the world's least annual average rainfall occur?
At Arica, in the northern desert of Chile, rainfall for the year averages only 0.02 inches over a 43-year period. The Kharga Oasis, in

Egypt, may be even drier, with mere traces of rain occurring throughout the year.

789. Where are the areas of greatest annual average rainfall in each continent?

Continent	Amount (inches)	Place
North America	251.30	Henderson Lake, B.C., Canada
South America	342.18	Buena Vista, Colombia
Europe	182.76	Crkvice, Yugoslavia
Asia	450.00	Cherrapunji, India
Africa	399.57	Dwbunja, Nigeria
Australasia	471.68	Mt. Waialeale, Hawaii
Oceans	57.81	Bermuda, Atlantic Ocean

790. Where are the areas of least annual average rainfall in each continent?

Continent	Amount (inches)	Place
North America	1.66	Greenland Ranch, California
South America	0.02	Arica, Chile
Europe	6.60	Astrakhan, U.S.S.R.
Asia	1.93	Aden, Arabia
Africa	Trace	Kharga Oasis, Egypt
Australasia	5.12	Charlotte Waters, Australia
Oceans	9.41	Sao Tiago, Cape Verde Islands

791. Where is the world's record rainfall for 24 hours? On July 14–15, 1911, 46 inches of rainfall was recorded at Baguio, Luzon, in the Philippine Islands. Other high 24-hour rainfall amounts are: 38.2 inches at Thrall, Texas, on September 9, 1921; 26.12 inches at Hoegees Camp, California, on January 22–23, 1943. The 24-hour rainfall record for Alaska is 14.13 inches at Cordova on December 29, 1955. In the British Isles, a 24-hour fall of 9.56 inches was recorded at Bruton, Somerset, on June 28, 1917.

792. What are some great rainfall extremes in the United States? The greatest local average annual rainfall in the United States is 150.73 inches at Wynooches, Oxbow, Washington. Greatest rainfall for a calendar year was measured near Cougar, Washington—171.83 inches. Greatest rainfall for a calendar month was 71.54 inches at Helen Mine, California. Louisiana has the greatest annual average rainfall for a single state—55.11 inches (average annual precipitation for the entire country is about 29 inches). The foggiest place in the United States is the Libby Islands just off the coast of Maine, which has an average of 1554 hours of fog per year. On the Pacific Coast, Point Reyes, California, averages 1468 hours of fog each year. In 1907, 2734 hours of fog were reported at Sequin Light Station, Maine.

793. Which United States areas receive the least amounts of rainfall? Greenland Ranch in Death Valley, California, receives only an average of 1.66 inches of rain a year. One extreme minimum rainfall record in the United States includes a total fall of only 3.93 inches at Bagdad, California, for a 5-year period (1909–1913). Nevada has the least annual average of rainfall of all the states—8.8 inches.

794. What are some extreme snowfall records in the United States? Greatest average annual snowfall amount is 575.1 inches at Paradise Ranger Station, Rainier Park, Washington. At the same location, the greatest amount fell in one season—1,000.3 inches. The greatest amount of snowfall for a calendar month occurred at Tamarack, California, in January, 1911—390 inches.

One snowstorm, on April 14–15, left several records in its wake, including the greatest 24-hour amount—76 inches—at Silver Lake, Colorado.

In Alaska, the greatest snowfall ever measured for a calendar month was 204.2 inches at Thompson Pass in November, 1952.

795. Where have the lowest and highest pressures been recorded? The lowest barometric pressure ever recorded on a land station anywhere in the world was 26.35 inches (892.3 millibars) on the Florida Keys during a hurricane on September 2, 1935. Pressures of 26.16 and 26.30 inches have been reported from ships in typhoons in the Pacific and the Bay of Bengal, but cannot be classified as official because of questionable standards of instrumental and observational accuracies. Lower pressures have undoubtedly occurred in other hurricanes and especially tornadoes. Similarly, the highest wind of official record occurring at Mt. Washington, New Hampshire, in April, 1934 (a gust of 231 miles per hour), has undoubtedly been exceeded, but standard measurement is difficult.

The highest barometric pressure ever recorded in the United States was 31.29 inches (1,059 millibars) at Lander, Wyoming, on December 20, 1924. The highest sea level atmospheric pressure in the world was recorded at Irkutsk, Siberia—31.75 inches—on January 14, 1893.

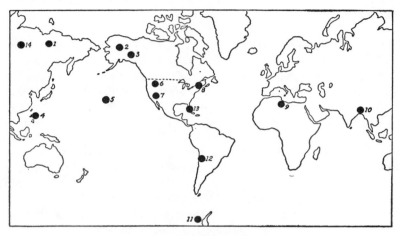

WORLD WEATHER EXTREMES

(1) Lowest temperature recorded on Earth: —90° F., Feb. 5 and 7, 1892, Verkhoyansk, Siberia; (2) Lowest temperature in Alaska: —76° F., Umiat and Tanana; (3) Lowest temperature in North America: —81° F., Feb., 1947, Snag, Yukon; (4) World's record rainfall in 24 hours—46 inches, July 14–15, 1911, Baguio, Luzon; (5) Greatest average annual rainfall in world—471.68 inches, Mt. Waialeale, Kauai, Hawaii; (6) Lowest temperature in U. S.: —69.7° F., Jan. 20, 1954, Rogers Pass, Montana; (7) Highest temperature in U. S.—134° F., July 10, 1913, Death Valley, Calif.; (8) Highest wind speed ever recorded—231 m.p.h., April 12, 1934, Mt. Washington, N. H.; (9) Highest temperature recorded on Earth—136.4° F., Sept. 13, 1922, El Azizia, Libya; (10) Rainfall record for one year—1,042 inches, Aug., 1860—July, 1861, Cherrapunji, India; (11) Lowest average annual temperature: —14° F., Franheim, Antarctica; (12) Least average annual rainfall in world—0.02 inches, Arica, Chile; (13) Lowest barometric pressure recorded—26.35 inches, Florida Keys, Sept. 2, 1935; (14) Highest barometric pressure recorded—31.75 inches, Irkutsk, Siberia, Jan. 14, 1893

VII. APPLICATIONS

Introduction. What do you do? Do you grow zinnias? Fly an airplane? Operate a fleet of trucks? Market fuel oil? Build houses, bridges or highways? Sell air conditioners, bread or furs? Ship bananas? Manage a department store? Command troops? Direct motion pictures? Sail yachts?

Whatever you do, the chances are that in some way, large or small, you are a weather consumer. Somewhere, obvious or hidden, is an area of your activity that is shaped to a degree by the factors of weather and climate.

A survey was recently made of the uses of weather information to business and industry in the United States. Quoting the Chief of the U.S. Weather Bureau, "There was an attempt to assign monetary values to the savings or profits realized through applications of daily weather reports, forecasts, storm warnings, and climatological data. The survey could not attempt to include all American business and industry, but the sampling was sufficiently broad to give an indication of the order of magnitude of weather service values. The total for the United States ran into 10 figures annually—a tremendous implication of what weather means to the economy of the nation and a reminder of how weather and climate affect our daily lives and means of livelihood."

It is not within the scope of the following section to illustrate more than just a few ways in which weather and climate are considered in relation to a few specific activities. A general word of advice might be offered to diverse groups of people to whom some serious weather study is necessarily a matter of vocation. Some may be interested only in temperature; others wind or ice or rain, weather in the air or on the sea. But—whatever the specific application—it is all *weather*—following basic general laws and principles. The trick is to first understand the principles and then to seek methods of application to particular problems so as to obtain beneficial or profitable uses of weather and climate.

796. What branches of industry are most dependent upon weather vagaries? If a list of industries were made whose operations were most directly affected by weather, the following would be outstanding: agriculture, transportation, communication, heating, cooling, lighting and the motion-picture industry.

797. What is agrometeorology? It is the study of the relationship of weather and climate to agriculture. The most important weather element of interest to the farmer is, of course, precipitation—either present moisture in the form of rain or moisture which is deferred in the form of snow. Despite the fact that agricultural climatology is not a new science (Its broad outlines were marked out more than 200 years ago), a serious lack of scientific climatological studies exists as applied to agriculture. The revolution brought about recently by new implements, insecticides, fungicides, maintenance of soil fertility, prevention of erosion, better seed and new strains and varieties of plants cannot be credited to meteorologists or climatologists. The general problem of agrometeorology is twofold: (1) to include plant measurements and phenological records along with detailed data of weather and climate environments in field experiments; and (2) a more comprehensive record of weather in the important American agricultural areas is needed and better methods of correlation of this information to agricultural needs.

798. What is agricultural microclimatology? It is an important and more recent study of weather and climate as related to farming operations. It includes special climatological observations made in *close proximity* to the growing plant with precise observations obtained in the layer of air near the ground.

799. How does soil form? Different kinds of soil form under different conditions. Typically, sun, rain, temperature changes and ice cause the crumbling of rock. Some of the minerals are changed and dissolved by water. At some time, plants appear and take hold—at first maybe only mosses and lichens; eventually trees. They die and form humus, their organic material being added to the crumbled sand or clay substances. Bacteria and other microorganisms cause the decay, and some "fix" nitrogen from the air and add it to the soil. Washed-down minerals and clay deposit under the surface and, gradually, a profile of soil with distinctive characteristics forms shaped by the climate and other conditions under which the soil is formed. In a chemical sense, soil is either acid or alkaline and physically, sandy, clayey or loamy.

800. What are some relationships between soil and climate? There are many variations. Vegetation, for example, modifies climate locally; wind, sunlight, rain, snow are different under a dense forest, under grass, under sparse desert vegetation. Steep slopes and gentle slopes, slopes facing the Sun and away, even small mounds and pits, all have different effects on the climate. At the surface and under the soil changes in air pressure at the surface are recorded instantly in the soil, even deep down, because of the pore spaces into which air enters.

Temperature of the soil is especially influenced by the amount of water and organic matter it contains; the less it contains, the more quickly the soil responds to changes in air temperature. The temperature of the rain has an influence. Evaporation has a cooling effect and greatly modifies the heat coming from the sun.

801. What weather elements affect plant growth most? The big three are temperature, moisture and light.

802. How does temperature influence growing plants? Temperature influences every chemical and physical process in plants and determines the great production belt for various crops. Though plant life as a whole is enormously adaptable (some algae thrive in hot springs at 200° F., and arctic plants survive —90° F.), most plants will grow only within a much narrower range. For each species and variety there is a minimum below which growth is not possible, an optimum at which growth is most rapid and a maximum beyond which growth stops. Damage from cold is a universal hazard throughout the United States, even in subtropical fruit-growing areas, because for economic reasons production is always being extended beyond the safe seasonal and geographic limits. One of the objects of plant breeding is to make this extension possible by creating hardier strains.

Plants vary in reaction to cold and many make surprising recoveries because not all their parts are equally effected. Nor is cold always harmful. Deciduous fruit trees, for example, go into a rest period during which no growth or injury occurs. Winter wheat requires a cold period in germination stages.

803. How does light influence plant life?
Light affects plant life in two major ways—both in a combination of opposite effects. For example, light is essential for the process of food manufacture within the plant, but the less light, the more a plant grows in length. Growth speeds up at night and slows down in the daytime. With many plants, day length rather than temperaure sets the time of maturity; they will flower and produce seed only when the days are of the right length.

The intensity of the light has different effects on different plants. Some reach maximum production with high light intensity—as in irrigated areas in arid regions. Others—sunflower, buckwheat, tobacco—produce more when slightly shaded.

804. How do vegetable growers get around weather problems?
At one time the native range of many vegetable plants was narrow with very exacting climatic requirements; yet, today they are grown over much of the Earth in climates quite different from those in their old regions. Seed is produced in the most favorable regions, often far removed from the places where the plants are grown for food. Short-season plants are grown to follow the seasons; northward in the spring, southward in the fall. Young plants are started in the South, and shipped to the North for transplanting weeks later. Greenhouses, hotbeds, cold frames and plant covers are used to lengthen the growing season. Irrigation is used to overcome droughts and make deserts productive. New plant varieties are bred to overcome climatic handicaps. Fresh products are hauled, flown or shipped long distances because of modern developments in refrigeration, transportation and packaging. Finally, the vegetable grower can afford to take more chances with the weather than other producers of crops—tree fruits, for example.

805. Where are the best climate areas in the United States for vegetable production?
The great market and truck-gardening areas in the United States are near large bodies of water, which reduce climatic extremes, or in protected valleys. Experts list these areas as: (1) a belt along the Atlantic and Gulf Coasts from Massachusetts to

Texas; (2) a broad area along the Great Lakes from New York into Minnesota; (3) certain intermountain valleys in Colorado, Utah and Idaho; (4) the Rio Grande Valley in Texas; and (5) the Pacific Coast and intermountain valleys of Arizona and California. The last three areas grow vegetables mostly under irrigation.

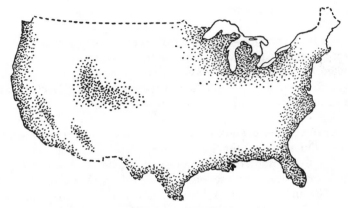

Best climate areas for U. S. vegetable production

806. What are the lowest temperatures citrus fruits can tolerate?

Of the citrus fruits, the Satsuma orange, which can stand 18° F., can be grown farthest north. The lime is injured at 28° F. Other citrus fruits are in between these limits. In general, the smaller varieties freeze first, followed by the larger fruits and, if the temperatures fall low enough, the trees themselves.

807. How do orchardists fight cold temperatures?

When temperatures start dipping critically (in the mid-30s), numerous orchard heaters are burned per acre. Wind machines in the form of powerful motor-driven propellers mounted on towers are turned on so as to circulate air around the orchard and possibly break up any shallow layer of supercooled air. Artificial light is sometimes applied on the theory that it will stir the tree sap circulation. Banking of soil above the bud union is another method of combating cold.

808. How do temperature and moisture affect citrus fruits?

High temperatures result in a green or yellowish-green rind color, cooler temperatures in a deep yellow or orange color. Normal production requires about 35 inches of water a year from rainfall or irrigation. High relative humidity apparently has a favorable effect on smoothness and thinness of skin, juiciness, richness; low humidity, with high temperatures, causes dropping of immature fruits. Both insects and diseases are related to climate; certain pests develop in humid areas, others in dry areas.

809. What is the growing season? The growing season of crops susceptible to frost damage—the so-called warm-weather crops—is restricted by the number of days between the last killing frost in the spring and the first in the fall. The length of the period between these dates is usually referred to as the *growing season.*

810. What are the growing seasons in different parts of the United States? On some of the Florida Keys freezing temperatures have never occurred, and, consequently, the frostless season covers the whole year. These are the only frost-free localities in the United States. Throughout most of Florida, along the coast of the Gulf of Mexico, and in favored localities in Arizona and California, the average growing season is more than 260 days. Along the northern margin of the Cotton Belt it is about 200 days and in the northern part of the Corn Belt from 140 to 150 days. In northern Maine and northern Minnesota, where hay, potatoes, oats and barley are the principal crops, it is about 100 days; and in higher altitudes in the West, about 90 days.

811. How deep does frost penetrate into the ground? Average depth of frost penetration varies greatly with latitude and persistency of cold weather. For example, the northern portions of Alabama, Mississippi, Georgia and Louisiana may average about one to three inches of frost penetration. The penetration of frost along the northern tier of states ranges widely from 12 to 72 inches. In some arctic districts shafts have been sunk more than 200 feet without going through permanent frost.

812. How does a dense forest modify climatic conditions? A dense forest has been compared to an enormous umbrella. Taking away or adding forests can influence climate locally to some extent. Temperatures are lower in a temperate-region forest, light is reduced greatly, the soil is several degrees warmer and freezing is much less deep. Humidity is greater and dew and fog form readily over adjoining fields. Evaporation from the soil under a forest is greatly reduced. Wind velocities are reduced. The water-storing capacity of the soil is greatly increased. The significance of this underground storage is indicated by the fact that a rise of six inches in the water table of the Tennessee Valley would mean an additional storage, in the ground itself, of four times as much water as the Norris Reservoir holds.

813. What effects are caused by forest reduction? Higher temperatures, much greater wind velocities and greater evaporation in denuded forest lands undoubtedly lead to soil destruction. Measurements made in various places also prove the great value of forest cover in preventing runoff and reducing flood hazards.

814. What weather conditions are linked to the spread of forest fires? When the forest leaf litter, dead grass, bush and dead wood are dry, a forecast of low humidity is a call for extreme precaution. If, at the same time, there is a forecast for enough wind to be called a breeze, the demand is for even greater care. Many forest fires are started by "dry" thunderstorms and under circumstances of low humidity and high wind; such fires are terrifically destructive. In mountainous areas in the West where dry warm Föhn winds blow across forests from the leeward sides of mountains, the fire risk is exceptionally great.

815. What are the special weather problems of commercial aviation? In charting a course for flight, the pilot of an airliner is concerned with several types of weather phenomena which may create hazardous conditions. At take-off and landing, visibility conditions and cloud ceilings are important considerations as well as the possibilities of extreme gustiness. During the flight, the possibilities of ice formation on wings, propeller blades, vertical fin, outboard stabilizer or in the carburetor may cause dangerous cumulative effects, involving a drag increase, a lessening of lift, a fall-off of thrust and, in general, a cutting of power demanded of the engines to maintain flight.

Severe air turbulence which goes beyond annoying bumps to cause structural damage is another important concern. Turbulence near

VISIBILITY

ICING ON WINGS

HAIL AND TURBULENCE

Weather hazards to aircraft

jet stream flows, in severe vertical currents within thunderstorms, and
in the vicinity of mountain ranges, deserves particular attention.

The punishing effects of a severe hailstorm is another aviation
hazard.

**816. What weather safeguards are employed by commercial air-
lines?** The main safeguard is a constant analysis of weather reports
received from the vast network of weather stations along the different
flight routes. This information of weather conditions at the surface
and in the upper air is studied, interpreted and delivered by means

of careful briefings to pilots and crews so as to provide a detailed picture of expected weather conditions along the flight path from take-off to terminal points. Most major airlines maintain complete crews of trained meteorologists who supplement the Weather Bureau's flight forecasting service.

Carrying many ingenious mechanical devices such as deicing equipment, storm-spotting radar, navigational and communication instruments, the airline of today is virtually a weather station in the air, with the meteorologically trained pilot the recipient of a constant stream of weather information. The amazing weather safety records of scheduled airlines reflect the elaborate and costly attention given to aviation weather problems by the airlines and many governmental agencies.

817. Do planes really fly above the weather?   Although commercial aircraft fly at heights which were once thought to be weather-free, they must still contend with effects of turbulence and clouds. In the main, modern aircraft are far less susceptible to the more complex and hazardous conditions which prevail at lower levels. Detailed weather reports allow selection of flight paths which present the least amount of trouble. In the short period of about 50 years after the first flight, the airways have become, virtually, a well-mapped, practicable highroad for much of the travel and commerce of the world.

818. What is an air pocket?   Air is in constant motion, like the sea; and an airplane is affected by that motion. There are no air pockets, for that would imply that there are areas of vacuum in the atmosphere. But there are winds blowing laterally and air currents moving up and down. These vertical currents give the flight vertical motion. An airplane does not *drop*. The currents occasionally cause the plane to lose some altitude so that it actually rides a downward current of air as a ship rides a downward current of water.

819. What weather conditions cause delays or cancellations of flights?   Weather conditions must be above certain standards at the destination and at alternate airports as well as at the point of origin of the flight. *Visibility,* the range of distance which can be seen along the ground, and *ceiling,* the distance between the surface and the bases of clouds which cover more than half the sky, are two important factors which determine whether or not flights can take place.

820. What are the minimum conditions which close an airport to all commercial air traffic? The minimum ceiling and visibility restrictions to flights vary greatly, depending upon terrain, end-of-runway obstructions, the kind of runway lighting facilities and similar factors, as well as different standards for night and daytime flights. For example, at Idlewild in New York, two-engined craft can take off or land when the ceiling is at least 300 feet and visibility one mile. Four-engined planes require only a 200-foot ceiling and one half-mile visibility at night and a 100-foot ceiling and one fourth-mile visibility during the day. At mountainous and valley airports, minimums are much hgher. At Wilkes-Barre, Pennsylvania, a minimum ceiling of 500 feet is required with visibility of one mile. At Rutland, Vermont, a ceiling of 2,500 feet and visibility of two miles is required, with no commercial traffic permitted at night.

821. What government agency is responsible for general flight rules and standards? In general, the Civil Aeronautics Board establishes broad-scale aviation regulations, and the Civil Aeronautics Administration has the role of recommending certain policies, administering and enforcing flight regulations and standards. In a sense, the CAA represents and protects the public, fosters and promotes an understanding of matters pertaining to aviation.

822. What precautions should all construction engineers take regarding climate? Engineering is one phase of industry which is deeply affected by climate rather than immediate weather. It is a regretful fact that dams burst, bridges collapse, housing projects suffer an assortment of ills because the designing engineers neglected to allow for maximum accumulation of water, highest winds or a host of climate extremes peculiar to a particular region. The collapse of the Puget Sound Bridge in 1940 stemmed from the fact that designing engineers neglected to study the greatest wind forces which might occur in the area.

823. When building a home, what are some climate and weather considerations? The new science of microclimatology—the study of small-size weather—reveals that a large amount of homes are built without regard to consideration of how local climates may be changed by such factors as slope of the land, orientation of the home to the Sun's path across the sky, the water-storing efficiency of surrounding land, exposures to prevailing winds, nearness of water bodies, etc. The wise home builder will plan ahead so that the home becomes truly functional with respect to a local climate, with the best possible effects of Sun, temperature and wind obtained. A consultant climatologist might well be initially far more important than an interior decorator!

824. How is railway transportation affected by weather? For rolling stock operations of railways, the three most important weather factors are visibility, precipitation and weather extremes. For railway maintenance a sharp weather eye must be turned on abnormal rainfalls that may result in washouts and floods, or collapse of bridges along routes. Heavy snow and icing are also fearful elements to the maintenance man. Track plowers must be fleeted and ready to operate. Heavy icing conditions cause havoc along miles of telephone and telegraph wires, disturbing important railway communications, not to speak of track and switch equipment that is made inoperative under coats of ice.

825. What are weather problems of highway transportation? On the highway, passenger and freight vehicle drivers are immensely concerned with the amount and type of precipitation along the road, especially if it is of the solid variety—sleet, hail or ice particles. The road surface becomes dangerously slippery. Papers record almost daily in the winter the gruesome details of accidents caused by iced roadways or poor visibility along the road because of pea-soup fogs, blowing snow, or again, the roadway condition as it changes under extremes of heating and freezing.

826. Which weather elements are of particular concern to the boatman and mariner? Like most outdoor people, boatmen and mariners are presumably interested in sunshine, cloudiness and rain—factors which have a bearing on the enjoyment of their pursuit. Two weather particulars, however, which may influence more than mere comfort, are of special concern. The first of these is visibility, particularly its restriction by fog. The second is the wind and the seas that rise with the wind.

827. How does the government recognize the weather needs of the mariner? The Marine Service is one of a number of specialized services of the U.S. Weather Bureau. It is concerned with providing weather information for all marine interests. Various publications of interest to the boatman or mariner are available for nominal costs from the Superintendent of Documents, U.S. Government Printing Office, Washington 25, D.C. These pamphlets and booklets include schedules of the principal marine weather broadcasts covering specific coastal and water areas. A recent bimonthly publication of the U.S. Weather Bureau, *The Mariner's Weather Log,* contains articles on marine meteorology and climate.

828. What weather problems face the communications industry? Windstorms and ice storms cause particular damage. In more open regions, telephone and telegraph wires which are not cabled and conduited underground are brought down, many miles at a time, by severe winds or by the heavy weight of glazed ice formation. In very severe cases of wind or ice, radio and television antennae and even transmission towers may crash to earth. A more indirect disturbance of radio communication channels is caused by the effects of solar activity on the radio-reflecting layers in the ionosphere.

829. What special weather problems concern the fuel-oil industry? When cutting crude oil for various needs, the amounts of oil needed for heating purposes must be anticipated long in advance of the heating season. If temperatures deviate very much from the expected normal, there are risks of oversupply or undersupply. If winter temperatures should average more than 10% colder, for example, the danger of serious shortages of supply may result. If the heating season is much warmer than normal, the fuel-oil market may be depressed financially by excess supplies. Long-range temperature forecasts are needed for application to this industry.

830. How is the production of steel influenced by weather? From the mining of raw materials to the shipment of finished steel products, weather exerts a definite, though generally a taken-for-granted influence in the production of steel. The severity of freeze-ups in the ore fields and on the lakes determines the length of the ore-shipping season. Harbor and docking operations and facilities are handicapped by strong winds, icing conditions, fog and high water. Snow and rain hamper the handling of raw materials needed in the manufacturing process. At the furnaces, differences of water vapor in the air alter the heating value of gases. High humidity in the summer may cause rapid rusting of unprotected steel in finished form.

831. How do sudden weather changes affect the lighting industry? Normally, the electric load in a large city varies with a minimum value around midday to one main maximum in the early evening and another maximum in the early morning. If a large cumulonimbus

cloud develops suddenly in the afternoon hours and darkens the city considerably, thousands of lights are switched on. This creates a large drain on power-plant generators already running, so that calls are made for more generators to be started as quickly as possible.

832. What are the special weather problems of a motion-picture director? Motion-picture production entails enormous amounts of costs related to the filming of outdoor scenes. When the cast of a production is on location, the requirements of the director for particular weather sequences are fantastically diverse. He demands accurate forecasts of natural light conditions, particular cloud formations as background, different types of sunrises or sunsets, particular kinds of winds; in short, the most tailored kind of weather information consistent with his needs for matching the film story.

833. How is meteorology related to the chemical industry? One important problem requiring special meteorological study is the control of wastes in the atmosphere. Meteorologists attempt to predict the dispersion of stack gases and assist in proper site selection and size of stacks for new plants.

834. How is weather related to department-store advertising and merchandising? The fact that weather affects sales has long been recognized by department stores. Many stores regularly write a brief description of the weather for the day immediately following the day's dollar sales figure in their records. The purpose of this is to help in determining the "planned sales" figure for the same day a year later. Although there are many weaknesses inherent in this method of correlation, it is becoming a springboard for far more elaborate quantitative measurements involving weather and sales in a department store. Some stores are applying advertising and merchandising techniques based on sales expectancy indexes as related to weather conditions.

835. What are the weather concerns of a gas utility? Weather is an important factor in the operation of a gas utility since temperature, wind, sky cover and precipitation affect the customer's demands for gas. The primary mission of the meteorologist working for a gas utility is to provide weather information required by the personnel who direct the production and distribution of the company's gas supply. Customer demands for gas must be anticipated 24 to 36 hours in advance, because of the time required to transport the gas from the field to the load centers. Temperature is very important. A one-degree fall in the mean temperature, when the mean is below 65° F., will increase the daily customer demand by 28 million cubic feet of gas, approximately.

836. Is there a connection between weather and coal-mine explosions? Weather conditions existing outside of 41 different coal mines which suffered major explosions were studied. In all but one, there had been sharp drops of barometric pressure, due to an approaching low pressure area. Most of the explosions took place a day or so after the pressure minimum. The explanation may be that lowered atmospheric pressure encourages the flow of gas out of the coal. If the increase is too much for the fans to handle, in poorly ventilated mines, the gas may accumulate until it forms an explosive mixture. Falling barometric pressure, therefore, should be a special warning to miners.

837. What is the importance of weather and climate to warfare? The story of weather importance in military operations can be developed into a heavy volume. The elements have turned tides of battle since armies first clashed. As one magazine expressed itself on the eve of World War II; ". . . weather, next to stomachs, is war's most basis consideration." If major wars should occur again, they will be fought in the upper air and space and may well be decided by the availability of the most expert meteorologists, geophysicists and astrophysicists. In the following questions, a few random examples of how weather affected military actions at different points of history will be considered.

838. How did climatology play a part in the Berlin Airlift? Climatology played an important part in Operation "Vittles," the Berlin Airlift, following World War II. Air Weather Service of the U.S. Air Force compiled detailed statistics in advance of the approximate frequency of specific types of ceiling and visibility conditions at each terminal. From these studies, an accurate calculation of over-all gas load and payload was made possible. The frequencies predicted on the basis of climatology approximated quite closely those actually experienced.

839. Why was the Korean invasion begun in June? Climatic conditions guided the choice. The North Korean commanders were undoubtedly aware that any opposition would be relying heavily upon air support. By scheduling the attack to begin early in the summer monsoon season, the chances of successful air support by the opposition were at a minimum.

840. How did climatology relate to the invasion of Poland by the Germans in September, 1939? The invasion of Poland, initiating World War II, was launched in September because German meteorologists had determined from historical weather information that Panzer tanks could expect the driest soil conditions and their Stuka dive bombers would be hampered least by cloudiness in this month.

Excellent weather conditions prevailed and the operation was completed in three weeks.

841. How did weather play an important part in the evacuation at Dunkirk? In June, 1940, the sudden surrender of Belgium placed the British Expeditionary Force of some 200,000 men, together with more than 100,000 French and Belgian soldiers, in a severely exposed position. They fought their way across northern France to Dunkirk where British air and sea power was concentrated. Despite the heroic action in the air and on the sea, it is doubtful if a tremendously successful evacuation (losses less than 10%) could have been made if not for a thick fog which rolled over the English Channel from seaward, screening the operation from the German level and dive-bomb attackers.

842. How did the Japanese take advantage of Aleutian weather? The Aleutian Islands area, one of the most consistently storm-covered regions, was invaded by the Japanese under cover of foul weather which kept American planes on the ground. Kiska was later evacuated under storm conditions to such a successful extent that the Americans found an empty island following a full-scale attack.

843. How did weather hurt the U.S. Navy in World War II? One of the most devastating blows received by the Navy in World War II was delivered not by the Japanese, but by a typhoon on December 17–18, 1944, following a withdrawal from a series of strikes against Luzon in the Pacific. The American fleet consisted of 20 carriers, 8 battleships and numerous small craft, besides two dozen tankers. Dangerously meager weather reports neglected to spot accurately the center of the vicious storm which struck with destructive wind force (up to about 150 miles per hour). A total of 790 men was lost with several ships, and heavy over-all damage resulted before the storm moved off.

844. What were the weather requirements for the Normandy invasion in World War II? Weather forecasters were presented with a complicated set of operational weather requirements by various units of the Armed Forces. The Navy did not want a heavy wind and swell condition in the Channel. For air transport, a ceiling of 2,500 feet and visibility of three miles were needed. Heavy-bomber operations required a ceiling not lower than 11,000 feet with clouds below 5,000 feet not covering more than half the sky. Lighter bombers called for ceilings no lower than 4,500 feet and visibility no less than three miles over the target. Fighter pilots required a ceiling of at least 1,000 feet.

845. What military group had the responsibility for the D-day invasion weather forecast? Climatologists working with commanders in chief had determined during the latter period of the winter that the month of June would be the best month consistent with the special weather requirements of air, sea and land-force units. Early and mid-June, 1944, periods were tentatively selected by the high command with choice of the exact date left to a specially selected forecast team. The forecast unit was made up of four British military meteorologists and two meteorologists from the U.S. Strategic Air Force Headquarters. They represented the final unit, although much of the interim weather analysis for Supreme Headquarters of the Allied Expeditionary Forces rested with the weather unit of the U.S. Strategic Air Forces in Europe.

846. How did weather influence the Battle of the Bulge? The German's final desperate bid for victory in 1944 took the Allies completely by surprise in the Ardennes sector. Allied air strength was reduced to an almost helpless condition because of thick weather that acted as a cover for the German penetration. Clearing finally took place and allowed strong air support of the U.S. First and Third Armies which destroyed Runstedt's forces, but not before 50,000 Allied casualties had occurred.

847. What spurred local weather forecast techniques in World War I? The widespread use of poison gas and smoke in the trench warfare of World War I demanded more effective means of forecasting surface and lower level winds; surface temperature of air and ground; and the kind and amount of precipitation. In many cases where gas was employed, poor wind forecasting resulted in gas being wafted back to the attacking group. The heavy increase in aircraft operations over France in the later war period also forced developments in forecasting techniques of upper air wind and cloud conditions.

848. How did weather hurt the British at Gallipoli?   British Expeditionary Forces attacked the Turkish peninsula at Gallipoli twice in 1915–16 in the hope of capturing Constantinople. They never made it, with climate bedeviling the British as much, if not more, than well-hidden defense gunners. The operation was planned without any consideration given to the desertlike summers in Turkey. As a result, provisions for water supply were dangerously inadequate in the first summer campaign. In the winter campaign, blizzard conditions helped to decimate troops because of exposure and frost bite.

849. Why has weather been called the reason for Napoleon's downfall?   The deep freeze of Russia in 1812 turned Napoleon's Grand Army into a pitiful few thousand emaciated warriors who staggered back to France, defeated by relentless cold. Hitler's supermen were similarly cut down by the icy Russian premature winter 129 years later. But it was at Waterloo, the greatest battle of the nineteenth century, where weather literally shaped history by is effect on Napoleon's campaign and caused his downfall.

850. How did weather help to defeat Napoleon at Waterloo?   On the night of June 17–18, 1815, torrential rain, lightning, thunder and high winds marked the area just south of Brussels where Napoleon's army faced the English under Wellington, with a Prussian army under Bluecher a few miles away hurrying to the aid of Wellington. Only a few thousand yards of rain-pelted mud separated the English and French. Napoleon was to have attacked at 6:00 A.M. but so desperately wanted drier ground for his rapid and effective concentration of artillery that he decided to wait a few hours in the hopes that the Sun would emerge. The Sun did not come out and the battlefield remained a mass of mud. It was not until noon that the French assault was ordered. As the armies locked in bitter combat, Bluecher's Prussian army arrived in time to execute a flank attack on Napoleon's army which resulted in a hasty French retreat and the end of the Hundred Days. Of this historic battle, Victor Hugo wrote: "A few drops of water . . . an unseasonable cloud crossing the sky, sufficed for the overthrow of a world."

851. How did weather play a part in an important Revolutionary War engagement? In March of 1776, Lord Howe, Lord Percy and Admiral Shulham decided to await the nightfall before counter-attacking General Washington's troops who, in a surprise movement, had moved into Dorchester Heights near Boston. While the British waited, a violent windstorm and torrential rain moved in from the south. Vessels were driven ashore and troop movements so hampered that it became evident that the assault attempt would be ruinous. Howe called a council of war, and its members advised the instant evacuation of Boston.

852. How did weather hurt the Spanish Armada? August of 1588 was a stormy month on the Atlantic. The huge Spanish galleons were harried by western ocean winds which reduced maneuvering ability against the small and swift English craft. The much-touted armada limped back to Spain, half the fleet victims of the intrepid English and the weather.

853. How did weather hurt Alexander the Great? History records that weather dealt the great military strategist a severe blow in one of his campaigns at around 325 B. C., near the end of his career. Traveling from India to Mesopotamia on his way home, Alexander's weary army encamped at Baluchistan in the western part of Pakistan, the region then being known as Dedrosia. The weather was hot and extremely dry and the men camped in the arroyos peculiar to the area. A sudden development of severe thunderstorms and cloud-bursts caused an unusually severe flash flood which resulted in the death of thousands of men by drowning, along with the loss of much valuable equipment.

854. How did weather decimate a Persian army? One of the most catastrophic losses in history due to weather occurred around 500 B. C. Cambyses, the Persian king, had just conquered Egypt with a huge army and was on his way through Libya to the oasis of Sieva. A violent sandstorm engulfed the army with fierce winds driving the sand in thick, punishing clouds. It is estimated that only a trickle of survivors remained out of an army of about 20,000 men.

VIII. MEDICAL METEOROLOGY

Introduction. About 2,500 years ago, Hippocrates, the father of medicine, wrote a treatise entitled *Airs, Waters and Places.* He showed deep concern and interest in the effects of climate and weather on man's comfort and health. For a long period of history his original thoughts about environmental factors were the only ones that man took into consideration in attempting to explain disease phenomena. In the latter part of the nineteenth century, the impact of Pasteur's discoveries diverted interest from climatic factors as the cause for sickness. Medical studies veered suddenly from the important study of man in his environment to that of bacteriology and the allied sciences.

Aided by the tools of modern technology, the study of the influence of climate and weather on human comfort and health is emerging from its long-dormant state. To be sure, this old-yet-new science is plagued by growing pains. A critical survey of the entire field is lacking and the horizons of study seem frustratingly wide. But many individual physicians, while exploring other fields of medical endeavor, have been drawn irresistibly to the obvious link that connects the health and comfort of man to his physical environment.

The following section is derived mostly from the findings of physicians, physiologists and meteorologists who have made special studies of the effects of climate and weather on man. In no sense is this chapter intended to be definitive. The intention is to outline a few of the areas of interest in medical meteorology.

855. What is bioclimatology? In its broadest sense, bioclimatology may be considered a science which deals with the effects of the complete physical environment on all living things. It includes the reaction and behavior of plants and animals as related to many individual factors that comprise climate and weather such as heat, humidity, atmospheric pressure, sunlight, winds and precipitation. This section concerns a few fundamental relationships of the effect of climate and weather on man.

856. How does the human body adjust itself to weather extremes?
The body is continually working to keep up a balance between heat
it produces internally and the heat it loses to the outside environment.
In many respects it is like an automatic heating and cooling system.
The food we eat represents the main source of heat. Body heat is
generated and stoked by the oxidative processes of food ingestion.
Heat is also gained from the surrounding environment by conduction,
convection and radiation. But the body cannot keep storing heat. It

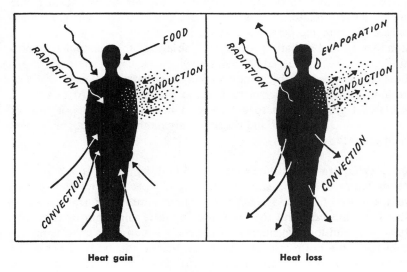

Heat gain   Heat loss

HEAT BALANCE

is essential for it to get rid of excess heat and there are four ways
to do it—by evaporation (cooling when sweat evaporates), radiation,
convection and conduction. Heat is also lost in a small way through
excreta and unevaporated sweat.

857. What part of the brain acts like a thermostat? The body
possesses thermostatic equipment which makes it possible for man
to maintain the balance of heat gain and loss and to adapt himself to
varying conditions of weather and climate. The *hypothalamus,* next
to the pituitary gland at the rear part of the brain where it joins the
spinal column, functions pretty much like the thermostat on the

living-room wall. This sensitive heat center reacts to heat or cold, alerting the neighboring pituitary gland to send the alarm to the body's entire glandular system.

This alert by the heat center results in a series of intricate physiological processes, closely bound with one another, which allow the body to continue functioning.

858. How does the body respond to heat? When the body is physically active or the temperature of the air above normal, the blood vessels dilate. This brings more blood to the surface which raises the skin temperature and facilitates heat loss by convection, conduction and radiation to the outside air. If the external temperature is further increased, a point may be reached where sufficient heat cannot be lost by conduction and radiation because the surrounding air will be warmer than the skin temperature. At this point, the sweat glands begin to secrete and heat loss by evaporation of sweat becomes the chief and possibly only method of removing excess heat from the body.

859. Why are excessive heat and humidity dangerous? If the temperature of the air and surrounding objects is above the surface temperature of the body, and if the air is saturated, no heat can be lost from the skin and the body temperature will begin to rise. The heat gain and loss mechanism of the body is supset. As the temperature rises, the pulse rate increases. An increase in body temperature stimulates the metabolism and this leads to a further increase in heat production. The effects accumulate until sometimes unbearable conditions are reached, depending upon whether the body is at rest or at work to different degrees.

860. What diseases may result from excessive heat? The two most disastrous results of intolerable heat are heat exhaustion and heat stroke.

861. What is heat exhaustion? This term applies to a multitude of cases which exhibit symptoms from a sensation of fatigue to complete collapse in hot weather. It occurs more frequently after a prolonged period of hot, moist days and nights. Symptoms usually begin with dizziness, fatigue and headache, sometimes with nausea. It is generally assumed that there is a failure of the circulatory system to compensate for the dilation of blood vessels and other adjustments to high temperature, with a resulting inadequate supply of blood to the central nervous system.

862. What is heat stroke? Heat stroke is far more dangerous than heat exhaustion. It comes on rapidly and often with drastic consequences, the outstanding symptom being a very high body temperature. In these cases, for some inexplicable reason, the heat-regulating center in the brain ceases to function and the body rapidly loses its ability to get rid of excess heat. Treatment consists in reducing the body temperature as rapidly as possible by ice-water baths, ice packs, cold sprays or ice-water enemas, until the temperature is reduced to about 102° F. Cold sponges may then be applied. The attention of a physician is demanded as quickly as possible.

863. What is prickly heat? Common in hot and humid weather, prickly heat is due to excessive sweating with skin congestion. Inflamed sweat glands cause tiny irritated red pimples to form, especially in the folds of skin where there is less chance of skin exposure to moving air which would lead to evaporation or drying. The affected skin pricks, stings, burns and itches. Remedial measures include, generally, exposure to drier air, sponging with a mild alkaline solution or the use of a good talcum powder.

864. What is the highest internal body temperature ever tolerated by a living human being? Many records of high body temperatures are poorly documented and must be considered as extremely questionable, especially those concerning malingerers. Unauthenticated records viewed with skepticism by the medical profession show that one patient in London in 1875 survived a body temperature range from 110° F. to 120° F. Another patient in Dublin in 1880 was supposed to have survived a temperature of 130.8° F. Under more exacting clinical circumstances, temperatures varying from 108° F. to 112° F. are considered very unusual. Few patients have survived these extremes.

865. What is the lowest internal body temperature ever tolerated by a living human being? Surgical experiences reveal that if a patient's temperature falls below 85° F., the chances of survival are doubtful. Industrial medicine studies of body exposure to low environmental temperatures report that when the body temperature reaches 80° F., coma sets in, although shivering may continue intermittently even as low as 75° F., with death occurring usually at 70° F.

An amazingly rare exception occurred in February, 1951, when a 23-year-old woman was found in Chicago, Illinois, in a severely frozen state, unconscious, but alive. When she was hospitalized, her body temperature was reported as being 64.4° F., (taken an hour and a half after admission to a hospital). She survived this low body temperature with subsequent amputation necessary just above the ankles. Medical researchers are studying techniques of controlled hibernation as a helpful method for the performance of certain therapeutic procedures.

866. What is the highest outside air temperature that can be tolerated by a human being?   Dr. Craig Taylor, associate professor of engineering at the University of California, Los Angeles, in the interests of science, subjected himself without serious results to an average temperature of 250° F. for 14 minutes and 32 seconds. According to a British publication listing unusual medical experiences, one human subject was reported to have survived a temperature of 270° F. for a 15-minute period and a temperature of 364° F. for one minute. These are unauthenticated reports.

867. What is meant by acclimatization to heat?   It is a shifting of the comfort zone to higher limits of temperature and humidity. This is a process whereby the body adapts itself by a series of complicated physiological changes to conditions of heat which are not normally encountered.

868. How effectively can the human body acclimatize to heat?   U.S. Army studies in World War II revealed that when suddenly exposed to hot environments (120° F. temperature and 20% relative humidity or 91° F. and 95% relative humidity), even fit young men were unable to work long or strenuously. Their working capacity was tremendously decreased by the sudden change from a temperate to a hot environment, and they exhibited a wide range of symptoms associated with excessive heat. After from four to ten days of work in the heat, they adapted themselves by acclimatization to the stresses of a hot climate and their working performance was nearly as effective as in temperate climate surroundings.

869. Does man acclimatize to cold conditions as readily as to hot environment? Man's ability to survive in extremely cold regions depends more on protective clothing and experience than upon bodily adaptation. Very little seems to be known about the long-term adaptation of man to a cold environment. Certainly the body does not shift its mechanism to meet the rigors of a cold climate as readily as it copes with a hot climate. There is even some doubt whether true acclimatization to cold actually occurs.

870. What is spring fever? Spring fever is a common set of symptoms occurring with many people in the Temperate Zones around mid-April, particularly with the appearance of a sudden warm spell following a long cold period. It is generally marked by a feeling of lassitude and is associated with physiological changes going on in the body in order to meet the changing conditions of the external environment.

When the temperature rises, the body has to get rid of heat. Dilation of the blood vessels occurs so that blood can be carried nearer to the body surface where heat can be lost more quickly to the outside air. In the early process of this blood-circulation change, plasma, a watery substance in the blood, increases in amount and is the basis for the long-held idea of "thinning blood" in the spring. The general enervated feeling that takes place is simply the result of the body's reaction to the large amount of work going on in the body process of shifting its blood circulation and production to cope with a warming external environment. For a period of a few days until the acclimatization takes place, the familiar spring fever symptoms may occur.

871. How does the body respond to cold? Cold evokes an opposite series of reactions in the body from heat. When the environment begins to cool below comfortable levels, the body endeavors to decrease the body heat loss by contraction of the blood vessels near the surface of the body. This tends to reduce the blood flow through the exposed skin and heat is conserved inwardly in the vital organ areas. If colder conditions develop, more heat is produced by an increase in muscle tone, shivering and voluntary movements. Because of these adaptive mechanisms, the body temperature does not usually fall more than 2° to 3° F. If, however, the exposure is intense and prolonged or if the body is unable to compensate due to other factors, such as alcoholism, the body temperature may continue to fall to dangerous levels.

872. What are the chief injurious effects of cold on the body?
The most damaging effects of cold are usually not the result of a fall in body temperature but of local changes which occur in the skin and subcutaneous tissue of the exposed parts of the body, chiefly the fingers, hands, toes, feet, ears, nose and cheeks. Different durations of exposure to coldness and wetness may lead to varying degrees of damage to the extremities, such as chilblains, trench foot or frostbite. Extreme or protracted exposure may lead to gangrene of the exposed part of the body and consequently to death if severe.

873. What is trench foot?   If legs are constricted or if held in more or less stationary position while exposed to coldness or wetness around freezing temperatures, injury may occur to the tissues. Pain, then numbness, may be followed by infection or gangrene, or the tissues may survive when circulation becomes re-established. Trench foot of World Wars I and II, shelter foot and immersion foot are all variants of a single disorder, modified by duration of exposure, wetness and chilling.

874. What is frostbite?   Exposure to air which is dry and well below freezing may lead to frostbite which is literally a freezing of skin tissue, usually in the extremities. If the affected part is thawed promptly, no serious consequence will develop. If freezing of the tissue is prolonged at very low temperatures, death of the tissue cells, gangrene and formation of clots in the blood vessels may develop.

875. What is chilblains?   Prolonged exposure to damp cold may cause this injury. The hands and feet are reddened, with burning and itching, sometimes with chapping and ulceration.

876. What is the treatment of injuries from cold?   In extreme degrees of exposure of the whole body to cold, it has been found that rapid application of external warmth is life-saving, just as the opposite procedure is essential in heat stroke. This represents a change of the older notion of very *gradual* application of heat.

877. Will exposure to cold precipitate a heart attack? When people whose hearts are generally under par are suddenly exposed to a much colder environment, an added and sometimes dangerous burden is placed upon the heart. The swift change of environment demands a greater expenditure of energy on the part of the heart to match the increased combustion rate of the body. This need for sudden adjustment of the body to cope with a low outside temperature is readily answered by the fit person, but may prove an excessive and fatal circumstance for the abnormally sensitive heart.

878. Why does one's nose run in the winter? When very cold air is inhaled and strikes the mucous membrane of the nasal cavity, a constriction of a network of capillaries which is imbedded in the membrane takes place. After a short period of constriction, the capillaries dilate again by reflex action. During this dilatation, there is a stasis (a pooling or slowing down) of the blood and a relative increased permeability of the cells of the blood vessels to fluids. This permits an outpouring of serum or mucous from adjacent mucosal glands. The release of liquid occurs more frequently when there is a sudden change from breathing warm air to cold. After the disturbance, a steady state ensues in which normal physiological functions are resumed and an excessive accumulation of fluid does not reoccur.

879. Does cold increase the appetite for fat? Food requirements are clearly altered by cold exposure. Studies have shown that a basal tropical diet in the neighborhood of 3,000 calories must be increased to about 4,900 calories for arctic conditions. Despite a general notion, it has not been clearly established that cold exposure favors a preference for a high fat content in the diet. A high calorie level is required and frequent small meals rather than a few concentrated large meals appear to be advantageous in a cold environment. But there is no positive evidence that a large effect on cold tolerance may be obtained by a qualitative change of diet.

880. What is the desirable combination of temperature and humidity for office working conditions? Studies by the American Society of Heating and Ventilating Engineers reveal that a temperature of 75° F. with relative humidities between 55–60% and a light air movement (about 15 to 25 feet per minute) represent optimum conditions of comfort. This comfort level applies to general office work conditions for all seasons of the year. For more strenuous physical work, the optimum temperature level is lower—from 60° F. to 68° F., depending on the degree of strenuous work. At higher temperatures or when the work is heavier, air movement should be increased

to facilitate heat loss, but velocities in excess of 150 or 200 feet per minute are considered undesirable. Strict attention must be given to the parts of the body exposed to strong air movement. For example, some ill effects may be encountered if a strong air movement is directed to the neck or back, rather than to the face.

881. How does summer air conditioning relate to comfort or health? One problem of air conditioning has been to determine whether the sudden changes between high outdoor temperatures and lower indoor temperatures are harmful, especially if skin and clothing are wet with perspiration. Detailed studies conducted by the American Society of Heating and Ventilating Engineers in Florida have shown that if the optimum comfort level of temperature, humidity and air movement, as described in the preceding question, are maintained indoors, no harmful effects have been noticed by people moving in and out. Other studies have compared illness records of employees who work in air-conditioned buildings and those in nonair-conditioned buildings. Harmful effects have not been revealed.

882. What precaution should be taken when heating the home in winter? Inside a house in winter, when dry, cold air is heated to comfortable temperatures, the relative humidity may fall well below 20%, and the air may be drier than the air over the Sahara Desert.

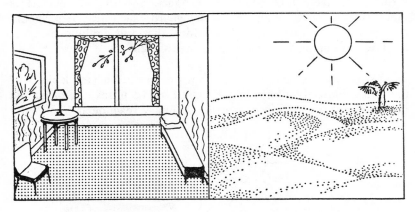

This is not healthful because very dry air irritates and dries the mucous membranes of the nose and throat. Precaution should be taken to increase the moisture in the air to comfort levels that match the temperature.

883. What is the ideal climate? The ideal climate for the mental and physical health and comfort of most people is one that is marked by frequent but moderate changes in weather, variations in temperature from night to day, and gradual seasonal changes. Such a "middle" climate has been found bodily invigorating and stimulating.

884. What geographical region meets the general description of an ideal climate? A climate that is neither invariably hot or continuously cold, that is neither monotonously rainy nor foggy, nor arid and cloudless—in other words a climate similar to that of the central region of North America—is probably most conducive to exercising the body's power of adaptation and reaction.

885. How is comfort affected by traveling suddenly from one climate to another? Rapid modern air transport often results in shifting suddenly from one climate to another. Until adaptation takes place, discomfort may occur. When traveling from the northern cold to a hot southern climate, people arrive with a bodily production of heat geared to a cold environment. The body must change its heat regulation and in the initial period of change, the discomfort is heightened. Conversely, southerners traveling to a cold climate take with them a relatively low body-heat production and find themselves more susceptible to chill. As a safeguard, such travelers should change their general tempo of activities, diet and clothing to assist the body in coping with a changed environment.

886. What causes sunburn? Sunburn or erythema (reddening) of the skin is basically a burn injury caused by the ultraviolet or actinic rays of the Sun. The most important primary reaction is apparently

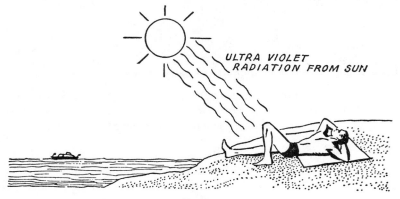

Cause of sunburn

an injury to the cells of the lower of two skin layers which make up the epidermis or superficial skin layer. The affected layer is called the malpighian or living layer; it is just below the corneum which is the outermost layer of the epidermis.

887. What serious skin disease is linked to sunlight? There are several skin conditions which are definitely related to sunlight. Most injurious is a condition known as *lupus erythematosus*. This may develop, following a severe sunburn. It is a constitutional disease affect-ing the skin, the vital organs and the blood stream. The recent use of cortisone is helping in the treatment of this once fatal disease.

888. What other diseases are caused by sunlight? Many skin conditions have been traced to the ultraviolet radiations of the Sun and most of them seem to be aberrations of the normal sunburn mechanism. These conditions appear in various forms and are com-monly classified as *polymorphous light irruptions.* Some rare condi-tions, classified as *urticara solare,* develop in people who are particu-larly sensitive to sunlight. When exposed even briefly to sunlight, they may develop a large itching but transitory wheal covering the skin area exposed.

889. Can the Sun's radiation cause cancer of the skin? Wave lengths of the same energy band that produces sunburn and causes the activation of vitamin D have induced cancer of the skin in experi-mental animals. Evidence is mounting that sunlight is also one of the principal causes of skin cancer. The type of skin cancer caused this way, however, is not highly malignant and may be treated with relative effectiveness if attended to early enough.

890. How can the eyes be damaged by the Sun's energy? Fre-quent exposure to high intensities of both the heat and ultraviolet rays of the Sun without sufficient protection may induce cataract of the eye. Looking into the sun for too long a time may literally cause a hole to be burned in the retina of the eye, resulting in a focus-point type of vision loss known as *eclipse blindness.*

891. What are the beneficial results of exposure to the summer Sun? Though literature contains much concerning the effects of sunlight on human health, relatively little is known about the subject. There is one clearly established therapeutic effect. Ultraviolet radia-tion of the sun produces vitamin D in the skin. This vitamin is useful in the prevention of rickets. However, the supply of vitamin D can be better controlled by giving it internally. Exposure to ultraviolet radiation has been claimed to have a preventive effect on the com-mon cold, but controlled studies are lacking.

There is no doubt that sunbathing generally produces a feeling of well-being, but it may be largely due to the heating effect of the Sun rather than the effects associated with ultraviolet radiation.

892. Why does looking at the Sun cause one to sneeze? Many allergy-prone people are susceptible to the ultraviolet radiation of the Sun. Such persons, for example, may be afflicted with allergic conjunctivitis (an inflammation of the inner eyelid). An allergy reaction to the Sun's radiation may cause sneezing in these cases of particular sensitivity. Antihistimines are usually recommended as well as the use of good sunglasses which screen the ultraviolet radiation of the Sun's energy.

893. When do most common colds occur? The common cold, along with many diseases, exhibits a seasonal pattern. In the United States the greatest number of cases occur during the winter months of January, February and March, and the least during July and August. Studies seem to support the fact that more colds occur during the colder portions of the peak months (following sharp cold front passages) than during the relatively mild periods.

894. Does weather actually cause the common cold? Weather itself cannot produce a germ. The upper respiratory tract and other parts of the body of the healthiest person are permanent lodgings for bacteria and a host of other parasites and microorganisms. The precise trigger mechanism of environmental change which is associated with lowering of bodily resistance leading to the common cold is not known. Local or sometimes total chilling of the body seems to produce some changes in the mucous membrane of the upper respiratory tract which favor invasion of the tissues by the dormant organisms.

Exactly how an environmental circumstance interferes to allow an infection to occur is a most vexing problem. To illustrate the many unsolved questions regarding the linking of an infective agent and environment, it is interesting to consider Pasteur's experiment in trying to cause a hen to develop hen cholera. Injecting or feeding the cholera bacteria had no effect until such time as the environmental temperature was lowered to a point where the hen's body temperature decreased by almost 3° F. This caused the infection to develop, although the exact relationship of the lower environmental temperature to the process of infection has never been explained.

895. What is climatotherapy? It is the treatment of disease by migration to a region having a particular climate which is assumed by some to help in successful treatment. Climatotherapy is sometimes suggested to the following groups of people: those with heart ailments, those with chronic respiratory infections (asthma, bronchitis, etc.), those who are allergy affected, those with rheumatic and arthritic diseases, and the aged.

896. What is the essential basis for climatotherapy? Physicians who recommend climatotherapy point out that the stresses and strains imposed on certain patients by the changeable weather patterns of the Temperate and North Temperate Zones are reduced considerably in regions where more equable conditions prevail. The movements of high- and low-pressure areas in middle latitudes such as the North Central and Northeast States are attended by passages of cold and warm fronts. Sudden changes of temperature, humidity, atmospheric pressure, precipitation and winds go hand in hand with these frontal movements. Some physicians cite the sudden environmental changes as providing excess burdens on certain patients whose physiological responses are not adequate to the demands made upon the body by these changes.

897. How important is climatotherapy? Many physicians agree that climate may strengthen or weaken the individual's resistance to many diseases by influencing the level of activity, metabolic rate, physical and mental state. However, they feel that climatic factors represent only a small portion of a complex pattern. It is believed that rest, relaxation, social adjustment, proper nutrition, relief from stress and strain, worry regarding income and medical care, together with adjustment of psychosomatic factors, should be given due consideration when evaluating climatotherapy. These human and social factors as well as biological and environmental considerations are more important than the most favorable climatic conditions.

898. What is the effect of climate on the recurrence of rheumatic fever? The effect has not been definitely established. There is no geographical region in the United States where rheumatic fever does not exist, although a greater incidence is reported in the Northern States as compared with southern sections. Areas with lowest incidence of rheumatic fever are southern California, Florida, Texas, Arizona and any of the other states along the Gulf of Mexico and the Mexican border. Highest incidence occurs in the Rocky Mountains and Great Lakes area.

899. How does climate relate to "heart trouble" or arteriosclerosis (hardening of the arteries)? No evidence exists that climate is in any way a direct causative agent in these diseases. But climate is related to their progress and they are more common in regions of strong frequent weather changes. Statistics show that people suffering from these conditions do much better in Florida, the Gulf States and southern California areas which do not demand the level of activity and rapid circulation necessary in colder northern regions.

900. How does climate relate to asthma? The effect of meteorological conditions on asthmatic patients has been studied recently in some detail. In treating large groups of patients, physicians have been impressed by the frequency with which asthmatic attacks are associated with or occur prior to or after changes of weather and climate. Yet repeatedly such climatic and weather changes occur without inducing symptoms. The evidences produced by different doctors contrast sharply. Some report adverse effects from lowering atmospheric pressure; others the reverse. A lower incidence of symptoms was reported during the periods of the year of high humidity and rainfall. One authority has pointed out that many patients attribute their improvement in asthma to a change of climate when, as a matter of fact, the benefit derived is often due to the avoidance of contact with environmental allergens. All in all, the role played by the many weather elements in relation to the incidence of symptoms in the asthmatic patient is still subject to continued research.

901. Why is smog more than just a nuisance? Since the acute and damaging smog episode at Donora, Pennsylvania, in October, 1948, smog has been recognized more realistically for the dangerous consequences it may have on an exposed populace. Beyond reduced visibilities, irritation to the eyes, nose and throat; damage to vegetation and local nuisances, acute smog collection exerts a heavy stress upon the respiratory organs and secondarily upon the circulatory system. Those already suffering from chronic respiratory or cardiac ailments may be seriously affected. (See Question 192)

902. What are some outstanding smog disasters on record? Acute smog disasters occurred in the Meuse Valley, Belgium, in December, 1930; in Cincinnati, Ohio, July 8–17, 1936; and at Donora, in the Monongahela Valley, Pennsylvania, October 25–31, 1948. In the polluted air near the factories at Liege in the Meuse Valley, 60 deaths

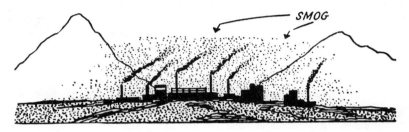

resulted. The "Big Smog" at Donora caused 20 fatalities with over 2,000 people affected in different degrees. At Cincinnati 16 persons died. Frequent smog situations, bordering on acute disaster proportions, have occurred in industrial valleys in England and Los Angeles, and pose a constant threat in those areas.

903. What chemicals were involved in the Donora incident? A wide variety of chemicals were involved, including oxides of nitrogen, halogen acids, zinc, lead, cadmium and other metals, sulphur, carbon monoxide and carbon dioxide. Spewed forth as waste products of factories, refineries, smelters and automobiles, the irritant gases were enhanced in effect by the presence of condensed water vapor (fog).

904. Who were most seriously affected in these smog disasters? Both incidence and severity revealed a direct relationship with increasing age. Generally the victims were the old people, asthmatics, cardiacs and the physically weak. At Donora, over 60% of persons 65 years of age and above reported some affection of the smog, and almost one half of these were in the severely affected group. Among the fatally ill, a significant factor at Donora was the pre-existing disease of the cardio-respiratory system, although in four cases no history of any chronic disease prior to the smog was obtained.

905. What steps are being taken to reduce smog? Many civic and industrial studies are being made regarding smog formation in local areas. Techniques of cutting down pollution dispersion in the air are being developed. Mechanical devices have been installed in stacks to purify wastes before they are discharged into the atmosphere.

At Los Angeles County, special wind studies are being made as well as a detailed study of noxious gases released from over two million automobiles in the area. County authorities there are attempting to design special mufflers for automobiles which will reduce the contaminating effects of exhaust fumes. Many local legal ordinances have been planned against violations of practices involving excessive pollution formation.

906. What group of people are exposed to abnormally high atmospheric pressures? Such exposure occurs in occupations where compressed air shafts and caissons are used, such as in the building of bridge piers, abutments, building foundations, mine shafts, traffic and sewer tunnels. High-pressure exposure also occurs in diving operations, such as in the construction of harbors, docks, piers and breakwaters, in the salvage and repairs of ships, in underwater explorations and in the pearl and sponge industries.

907. What is the rate of air pressure increase with descent below sea level? The atmospheric pressure at sea level (one atmosphere) is 14.7 pounds per square inch. The pressure increases at the rate of one atmosphere for each 33 feet of descent below sea level. At 33 feet below sea level, the pressure is 29.4 pounds per square inch. At 66 feet below sea level, the pressure is 44.1 pounds per square inch.

908. What are some mechanical effects of high atmospheric pressure on the body? If the Eustachian tubes and the openings of the air sinuses are blocked, pressure differences between the inside of these air cavities and the outside air may result in severe pain with congestion, edema and hemorrhage within the cavities. A large pressure difference may result in rupture of the eardrum. Normally, the pressure in the middle ear can be equalized by repeated swallowing, yawning or a strong attempt at exhalation with the nose and mouth closed (recommended for air travelers who experience discomfort when a plane is descending for a landing).

909. What are the bends? When a person who has been exposed to air under high pressure is decompressed, that is, brought back to normal pressure too rapidly, certain characteristic symptoms may develop. The most common symptom is severe pain in the muscles and joints of the arms and legs which is commonly called the *bends*. More severe symptoms may occur, such as itching and skin rash, vertigo, nausea, vomiting, epigastric pain, fatigue, dyspnea (chokes) and shock. These effects are apparently traced to the fact that decompression permits the formation and enlargement of nitrogen bubbles in the tissues. The condition is also known as *caisson disease, tunnel disease* and *diver's paralysis*.

910. What is aviation medicine? When man ascends to high altitudes, he encounters reduced atmospheric pressure, reduced temperature, and, if in an airplane, the problems of acceleration, gravity and motion. Changes in concentration of gases, especially oxygen and nitrogen, are also encountered. Aviation medical studies are concerned with physical and mental reactions to these various environmental factors.

911. How does the body react to low atmospheric pressure? At sea level, the body is adapted to equalize internally against an outside air pressure of about 14.7 pounds per square inch. In ascending to higher altitudes, the atmospheric pressure is reduced by one half this amount for every 18,000 feet of ascent. This pressure difference between the internal body areas and the external air results in an outward expansion of gases and liquids within the body. Capillaries may burst. If critically low pressures exist, explosive decompression takes place. Air rushes violently out of the lungs. Water vapor and blood literally boil.

912. How does the body react to lack of oxygen? Oxygen is called the *gas of life* for a very good reason. Every physiological function is guided proportionately by the supply of oxygen. When there is less oxygen to breathe, as in higher-altitude air, the blood which takes up oxygen from the lungs has less to deliver to other parts of the body. This shortened supply, which shows up in the blood, manifests itself readily in blueness of skin, lips, earlobes and fingertips, in turn, eventually affecting every part of the body. When the oxygen supply falls below a critical level and when the brain is not receiving enough oxygen, consciousness is lost. Unless the supply is restored immediately, death occurs.

913. What is mountain sickness? Visitors to various mountain communities which are located some 10,000 to 12,000 feet above sea level usually experience symptoms of what is called mountain

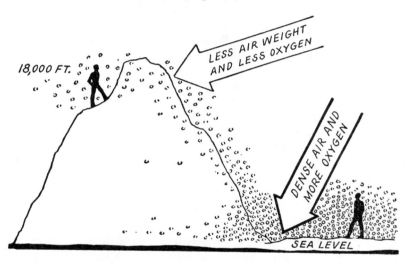

or altitude sickness, caused by a rather sudden exposure to much lower air pressure and a smaller supply of oxygen. Breathlessness and palpitation, loss of appetite and sometimes nosebleeding occur.

914. Where is the highest known community of human beings? By gradual acclimatization over periods of years, groups of people in the Himalayas of Asia and the Andes Mountains in South America have adapted themselves to live at high altitudes. One representative community is the little mining town of Cerro de Pasco in the Peruvian Andes at an altitude of 14,385 feet. Another fabulous small group of men and women live and work near the sulphur mines at Mount Aucanquilcha in Chile, about 18,000 feet above sea level.

915. How have these people adapted to such high altitudes? A virtually separate breed of men and women have developed under low pressure and reduction of oxygen conditions. Their hearts are long and thick and they have enlarged chests. Veins are distended with corpuscles more numerous. The heartbeat is slower. Their general work production is carried out with an energy that would cause a newcomer to fall unconscious.

916. What is the highest altitude at which atmospheric oxygen will sustain life? Depending upon physical condition and the dura-tion of stay, 20,000 feet represents the approximate altitude at which sufficient oxygen exists in the air to sustain human life for a period of time. A pilot *suddenly* exposed to the rarefied air at 25,000 feet without any acclimatization will lose consciousness within three or four minutes. At 30,000 feet, consciousness would only last one minute and at 50,000 feet, from 11 to 18 seconds.

917. At what altitude does water vapor in the human body start to boil? At about 55,000 feet above sea level, atmospheric pressure is low enough to permit water vapor in the body to boil, causing the skin to inflate like a balloon. At about 63,000 feet, blood will begin to boil at normal body temperature. By the protection of a pres-surized cabin and specially designed suit, a U.S. Air Force pilot was able to reach about twice this altitude in a rocket-powered plane in 1956.

918. How does acceleration affect the human body? Like any other object, the human body opposes any accelerating force because of its own inertia. A motorist gets an inkling of these forces when he steps on the accelerator. As the car surges forward, he is gently pressed against the seat. This stress is tremendously increased and becomes a crushing force in a rocket ship where the pilot will gradu-ally feel the weight of his own body increasing during the ascent when acceleration must be extremely fast. Because these stresses occur frequently in present-day fighter aircraft, detailed studies have been made on the tolerance of human beings to increased body weight.

919. How are studies of acceleration stresses made? Huge *centrifuges,* consisting of a bridgelike structure which rotates about a vertical axis, have been built for aviation medical studies. The subject is strapped into a cockpit-like chair and whirled around at different speeds. This whirling results in the generation of centrifugal force which pulls the subject outward and presses him down in his seat. Any desired force can be produced on the human body by controlling the rotational speed of the centrifuge.

920. How are acceleration forces expressed? They are expressed in units of a subject's normal weight. For example, when a person sits in a chair under ordinary circumstances, he is at "1g." When being whirled in a centrifuge, the person's weight can be increased to 2g, 3g, etc.

921. How does the body react to many "g" forces? Below 3g, no particularly adverse affects beyond some unpleasantness are experienced. At 4g, the head cannot be supported by the neck muscles and the arms can hardly be raised. At 5g to 6g breathing becomes difficult. The average subject generally blacks out after a few seconds because the blood rushes toward the abdomen and lower extremities, depriving the brain of oxygen.

922. What is the maximum of "g" forces that the human body can take? The findings in the preceding question were obtained with test subjects sitting erect. However, when the subject is placed in a supine position, the g forces act from chest to back rather than head to foot, not disturbing the circulatory system as much. Under supine conditions, some exceptionally adaptable test subjects have actually taken 17g, although not many, on the average, remained conscious over 10g.

923. What is space medicine? Space medicine is a logical extension of aviation medicine. It is based on the anticipation of many hazards which man will face when he travels beyond his protective aerial environment to the hostile depths of space itself.

924. What are the special problems of space medicine? Lack of oxygen, dangerous radiations, the blackness of space, meteoric bombardments, weightlessness—these are some of the extraordinary environmental conditions which are the concern of space medicine. It is interesting to note that in the development of aviation, medical science has always trailed engineering progress. In the case of space medicine, it is absolutely necessary that medical studies be the forerunner of space flight. It is the task of the space medical researcher to probe the limitations of man in a space environment. Man and not engineering is the weakest link in the conquest of space.

IX. WEATHER LORE—
FACTS AND FANCIES

Introduction. Earliest man emerged from the animal kingdom when he began to recognize the implications of time. He had to manipulate his time for hunting, fishing, planting and harvesting. Today's needs required information about tomorrow's possibilities, and so he began to observe his environment for clues about the future. One of the most obvious ways of anticipating the future was to observe the recurrence of weather phenomena.

Gradually he noticed the sequences and rhythms of climate and weather. While he could not establish cause and effect relationships, he was able to make some deductions about recurrent phenomena. These early observations about weather were gathered together into parts of local history and passed down from generation to generation, altered by the wisdom of the times. Thus, weather lore, an enormous body of proverbs and adages, has been carried down to the present day.

Much of it is nonsense, products of ignorance, imagination and whimsy, and contradictions are numerous. But some weather lore shows connections to factors now considered scientific. Apparently only some deeply ingrained universal tendency to speculate and philosophize about the weather has kept weather lore from receding into oblivion.

925. Which types of weather proverbs are worthy of some consideration? Proverbs pertaining to the condition of the atmosphere, the appearance of the sky, the character and movement of the clouds, and the direction and force of the winds are, generally speaking, all that are worth testing out for any particular locality. Proverbs regarding the actions of birds, animals and insects are of little value and sayings which pertain to the Moon and planets are useless. Those which apply to forecasts for coming seasons are entirely without foundation.

926. Why do so many weather proverbs concern animals? Earliest man believed that animals were supersensitive to the changes of weather because they lived closer to nature than man. Extreme winds or precipitation, for example, often threatened the survival of many animals and so they were observed carefully in the belief that their appearance, actions and habits would reflect the weather of the future. While it is quite true that changes in atmospheric conditions are responsible for many peculiar actions of animals, the mistake is to assume that these actions foretell future weather conditions.

927. Why is the figure of a cock used in many weather vanes? By a papal enactment made in the middle of the ninth century, the figure of a cock was set up on every church steeple as the emblem of

St. Peter. It is in allusion to his denial of Christ thrice before the cock crew twice. A person who is always changing his mind is, figuratively, a *weathercock*.

928. Can the severity of winter be judged by the manner in which a muskrat or beaver builds his home? The character of the muskrat's house or the beaver's dam is the direct result of the stage of the water at the time the structures were made and in no way offers a clue to the approaching winter.

929. Will studying squirrels provide a clue to the coming season's weather? Although people have watched squirrels "prepare for the winter" for centuries, animal-behavior scientists have proved that their activities in the fall season are motivated by many factors, none of which have to do with the weather of the future. They suggest that the number of nuts which squirrels store away in the fall depends upon how great the nut harvest is.

930. Do albatrosses indicate good weather at sea? Sailors for centuries have believed that it is fatal to kill an albatross. In Coleridge's "Rime of the Ancient Mariner," the story is told of a vessel which encountered bad weather after a crew member had killed a black albatross. The appearance of an albatross at sea is a harbinger of good weather. Similarly, the lively antics of the porpoise at sea were supposed to have foretold favorable winds. There seems to be no scientific bases for these beliefs.

931. Do migrating birds steer clear of storms? It is a common belief that birds can foretell severe storms during their migrations, and, therefore, they are able to fly safely for thousands of miles. In actual fact birds often fly directly into dangerous storms. In following definite routes of migrations, they frequently encounter storms which have destroyed whole flocks.

932. Is the ground hog a reliable forecaster? According to tradition, the ground hog (or woodchuck) comes out of its hibernation on February 2. If it sees its shadow, that is, if the day is sunny, it goes back for six weeks more and winter will be prolonged. In medieval Europe, Ground Hog Day was known as Candlemas Day and this particular proverb took the following form:

> If Candlemas Day be fair and bright,
> Winter will have another flight;
> But if Candlemas Day brings clouds and rain
> Winter is gone and won't come again.

The presence or absence of sunshine on any one day can in no way determine the length of a winter.

933. Can weather be predicted from the stripes on a woolly-bear caterpillar? It is an old superstition that the severity of the coming winter can be predicted by the width of the brown bands or stripes around the woolly-bear caterpillar in the autumn. If the brown bands are wide, says the superstition, the winter will be mild, but if the brown bands are narrow, a rough winter is foretold. Meteorologists hope that recent setbacks to this belief as the result of a series of annual woolly-bear studies at the American Museum of Natural History in New York will put the superstition in moth balls.

934. Can the air temperature be estimated by a cricket's chirp? The chirping of a cricket has been shown to provide a rather close indication of the air temperature. By counting the number of cricket chirps in a 14-second period, and then adding 40, the total will equal the air temperature to within one degree, three times out of four.

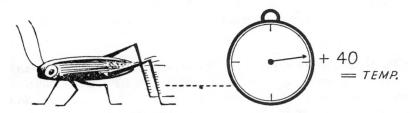

935. Do flies bite excessively before stormy weather? This bit of weather lore comes down from Theophrastus, a student of Aristotle, about 300 B. C. The origin of the belief is actually lost in antiquity. The only thing wrong about the belief is the fact that flies bite just as viciously in advance of a fine, dry weather period! Since weather in the Temperate Zones is continually changing, it is easy to find correlations of this type.

936. Is excessive croaking of frogs or quacking of ducks a forecast of rain? Frogs and ducks, perhaps because of their close association with water, have long been considered providers of weather clues. In Europe, especially, the popular adage is that excessive croaking or quacking is a sure sign of rain. To this day, on the Continent, a small glass containing a tree frog can be purchased as a personal weather forecaster. Scientists do not accept the idea, pointing out that the frog's croak, far from being a weather forecast, is a mating call!

937. How is the leech supposed to forecast weather? In the early nineteenth century in England, the ordinary medicinal leech was hailed by many people as a worm of great forecasting talent. One enthusiastic English doctor perfected what he termed a *tempest prognosticator* which consisted of several leeches housed in an enclosed bottle with the leeches attached by a pulley system to a bell. The doctor (appropriately named Dr. Merryweather) claimed that the leeches always remained quietly at the bottom of the bottle during calm fair weather, but that they became restless in advance of stormy weather and moved to the top of the bottle, ringing the warning bell. He added that the degree of restlessness of the leeches was a tipoff to the severity of the storm.

Many people still consider the leech to be a reliable weather forecaster and no amount of scientific repudiation sways them from their belief.

938. How did the expression "it's raining cats and dogs" start? In northern mythology the cat is supposed to have great influence on the weather. English sailors still say *the cat has a gale of wind in her tail.* The dog is a signal of wind, like the wolf, both of which animals were attendants of Odin, the storm god. In old German pictures the wind is often shown blowing from the head of a dog. From this mythology, the cat may be taken as a symbol of the down-pouring rain and the dog of the strong winds that sometimes accompany a rainstorm.

939. How valid are the following sayings? Red sky in the morning, sailors take warning. There are many proverbs which warn of rain in the event of a morning red sky. Shakespeare wrote "A red morn that ever yet betokened, wreck to the seaman, tempest to the field . . ." According to Matthew, Christ was supposed to have said to the Pharisees in answer to a call for a sign: "And in the morning, it will be foul weather today; for the sky is red and lowering." There is an element of scientific basis for the proverb because a red sun commonly indicates the presence of essential rain elements, dust and moisture. There are, of course, many other factors, but it is sufficient to give the proverb more than just a chance basis.

940. Mare's tails and mackerel scales make tall ships take in their sails. The familiar mackerel sky, consisting of tiny ripplelike formations of cirrocumulus clouds, often precedes an approaching warm front, with veering winds and precipitation impending. The proverb has a definite scientific basis of explanation, especially if the clouds should fuse and thicken.

941. In like a lion, out like a lamb. This proverb is one of many which seem to stress a comparison of contrasts. It usually refers to the month of March and probably reflects the fact that the early part of March often retains the bluster of winter while the latter part of the month tapers generally into occasional mild periods. It is scientifically inaccurate to say, however, that if the first day of March is stormy, then the last day will be pleasantly mild.

942. Rain on St. Swithin's Day, 40 days of rain to stay. So many control days have found their way into the folklore of various nations that they have been given a name. They are known as *key days* —days whose weather is popularly supposed to be a sign of the weather to come. Perhaps the most commonly found key days in weather lore are the Saints' days. There are approximately 58 Saints' days which, in one land or another, are thought to give a sign of the future! St. Swithin's Day, on July 15, is one of the outstanding examples.

Swithin, the Bishop of Winchester in England, died in 862. A century later, it was decided to remove his remains from outside the church (where he had requested burial) to the church interior on July 15. By way of protest, the Saint arranged for a 40-day deluge, whereby the monks were persuaded to abandon their project.

A British Royal Navy officer once tallied up all the key day forecasts and ended with a year of continuous rain!

943. Clear moon, frost soon. If the Moon is clear, that is, the atmosphere clear and cloud-free, the surface of the Earth will cool rapidly by radiation, and if no wind exists and the temperature is low enough, frost may well form. It is one of the proverbs that has more than a strand of scientific basis.

944. Rain before seven, stop by eleven. It has the same kind of validity that is implied in a statement attributed to Mark Twain concerning New England's weather, "If you don't like the weather, wait awhile!" Most regions of the middle latitudes lie in the prevailing-westerlies belt which carry transient high- and low-pressure areas from west to east, attended by cold and warm fronts, each with a different pattern of weather. These weather patterns are generally in motion and stagnation of weather is not typical. Change of weather, therefore, is quite normal, often within a few hours. To that extent, the proverb is somewhat valid, although the specific hour of day reference is meaningless.

945. A year of snow, a year of plenty. There are a few seasonal proverbs that have evolved from the history of farming that are rationally founded. This is one of them and is similar to *frost year, fruit year* or, *year of snow, fruit will grow.* These and similar statements are true as evidenced from the fact that a more or less continuous covering of snow, incident to a cold winter, delays the blossoming of fruit trees till after the probable season of killing frosts. It also prevents the alternate thawing and freezing so ruinous to wheat and other winter grains.

946. Halo around the Sun or Moon, rain or snow soon. The presence of lunar or solar halos is evidence of a layer of cirriform clouds —high ice-crystal clouds, perhaps 30,000 feet above the Earth's surface. These clouds are often indicative of an approaching warm front associated with a low-pressure area. Rain or snow will not *always* follow the appearance of a ring around the Sun or Moon, but there is a higher probability of precipitation after a halo is seen and the probability is greater if the circle is brighter. A rough rule of thumb is that rain or snow will come within 12 to 18 hours of seeing a halo on two out of three occasions.

947. Rainbow in the morning gives you fair warning. If two factors are considered, it can be seen that this proverb has a sound basis. First, weather in the middle latitudes generally travels from west to east. Also, a rainbow is seen when the observer's back is toward the Sun while he is looking at a rain shower. Consequently, in the morning, when the Sun is in the east, the shower and its rainbow are in the west. As the weather moves from west to east, the morning rainbow is a promise of rain moving toward the observer from the west.

Conversely, a rainbow in the evening is seen in the east, and this means that the rain has passed and will continue to move eastward. But weather backs up once in a while, so that although the rhyme has a scientific basis, it is by no means an infallible guide.

948. The bonnie Moon is on her back, mend your shoes and sort your thack. Many people have supposed, and some still hold, that the Moon appreciably controls the weather. Numerous proverbs are based on this assumption. But careful study of the records shows that the Moon's influence on the weather, beyond a very small tidal effect on the atmosphere as indicated by the barometer, is negligible, if indeed it has any influence at all. However, the *appearance* of the Moon depends on atmospheric conditions and proverbs relating to these appearances may have some basis, whereas Moon *phase* proverbs are completely without basis.

949. When the stars begin to huddle, the Earth will soon become a puddle. This is one out of very few proverbs about the stars that has decided merit. It furnishes, in general, a correct forecast. When increased cloudiness develops, whole areas of stars may be hidden by clouds while groups of stars, still in clear sky conditions seem relatively to *huddle together*. When the last group of stars is hidden, the chances for precipitation are logically increased.

950. Does the Sun draw water? It is not possible to see the Sun "drawing water" because the water vapor evaporated from the Earth's surface into the atmosphere is invisible. Actually what we see is sunshine and shadows in a normally dusty atmosphere—sunshine through the clear spaces between broken patches of cloud, and shadows caused by the scattered cloud fragments. These shafts of sunlight shining through cloud openings seem to radiate from the Sun like giant drinking straws (See Question 608).

951. Is it ever too cold to snow? No matter how cold the air gets, there is still some moisture in it, and this can fall out of the air in the form of very small snow crystals. The reason why we associate very cold air with no snow is because such invasions of air from northerly latitudes are associated with general clearing conditions of strong high-pressure areas behind cold fronts. Heavy snows, on the other hand, are associated with modified, relatively mild air in advance of a warm front.

952. Does a hot summer mean a cold winter to follow? There is no evidence to show that this belief has any factual basis. Many weather proverbs seem to be based on the idea of contrasts in seasonal weather. It has been suggested that if a summer has been hotter than normal, the winter, relatively speaking, is bound to *seem* colder than normal.

953. Is there such a thing as yellow or red snow? Sometimes snow crystals are formed on microscopic dust particles which may result in a freakish dark brown color, as happened to startled Chicago residents in 1947. Snow has also been given a golden or yellow appearance by the presence in it of pine or cypress pollen. In some mountainous and polar regions, pink or red snow is not uncommon, and when such specimens of melted red snow have been examined, they were found to contain tiny colored plants which caused the coloration. A Hungarian scientist exploring the mountains of Alaska came upon some 50 varieties of such plants which sometimes seem to "grow" in snow, imparting different colorations.

954. Does salt water ever freeze over? It is a fact that salt water will freeze. The temperature at which salt water will freeze depends on the amount of salt present in the water. This quantity varies considerably, depending on the amount of rainfall and the supply of fresh water from rivers. The figure usually quoted for the temperature at which sea water freezes is 28° F., or 4° F. lower than that required to freeze fresh water. Normally, the open ocean is rough enough to prevent freezing. But sheltered coves and harbors freeze over frequently.

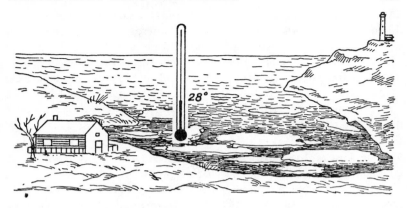

955. Can some people actually smell an approaching storm?
There is some basis of fact to this type of claim. Because of the decrease of atmospheric pressure that ordinarily precedes a storm, various types of decay odors such as in ditches, marshes or swamplands
normally contained close to the ground by high air pressure are re-

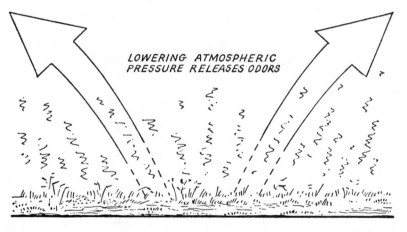

leased and penetrate over a wider area. This same lowering of pressure may be marked by the rising of water in wells, by more abundant
flow of certain springs and by the bubbling of marshes. An old rhyme
is quoted: "When the ditch and pond offend the nose, then look for
rain and stormy blows."

956. Does a thunderstorm turn milk sour? Lightning and thunder do not affect the milk at all. It appears, rather, that hot moist weather which produces so-called heat thunderstorms also favors the growth of bacteria and the spoiling of food. The process of souring, then, is linked to the same conditions of heat and humidity which cause the thunderstorm, and not to the storm itself. This fanciful belief has been disappearing in the face of modern refrigeration which keeps milk quite fresh, despite raging thunderstorms.

957. If leaves turn up, will a storm brew? Many people are adamant in their belief that just before a storm, leaves tend to curl or cup upwards as if anticipating needed rain. Botanists say that some types of plant leaves, especially of the bean family, react to changes of light by the turning together of leaflets. Since cloudiness (less light) often precedes a storm, there might be some validity with regard to leaves of the bean family type. Curling or cupping of other type leaves might possibly represent a constriction of the leaf caused by very dry air. This would be contradictory to the belief that it indicates rain (an increase in moisture).

958. How does the scarlet pimpernel relate to weather? The scarlet pimpernel is supposed to close its petals at the approach of rain. This flower has been called the *peasant's weather glass* or the *poor man's warning*.

959. Does a long falling star indicate rain? Beyond the obvious relation to a clear sky, the appearance of a meteor streak has absolutely nothing to do with the coming weather. People of the earlier civilizations who developed astrology constantly searched the heavens for portents of future weather. Comets were said to bring cold weather. The brightest edge of the Milky Way, in weather lore, pointed towards the direction from which to expect a storm. They ascribed to a well-defined shooting star the sign of an impending rainstorm.

960. What are the dog days? The term is used to describe an excessively hot spell, usually in August. It can be traced to the early Egyptians who held the theory that the appearance of Sirius, the Dog Star rising with the Sun, added to its heat, and the dog days (about

July 3 to August 11) bore the *combined* heat of the Dog Star and the Sun. (See Question 1076)

961. What is Indian summer?   Indian summer usually refers to a period of calm, hazy and unseasonably warm weather following the first frost of autumn, usually in late October or early November. The weather conditions which give rise to these conditions is the establishment of a stagnant high-pressure system over eastern North America. The air layers become stratified and stable. Calm, mild and hazy days and cool nights occur. The haze results from the stable atmosphere's inability to carry away smoke or dust from the surface.

In the United States, the term dates back only to the eighteenth century and its origin is uncertain. One explanation is that it was coined by the North American Indians who recognized it as a season in which final preparations for the winter could be made. There are equivalents of Indian summer in some European countries, but there they are known by different names.

962. What is an equinoctial storm?   The long-standing popular belief is that the occurrences of the autumnal and vernal equinoxes are marked by severe storms. Certainly, a severe storm may very well accompany the arrival of autumn or spring. But these storms which may occur on the equinoxes (or a few days before or after) have nothing to do with the crossing of the equator by the Sun. Stormy weather at any time of year, including equinoctial times, depends on atmospheric conditions which are quite independent of a precise calendar date. It is true that in March and September the middle latitudes are invaded by air masses which often are at opposite extremes of temperature and humidity and this generally creates frequent storminess. But it is incorrect to pinpoint the day of the autumn or spring equinox as the basis for these storms.

963. Do spring tides always occur in the spring?   Spring or unusually high tides are those which occur near the times of full Moon and new Moon, when the range of tide tends to increase. Therefore, they occur, on the average, twice each month at *all* seasons of the year as well as in the spring. *Neap* tides are those occurring near the times of first and last quarter of the Moon, when the range tends to decrease.

964. Are atomic bomb explosions affecting the world's weather?
Because of fears and speculations concerning the possible effects of
atomic explosions on large-scale weather patterns, special studies
have been made regarding the matter. A committee of meteorologists
appointed by the National Academy of Sciences reported that it is
unlikely that atom bombs have caused any important weather changes.
Some members of the World Meteorological Organization concur with
these findings, seeing no reason to conclude that any large-scale effect
has taken place. The general opinion of most scientists is that the
effects and energies of an atomic explosion are dissipated or swallowed up by the vaster energies and motions of the atmosphere to
such a degree that the broad weather patterns are unaffected. The
dangers inherent in the increase of atomic explosions do not relate to
weather changes so much as to the perils of radioactive fall-out on
plant and animal life. To this degree, many science groups and individuals have expressed grave concern about widespread atomic
tests.

965. Can atomic bombs be used to divert or modify hurricanes?
It is extremely doubtful that the energy of an atomic bomb, expended
as it is within a relatively small area, would alter the course or greatly
modify a hurricane. It has been computed that the energy released by
a typical hurricane in one second is greater than that produced by
several atomic bombs of the Hiroshima type. Another consideration is
that the contamination danger because of the radioactive material
acting in concert with heavy rainfall would be seriously increased.
These are just two basic reasons out of many which would put such
an experiment beyond consideration at the present time.

966. What do meteorologists think of flying saucers?   According
to the U.S. Air Force which has been assigned the task of evaluating
all saucer reports, about 80% possess natural explanations with
about 20% defying classification. Most meteorologists submit that
observers of flying saucers have been untrained in atmospheric or
meteorological optics, an area of study that scientists point to as containing the explanations for those observations which await clarification (See section on Color in the Sky). Flying-saucer reports, while
sometimes accurately described, are nevertheless based on many ig-

norances of atmospheric conditions which may lead to a wide assortment of startling visual phenomena.

It seems rather remarkable that out of thousands of meteorologists and astronomers whose very work is based on an understanding of sky phenomena, not one reputable and authoritative representative has ever seen or verified in any way the existence of a flying saucer as it is *popularly conceived*.

X. THE WEATHER SERVICES—
CAREERS AND EDUCATION

Introduction. The weatherman has been the butt of jokes and cynicism from the time of the first weather wizards of tribal communities thousands of years ago. Recently, an anthropologist pointed to the fact that one of the major jobs of primitive medicine men was to *insure* the right kind of weather for the growth of food, hunting or fishing and the comfort of the people. If the weather deviated to a harmful degree, the weather magician was often condemned to a quick death. The anthropologist suggested that present-day jokes about the weatherman might be traced to those early excesses of grim public censure. Nowadays, instead of being put to physical death, the weatherman's reputation is at stake.

This section concerns chiefly the organizations of weathermen. These groups are involved in a booming science with a tremendous impact on every level of our society. Not too long ago the weatherman was caricatured as an ineffectual, bearded, finger-wetting and preoccupied sky gazer. Today's weatherman is much more impressive. He is part of a complex corps of specialists who are constantly laboring to expand and apply their science. Industries and armies have come to rely upon him for guidance in their planning. The expectancies and demands made upon the meteorologist are sometimes incredible. Like Moses of old he is expected to bring forth water in the desert or to explore the frontier of space with the same detail as he would an earth-bound fog. In World War II, many of the bitterest complaints by commanding officers were not about the forecasters but about the presence of foul weather. General Patton, hampered by poor visibility, turned from the meteorologist to the chaplain and shouted, "God damn it, get me some good weather!"

967. When did the U.S. Weather Bureau come into existence? In 1870 the Congress of the United States passed a joint resolution that set up a national weather service as a part of the United States Army Signal Corps. In 1891 the service became known as the Weather Bureau and was administratively transferred to the Department of

Agriculture. Cleveland Abbe of the Cincinnati Observatory became the first meteorologist of the Weather Bureau when it was transferred to civilian operation.

968. What were its earliest functions? The work assigned to the Army Signal Corps on November 1, 1870, involved the issuance and collection of weather reports as an aid to navigation along the seacoast and in the Great Lakes. Besides the issuance of storm warnings, the responsibilities of the national weather service included the maintenance of climate records of the United States. Shortly thereafter, the weather service was authorized to measure the heights of water in rivers and issue warnings to protect lives and property from floods. In 1891 responsibilities were expanded to include weather information for a wide range of agricultural pursuits from planting and growing through harvesting and processing. Government forecasting for aviation started in 1918 for airmail services.

969. Under which government department does the bureau operate? In 1940 the U.S. Weather Bureau was transferred from the Department of Agriculture to the Department of Commerce under which administration it operates today.

970. What is the main function of the Weather Bureau? The primary job of the Weather Bureau in its capacity as the national weather service is to provide the basic data of weather and climate, the daily forecasts, and the warnings of severe weather conditions required for general public welfare.

971. Why is it best for the federal government to operate our weather service? Weather moves across continents and seas without regard for national boundaries. The very basis of a weather service is the necessity for exact procedure and scheduling methods over world-wide areas. Measuring, observing and reporting weather elements have to be synchronized to such exacting standards of timing in conjunction with the flood of reports from other national weather services, that a division below the national level would involve impossible deviations of procedure. Because so much widespread data must be gathered and used in forecasting and in issuing warnings, with much of it coming from foreign countries, it has been found to be most economical and best for the public welfare for the weather service to be operated by the federal government.

972. Do U.S. Weather Bureau stations exist outside of the 48 states? Weather Bureau meteorologists operate from stations in Alaska, islands in the Caribbean and the Pacific Ocean, from ships in the Atlantic, from stations as far north as Thule, Greenland, and from weather stations in the arctic within eight degrees of the north pole.

973. What kinds of U. S. Weather Bureau stations are there? There are three basic classifications of weather stations of the U.S. Weather Bureau; first order stations, second order stations and substations.

974. What is a first order station? First order stations are those at which only regularly commissioned civil service status employees engage in the operation of providing a weather service for the public.

975. What is a second order station? Second order stations are wholly or partially staffed by other than duly commissioned Weather Bureau personnel, such as Civil Aeronautics Administration or airlines personnel who do not deal with the general public from the standpoint of providing a weather service. The basic function of these stations is to supplement or round out the network reports. Their main responsibility is to take weather observations.

976. What is a co-operative substation? Substations of the Weather Bureau include co-operative observers who perform their services without pay. The main function of a co-operative observer is to maintain climatological records such as regular measurements of rainfall amounts, maximum and minimum temperatures, etc. These climatic observations are usually mailed to a Weather Bureau office at periodic times, normally once a month, and are added to the total collection of climate data assembled by the national weather service. Approximately 10,000 co-operative observers carry out this extremely useful function in the United States. Weather Bureau officials are highly complimentary concerning these unpaid and devoted people, pointing to the fact that without this assistance, the detailed assembling of valuable climate information would be a virtual impossibility.

977. How are co-operative observers selected?   Because of interest in meteorology and civic affairs, many people in local communities throughout the nation offer their services voluntarily to the U.S. Weather Bureau. If the community is geographically located so as to fill a needed gap in the collection of climatic data, and if the observer is shown to be a responsible member of the community, his services may be gratefully accepted. In most cases the Weather Bureau supplies the necessary weather instruments for taking climate observations as well as standard mailing forms for the compilation of statistics.

Many co-operative observers have devoted a lifetime to this public-welfare service. Sometimes entire families participate in the observation program, from one generation to another.

978. How many Weather Bureau stations are in operation?   In the continental United States and Alaska, there are a total of approximately 315 first order stations. Added to this total are about 600 second order stations. Other Weather Bureau stations in the ocean areas, Caribbean Islands and elsewhere, added to these, make a total of approximately 950 stations, excluding co-operative observing stations.

979. Where is the central office of the Weather Bureau located? The central office of the bureau at which most major administrative functions are performed is located at 24th and M Streets, Washington, D.C. The buildings include an older structure erected early in Weather Bureau history and a newer building built in 1935. The total area including the buildings, parking space and grounds take up one fourth square block. It is hoped that modern quarters will be built which will allow for expanded administration work and consolidation of some activities which are now spread out.

980. What is a regional office?   It has been found most practical to designate certain areas as focus points of administration for a multi-state region. These regional offices are located in key cities. Their function is, generally, to follow directions and policies of the central office so as to administer to the needs of service field stations in their particular region (personnel requirements, equipment, clarification of policies, etc.). They also keep the central office in Washington appraised of the status of assigned programs of the Weather Bureau stations in their regions. There are four regional offices in the continental United States; at New York City, Kansas City, Salt Lake City and Fort Worth. A regional office is also located at Anchorage, Alaska, and a Pacific supervisory office in Honolulu. The activities of the regional offices are extremely important in the successful administration of the Weather Bureau. Regional office directors are on the chief's staff.

981. How many persons are employed by the Weather Bureau?
There are approximately 5,000 positions listed for regularly commissioned Civil Service grades. Salary grades range widely depending upon qualifications and assignments from GS-3 to the chief's grade which is GS-18. New appointees with basic qualifications in meteorology are usually assigned as meteorologist trainees at a forecast center or active airport station with a rating of GS-5. Here they acquire practical experience in weather analysis and forecasting in addition to performing observational duties and furnishing meteorological information to newspapers, radio and TV stations and the general public. After a year's experience, trainees are eligible for promotion to general service meteorologist positions at higher grades.

982. What are the qualification requirements for meteorologist positions? Candidates must have a bachelor's degree in meteorology; or four years of progressive professional experience in meteorology; or an equivalent combination of academic training and professional experience. Appointees with these basic qualifications are usually assigned as meteorologist trainees (See preceding question). New appointees who have graduate degrees in meteorology or an additional year of professional experience in addition to basic qualification requirements, are appointed directly to positions as general service meteorologists, grade GS-7.

983. How can a person apply for a position in the Weather Bureau?
New appointees in the Weather Bureau are selected from registers of eligibles established by competitive Civil Service Commission examinations. Appointments are made from the top of the registers in accordance with Civil Service rules. Information regarding Civil

Service examinations for meteorological positions may be obtained by writing to the U.S. Weather Bureau, Washington 25, D.C., or by contacting the local Civil Service Commission representative.

984. What are the special services of the bureau? Although the bureau cannot provide specialized services for individuals or individual business concerns, it endeavors to include in its daily reports and forecasts those weather elements of special interest to major branches of the local business economy. Certain major fields have been designated for which meteorological services are to be provided by the

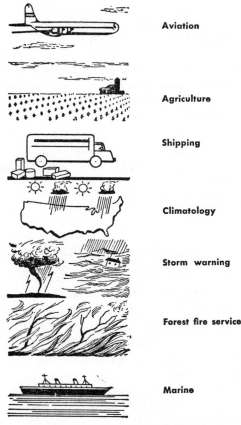

Aviation

Agriculture

Shipping

Climatology

Storm warning

Forest fire service

Marine

Special services of the U. S. Weather Bureau

bureau. These include agriculture, aviation and shipping. In addition to the general weather service for the public, the special services are designated as follows: Aviation Weather Service, Climatological Service, Hurricane and Storm Warning Service, Forest Fire Weather Warning Service, Marine Meteorological Service, Agricultural Meteorological Service and River and Flood Service.

985. In what type of work are most bureau meteorologists engaged? The Weather Bureau is described essentially as a service function of government. Most of the work may be compared to a production-line activity. The heaviest percentage by far of the bureau's funds and manpower are employed in taking weather observations, plotting maps and charts, preparing and issuing forecasts and processing weather information in various ways.

986. What is the average number of storms and weather changes reported by the bureau each year? The following tabulation shows the *average* number of storms and weather changes reported each year. It reflects the enormous scope of the bureau's operation.

8	hurricanes
28	coastal and Great Lakes storms
30	cold waves, or blizzards, heavy snowfalls, severe freezes
70	disastrous floods
300	tornadoes or other severe local storms
5000	forest fires of serious proportions
500,000	more or less local weather changes, that is, daily and sometimes hourly changes in weather in the many thousands of different localities in the United States.

987. Approximately how many weather observations, reports, forecasts and warnings are issued each year? In order to provide forecasts and warnings for the public and business community in general, the bureau's observations, reports, forecasts and warnings number approximately as follows each year:

800,000	basic synoptic weather observations and reports
8,000,000	hourlies and specials for aviation
22,000,000	climatological observations
190,000	specific forecasts for agriculture, marine shipping, etc.
650,000	local weather forecasts for general public
600,000	aviation weather forecasts

988. What is the NAWAC? A master weather analysis unit is maintained in Washington known as the *National Weather Analysis Center*. Its duty is to gather together all weather observation material, to plot maps and charts, to analyze the weather data and to disseminate this information to all parts of the nation via facsimile machine and coded teletype transmission. This serves to provide field offices with the essential meteorological information needed for their basic forecast responsibilities.

989. How is weather information gathered and transmitted? At scheduled times throughout the day, simultaneous observations and measurements of different weather elements at the surface and aloft are taken by hundreds of weather stations. This information is relayed along the entire network of Weather Bureau stations by teletype circuit transmission in code form so as to save time and space. When the information is received at each Weather Bureau office, it is decoded and assembled into various maps and charts which reveal a composite picture of surface and upper-air conditions over the entire country. These maps and charts prepared from the reports serve at most offices as the principal means of making weather forecasts for the general public which are disseminated by means of public automatic telephone systems, commercial radio and television, newspapers and individual telephone calls, etc.

990. What are the two main types of forecasts issued by the Weather Bureau? The two main types are (1) generalized forecasts for broad public distribution by newspapers, radio, television and automatic telephone and (2) forecasts for the aviation industry. The generalized public or district forecast is made by a senior forecaster and is generally valid for 24 to 48 hours. There are 14 district forecast centers covering the 48 states. The airways forecast is of shorter range and more specialized. It is usually valid for periods of 12 to 24 hours and is issued from 25 airway centers in the United States. Both main types are issued at standard intervals, four times daily.

991. What are some specialized Weather Bureau forecasts? In addition to the two main types described in the previous question, specialized predictions are made by district forecast centers of severe conditions such as tornadoes, hurricanes, frost conditions, fire weather and flood warnings. In addition to these special types of forecasts at local levels, forecasts are also issued for the purpose of providing assistance for operating groups such as farmers, shippers, industry, recreation, etc.

992. What long-range forecasts are issued for the public? A long-range or extended forecast unit is stationed in Washington, D.C. Three times each week a forecast of the weather expected during the next five days is released. It is broadcast by radio and television stations and is printed in many newspapers. Twice each month, on or about the first and fifteenth, an "outlook" is published, giving the average conditions of temperature and precipitation expected during the next 30 days for the entire United States. A summary of the 30-day outlook is printed in most newspapers. It can be obtained by subscription from the Superintendent of Documents, Government Printing Office, Washington 25, D.C.

993. Will a local Weather Bureau office answer personal weather questions? As a public-service organ of the government, the Weather Bureau wants to help in every way that is possible. However, regarding telephone inquiries made directly to the weather station, a difficult problem arises. Frequent interruptions to the forecaster's routine due to the answering of telephone inquiries, particularly during bad weather, often creates an upsetting of a tight schedule. Weather inquiries, often of a minor affair, detract from the station's ability to devote the proper attention to the quality of its product, and to the primary responsibility to serve the public as a whole.

994. How is the recorded telephone weather forecast prepared? In those cities where automatic telephone forecast recordings are made, a private teletype hookup is made to the telephone company. The Weather Bureau office sends out a forecast or special warning periodically on the teletype circuit to the telephone company. The forecast is revised when necessary. The telephone company then completes the operation, cutting an automatic tape each hour according to the forecast. The voice on the telephone is that of a telephone company employee.

995. How many calls for the weather forecast are made? In

Washington, D.C., the "iron forecast repeater" started its first year by answering calls at a pace of 10,000 each 24 hours. At the present time about 65,000 a day or nearly two million calls a month are handled by the mechanical weather system in Washington. The forecast repeater has answered more than 390,000 calls in one day in Washington and has answered more than 350,000 calls in one day in Chicago and 290,000 in Detroit. During one bad-weather 24-hour period, over 405,000 telephone calls for the weather forecast were received in the New York City area alone!

996. Are all newspaper, radio and TV weather summaries and forecasts based on Weather Bureau information? All forecasts which appear in newspapers, radio and TV can be directly or indirectly traced to the national weather network of the Weather Bureau. In many larger cities, news departments of wire services, newspapers, radio or TV have direct teletype transmission facilities so that they can receive various types of district and local forecasts. Close telephone contact between the Weather Bureau and news desks or editors is frequently maintained.

997. Why is a newspaper forecast sometimes different from radio or TV? It must be remembered that a period of several hours lapse between the time a newspaper gathers its information until the papers are actually distributed. During this time, revisions of the forecast may be necessary and cannot be included until a new edition is printed. Radio, TV and automatic telephone forecasts, on the other hand, can include up-to-the-minute forecast revisions. Despite this time lag, the majority of forecasts are usually the same. Also, newspaper treatment of weather summaries and forecasts are usually more complete and cover wider areas.

998. Who determines the kind of words used in a Weather Bureau forecast? While the central office in Washington provides general guide lines of terminology, the forecaster is given considerable leeway. His usage of words must be adapted to the type of forecast and the area or community. As much as possible, clarity and brevity are desirable with special care taken to issue warnings when necessary without creating panic. Public forecast terminology is also shaped by the local needs of a community. In a cotton-belt or fruit-crop area, for example, the forecast terminology may be quite different than in a large industrial city.

The problem of word usage in forecasts is a difficult one. Some meteorologists feel that a forecast should be expressed in terms of *probabilities* as in a ratio. For example, instead of saying *rain this evening*, the forecaster might say there are *8 chances out of 10 for rain this evening*. Methods for expressing forecasts in such probable ratios are being studied carefully for possible applications.

999. Who determines the percentage of accuracy of Weather Bureau forecasts? At the present time, the Weather Bureau has a system of verification which is maintained and kept up to date in the central office at Washington, D.C. In general, a verification system matches the weather forecast against the actual weather which occurs during the forecast period. The percentage of accuracy is determined by the degree to which the forecast falls within established reasonable limits of tolerance. The public may check the accuracy of a Weather Bureau forecast by first noting what the forecast calls for early in the day and then reviewing at the end of the day to see if they were in reasonable agreement.

1000. How accurate are the day-to-day Weather Bureau forecasts? Despite some belief that Weather Bureau forecasts are often in error, the statistical analysis made by the Bureau itself as well as numerous informal groups reveals a national average accuracy of close to 85%. This is a remarkable achievement when one considers the nature of the science itself with so many unknown variables.

1001. Can the Weather Bureau be held legally liable for damages resulting from a forecast? Weather Bureau responsibilities extend up to the issuance of a forecast. The forecast is not couched in operational terms. The responsibilities of how the forecast is applied rest with the individual. Therefore, there can be no recourse to legal restitution for damages which may conceivably stem from the application of a forecast made by a Weather Bureau forecaster. Weather forecasting has not reached perfection because the wide range of variable factors inherent in the moving atmosphere have not as yet been contained by exact mathematics.

1002. What is the total annual budget of the bureau? For the fiscal year, 1957, the total appropriation of funds provided to the U.S. Weather Bureau by Congress amounted to 37.9 millions of dollars. Although this appropriation falls short of the bureau's needs, it represents an increase over preceding years. It permits immediate improvements in its services and instigation of research programs, some of which were recommended by the non-governmental George Committee and some of which have been in a planning stage in the bureau for many years.

1003. What improvements are made possible by recent appropriation increases? In the research area, large-scale investigations are being made of the hurricane, tornado and other severe storm problems. The electronic computer, with its ability to resolve mathematical problems of staggering complexity, is being used to investigate atmospheric problems. In observations, radar is being used more widely for the detection and tracking of severe storms. The use of modern radio equipment permits the measurements of upper-air conditions with greater accuracy and to greater altitudes. The quality of surface observations has been improved by the installation of improved instruments of standard design and by the development of new instrumentation.

1004. How are the costs for various Weather Bureau branches divided? Although the different weather service functions are so interrelated that clear separation of costs is impossible, the approximate division for different purposes is indicated as follows (based on 1956 statistics):

Meteorological observations	48%
All forecasts (public and specialized)	26%
Information services	4%
Hydrology	4%
Climatology	6%
Research	7%
Administration	5%

1005. Are booklets and pamphlets concerning the weather available to the public? Over fifty periodicals and publications are issued by the Weather Bureau. These deal with a wide range of subjects such as climatology, local weather maps, agricultural weather reports, aviation weather, clouds, hurricanes, tornadoes, thunderstorms, Weather Bureau operations, etc.

1006. How is this material obtained? These items are for sale at nominal charges only by the Superintendent of Documents, U.S. Government Printing Office, Washington 25, D.C. For information concerning a complete list of publications, prices and ordering, requests should be sent to the superintendent.

1007. Does the Weather Bureau have a photographic library? The bureau library maintains an extensive collection of photographs of various weather phenomena.

1008. What are the general duties of a weather forecaster? Forecasting is only one of the many phases of meteorological activity but perhaps best known to the layman. In general, forecasters perform the following duties:

Record temperature, humidity, winds, amount of rainfall and pressure with weather instruments.
Take observations outside in all kinds of weather.
Draw weather maps using codes and symbols.
Check data received over the teletype machine.
Produce final form of forecast from weather maps.
Administer and supervise activities of station.
Inform newspapers, radio, TV and the general public about weather conditions.
Alert special groups such as pilots and farmers about storms.

Forecasting work becomes quite specialized at large weather stations. Instead of performing all the duties listed above, a forecaster may be concerned with only one of the following: fruit-frost, fog, clouds, hurricanes, storm warnings, forest fires, etc.

1009. What are the duties of the non-professional meteorologist at a weather station? Weather observers, computers, and plotters are known as meteorological aids in the government service. Few opportunities for nonprofessional personnel exist outside of the government service. The duties are generally as follows:

Record and check weather observations.
Plot weather maps with codes and symbols.
Encode weather data for teletype transmission.
Run computing machines.
File and classify weather reports and charts.
Duplicate weather maps.
Assist in the forecast and related duties.

1010. What is a physical meteorologist?   The physical meteorologist studies the chemical composition of the atmosphere, the laws of radiation, absorption and scattering; and the optical, acoustical and electrical properties of the atmosphere. He is concerned about such things as the nature of the processes by which water vapor is transformed into the liquid and solid states, and the mechanism by which trillions of tiny cloud droplets combine to produce rainfall.

1011. What is a dynamic meteorologist?   The dynamic meteorologist is concerned with the mathematical theory of air motions in atmospheric systems. These are motions which range in size from the globe-encircling jet stream at high levels to the small turbulent eddies in a plume of smoke escaping from a chimney.

1012. What is a synoptic meteorologist?   To the synoptic meteorologist falls the task of co-ordinating hundreds of weather observations, taken at the ground and aloft all over the world, into a coherent picture of the day-by-day weather. He charts the movements of high and low pressures, of air masses and fronts, and of hurricanes and tornadoes. He tries to bring all these things together into a prediction of the weather for tomorrow, for next week and for next month.

1013. What is a climatologist?   He is interested in marshaling and analyzing the accumulated weather records of past years to provide a description of climate—including its changes over periods of decades, centuries, thousands and even millions of years.

1014. What does the specialist in instrumentation do?   He is the measurer of every conceivable characteristic of the atmosphere. The size and number of droplets in a cloud; the amount of solar energy received on a unit surface at the ground; the base and top of cloud decks as revealed by radar; the pressure, temperature and humidity miles above the surface of the Earth, or the turbulent fluctuations of air motions a few inches above the surface of the Earth—all these are the measurements which his instruments must make.

1015. What is an applied meteorologist? Since World War II, industrial, or applied, meteorology, has emerged as a major subdivision of the field of meteorology. These *weather engineers* or consultants are concerned with the relationship between weather and a broad range of human activities in business, industrial, agricultural and recreational operations.

1016. Why is the private meteorological consultant needed? The Weather Bureau has too many customers to take care adequately of individual needs. The Weather Bureau is not set up to provide operational advice for the individual. The very nature of the Weather Bureau function prohibits this specialized attention. The private industrial meteorologist consultant can step into an area where the Weather Bureau cannot operate—by providing a specialized service to fit the particular needs of business and industry. The private consultant, with only a few clients whose special wants are always kept in mind, provides a most helpful service. He can guide day-to-day operations and future planning of a business so as to make the best use of the weather at hand.

1017. Is there competition between the private meteorologist and the Weather Bureau? Some time ago, there were feelings among a few Weather Bureau employees that encouragement of private meteorology was not compatible with the growth of the national service. Events have shown, however, that the development of the private profession of consulting meteorologists is highly desirable and in many cases quite necessary to meet many individual needs which the bureau cannot cope with because of understandable limitations of policy, time and personnel. The bureau today strongly encourages and actively co-operates with the many efficient meteorologists who are engaged in the growing profession of private meteorology.

1018. How may the services of a private weather consultant be obtained?   In many cities throughout the country, the services of private meteorologists are available now. Any individual or firm interested in obtaining consultancy services which may range from a single effort such as for a special forecast in connection with a

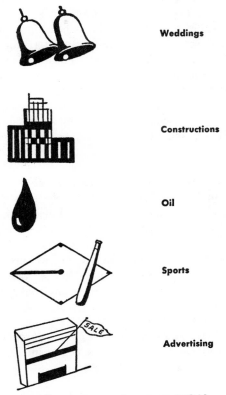

Weddings

Constructions

Oil

Sports

Advertising

Private weather consultants are available

wedding, picnic, painting job or sidewalk-laying through a long-term contract as for special fuel oil or gas heating temperature forecasts, should write to the American Meteorological Society, 3 Joy Street, Boston 8, Massachusetts. This society will be happy to recommend a bona fide consultancy firm or individual who may assist in solving a special weather problem. Consultancy fees vary widely, depending upon the type of service required.

1019. What is the function of the American Meteorological Society? The American Meteorological Society is a clearinghouse for weather. The society is the one and only agency in the United States through which the collective interests of the profession are represented. The society attempts to satisfy through its publications and services the professional and nonprofessional meteorologist as well as the amateur who is generally interested in weather. The objectives of the A.M.S. are the "development and dissemination of knowledge of meteorology in all its phases and applications, and the advancement of its professional ideals." The work of the society is carried on by its publications, by papers and discussions at national meetings, local branch activities, the library and through the central office of the executive secretary.

1020. What types of membership are available in the American Meteorological Society? *Corporation* membership is open to commercial enterprises and educational institutions interested in the advancement of meteorology. Dues are $100 per year. *Professional* membership is designed for those who have special qualifications in experience or training or who have made contributions to meteorology. Dues are $12 per year. *Regular* membership is open to those whose interests or activities in meteorology would make them desirable members of the society. Dues are $7 per year. *Student* membership is maintained for graduate or undergraduate students who are specializing in meteorology. Dues are $5 per year. *Associate* membership offers affiliation with the society to those whose interests are primarily of an amateur nature. Dues are $3 per year.

1021. What is the total membership of the American Meteorological Society? In 1956, total membership of the society amounted to 5,770. Included in this total were 2,650 professional members, 1,853 members, 1,067 associate members, 141 student members and 56 corporation members. There are 34 local branches or chapters of the society spread out through 24 states of the United States. Branches outside the United States are located in Alaska, Hawaii, Puerto Rico and Germany.

1022. What publications are printed by the society?   *The Bulletin of the American Meteorological Society* is the society's official publication. It serves the profession as a medium for original papers and contributions to meteorology with emphasis on practical application. It is published monthly except in July or August. It is included with most memberships and is $8 to nonmembers. *The Journal of Meteorology* includes results of original research. It is published bimonthly and is circulated to professional and corporation members. The cost is $9.50 to nonmembers; $5 to members. *Meteorological Monographs* are a series of original contributions appearing periodically when warranted. They are included with corporation members. *Weatherwise* is a popular, nontechnical magazine with appeal to all who are interested in weather. This well-illustrated magazine is included with all society memberships. Cost to nonmembers is $4. *Meteorological Abstracts and Bibliography* includes English abstracts of publications in meteorology and related fields from every country and in every language. Subscription is $8. *The Compendium of Meteorology,* published by the society in 1951, contains all phases of weather science from pole to equator, from microseisms to the ionosphere, in 108 articles covering 1,315 pages. Cost of this book is $12.

1023. What special guidance services does the society offer?   The society assists members in securing employment through the operation of the only guidance service devoted exclusively to weather personnel. Current job listings are mailed regularly to members using the employment service and a complete file of data on meteorologically trained personnel is available in the Boston office for prospective employers. A guidance service in education is also offered. Students wishing to pursue training in meteorology may receive free literature from the executive secretary's office.

1024. What colleges offer meteorology courses?   Colleges offering professional training in meteorology are listed below. Inquiries regarding course details should be addressed to the department listed after each school. A more complete list of meteorology courses offered throughout the United States is available from the American Meteorological Society, 3 Joy Street, Boston 8, Massachusetts.

Cornell University, Dept. of Agronomy, Division of Meteorology, Ithaca, New York.

Florida State University, Dept. of Meteorology, Tallahassee, Florida.

Iowa State College, Dept. of Physics, Ames, Iowa.

Johns Hopkins University, Dept. of Civil Engineering, Baltimore, Maryland.

Massachusetts Institute of Technology, Department of Meteorology, Cambridge, Massachusetts.

New York University, College of Engineering, Dept. of Meteorology, New York, New York.

Ohio State University, Dept. of Physics, Columbus, Ohio.

Oklahoma A&M College, Department of Meteorology, Stillwater, Oklahoma.

Oregon State College, Department of Physics, Corvallis, Oregon.

Pennsylvania State University, Dept. of Earth Science, University Park, Pennsylvania.

Rutgers University, College of Agriculture, Dept. of Meteorology, New Brunswick, New Jersey.

St. Louis University, Institute of Technology, St. Louis, Missouri.

University of California, Dept. of Meteorology, Los Angeles, California.

University of Chicago, Dept. of Meteorology, Chicago, Illinois.

University of New Mexico, Dept. of Physics, Albuquerque, New Mexico.

A&M College of Texas, Department of Oceanography, College Station, Texas.

University of Texas, Dept. of Aeronautical Engineering, Austin, Texas.

University of Utah, Dept. of Meteorology, Salt Lake City, Utah.

University of Washington, Dept. of Meteorology, Seattle, Washington.

University of Wisconsin, Dept. of Meteorology, Madison, Wisconsin.

1025. What are the occupational possibilities for a weatherman?
Since weather is a valuable ally and tool in practically every phase of our lives, the possibilities are unlimited for those with training, skill, imagination and ambition. Weathermen work for the government, universities, airlines, instrument companies, weather and engineering consulting firms, business and industrial firms, agricultural experiment stations and radio and television stations. Popular writing and speaking offer an interesting career for weathermen. Meteorologists are editors and also administrators of scientific foundations. Because weather has no national boundaries, opportunities for weather-

men exist everywhere. The Weather Bureau, some airlines and the military services maintain foreign stations.

1026. Who employs the largest number of weathermen? Of all organizations, the U.S. Weather Bureau provides the largest number of jobs for weather personnel. Since it is a field organization, personnel working for the bureau may be transferred from one station to another, in and outside of the country.

1027. What other government agencies employ weathermen? Besides the Weather Bureau, weathermen are employed in various capacities by the military services, the Research and Development Board of the Department of Defense, the Bureau of Reclamation, the Civil Aeronautics Administration and the U.S. Geological Survey. Units of government such as cities, counties and states are employing weathermen for many specialized weather services that are beyond the scope of the national weather service.

1028. Where is meteorological research conducted? Meteorological research is conducted by government agencies, universities, industries, foundations and societies. In recent years, the scope of meteorological research carried on by U.S. Air Force and Navy centers has broadened to a degree only rarely appreciated by the general public. Laboratories of the Air Force Cambridge Research Center at Cambridge, Massachusetts, and at the Office of Naval Research and Naval Research Laboratory in Washington, D.C., employ hundreds of civilians and military personnel who are concerned with a host of meteorological and geophysical research problems. Besides in-house laboratory research, centers like these let out contracts to scores of universities, industrial groups and societies for the conduction of meteorological and allied research. In the Weather Bureau, research is conducted by the Offices of Meteorological Research and Physical Research at Washington, D.C.

1029. What single agency conducts the largest amount of meteorological research in the United States? For magnitude of meteorological research operation, the Geophysics Research Directorate of the Air Force Cambridge Research Center, Cambridge, Massachusetts, is the largest single agency in not only the United States, but probably of the free world. This geophysical center employs a heavy percentage of civilian scientists as compared to active military personnel. The directorate is part of the Air Research and Development Command, U.S. Air Force, Department of Defense.

1030. How can an individual obtain information about career opportunities in meteorological research? Interested persons should request an announcement entitled "Research Opportunities for Meteorologists" which may be obtained from the Board of U.S. Civil Service Examiners, Air Force Cambridge Research Center, Cambridge, Massachusetts. For general information concerning the activities of Air Force and Navy meteorological research groups, inquiries should be addressed to: Office of Information Services, Air Force Cambridge Research Center, Cambridge, Massachusetts and Office of Information Services, Office of Naval Research, Washington 25, D.C.

1031. What is the Navy Aerological Service? The vast controlling influence of weather conditions on modern warfare has emphasized the role that the science of meteorology plays in the operations and planning of the U.S. Navy. The organization charged with the responsibility of meeting the over-all meteorological requirements entailed by the world-wide disposition of Naval forces is the Naval Aerological Service. (Aerology as used by the Navy is synonymous with meteorology.)

1032. Where is the U.S. Navy Aerological officer trained? Upon commissioning and completion of a tour of duty with the operating forces of the Navy, the potential aerological officer applies for aerological training at the Naval Postgraduate School at Monterey, California. The satisfactory completion of a course in aerological engineering, which is of 18 months duration, qualifies him as an aerological officer and leads to a 3-year tour in an operational billet with the Naval Aerological Service.

1033. What advanced training is available for the Navy Aerological officer? On completion of his 3-year tour, the aerological officer is eligible to apply for the advanced aerological engineering course at the postgraduate school for additional training. This 18-

months course includes advanced theoretical and applied meteorology, oceanography, climatology and higher mathematics. Officers who meet the special academic requirements of the school receive a master of science degree upon graduation. Officers demonstrating outstanding abilities are offered the opportunity to receive additional training in meteorology at selected civilian institutions leading to a doctorate degree.

1034. Where is the career Navy Aerological officer assigned? Having acquired a substantial background in meteorology, the aerological officer may apply for the designation Aeronautical Engineering Duty (Aerology), which classifies him as an aerology specialist and a career member of the Naval Aerological Service. He is then assigned to key operational and administrative positions. He also participates in applied research, operational weather analysis, and research and development of new meteorological methods.

1035. What training is available for the enlisted man in the Navy Aerological Service? Personnel wishing to become aerographer's mates must be high-school graduates or have satisfactorily passed the U.S. Armed Forces Institute high-school test. Following Navy indoctrination and basic training, enlisted men attend the Primary Aerographer's Mates School at Lakehurst, New Jersey. Applicants are chosen on the basis of individual aptitude, academic background, achievement in previous training and personal desires.

1036. What basic course is provided for the aerographer's mate in the Navy? The primary school at Lakehurst provides 16 weeks of instruction in operation and maintenance of meteorological instruments, weather codes, adiabatic charts, upper-air soundings, pilot-balloon soundings, weather and cloud observations, and practical work in the preparation of meteorological charts.

1037. What advanced courses are offered for the aerographer's mate? Advanced training is available to petty officers of this rating to augment their technical knowledge and to further their proficiency. In addition to the primary school there are two advanced schools for the aerographer's mate involving a total of 25 weeks of training. Subjects covered in these schools are: meteorological theory, map analysis, forecasting, and the operation and theory of current types of meteorological electronic equipment.

1038. Where may information concerning the Navy Aerological Service be obtained?   For details and additional information write to the Bureau of Naval Personnel, Department of the Navy, Washington 25, D.C.

1039. What is the Air Weather Service of the U.S. Air Force?   The Air Weather Service of the U.S. Air Force is charged with the responsibility of training and assigning commissioned and enlisted personnel for a wide range of meteorological duties as required throughout Air Force operations.

1040. Where are weather officers trained in the U.S. Air Force?   At the commissioned level, weather officers are trained in meteorological departments of outstanding universities throughout the United States. For entry into this program, applicants must hold a bachelor's degree with credit for mathematics through integral calculus and a year of college physics.

1041. What courses are given to the training weather officer in the U.S. Air Force?   Depending upon the university attended, courses include synoptic and dynamic meteorology, as well as climatology, observations and meteorological instruments. Possible electives include physical meteorology, thermodynamics, oceanography, physics, advanced mathematics, electronics, statistics and possibly a foreign language, in addition to micrometeorology and other special topics.

1042. Where are weather officers assigned after training?   New graduates are assigned to forecasting duties at one of the many weather stations in the world-wide Air Weather Service network. After basic experience with operational use of weather advice, weather officers are qualified for duties of greater responsibility in weather centrals. Further experience readies them for positions requiring the ability to direct the operation of weather stations and for later specialization in micrometeorology or climatology. After several years of experience in the field, many weather officers are assigned to command a weather station. This duty carries grade requirements varying from first lieutenant to lieutenant colonel. Administrative duties vary widely depending on the size of the station.

1043. What are some special activities of the Air Weather Service? There are a variety of activities in which the Air Weather Service engages, over and above the observing, forecasting and briefing services provided at each Air Force base. Aerial weather reconaissance is conducted over the arctic, the Atlantic and Pacific Oceans, and over land areas where ground stations cannot be feasibly located. Specialized meteorological assistance is supplied to military agencies engaged in rocket and guided-missile operations, chemical warfare and nuclear-physics projects. Climatological studies are prepared in support of research and development, military planning and daily operations.

1044. Where are enlisted weather personnel trained in the U.S. Air Force? Most weather courses for enlisted airmen are conducted at Chanute Air Force Base, near Champaign, Illinois. After graduating from basic training, an airman entering the weather career field may attend the Weather Observer Course to prepare him for observer duties. Upon completion of this course, he will most likely be assigned to a base weather station. There he will perform a multitude of duties such as observing and recording weather conditions, coding and decoding weather data, plotting weather data and operating teletypes.

1045. What additional training is offered the enlisted Air Force weatherman? After seasoning as an observer, the weather airman is ready for the Meteorological Technician Course. There he gains the necessary know-how to analyze a mass of plotted data, to prepare detailed weather information for forecasts and briefings, and to make use of upper air data, in addition to learning the mathematics and physics needed for an understanding of the behavior of the atmosphere. Following graduation, the new "met tech" will apply his know-how to operational situations at Air Force field stations and centrals.

1046. What special courses are offered to the enlisted weather airman? Graduates of the Meteorological Technician Course may attend the Weather Forecasting Superintendent Course to qualify as full-fledged forecasters. A few graduates may enter the Climato-logical Technician Course where students learn to collect and cata-logue weather data and to analyze past data, the procedure necessary to utilize historical weather data. Skilled operators are needed to use and maintain a wide variety of electronic weather instruments. Airmen entering the equipment division of the weather career field are enrolled in the Ground Weather Equipment Operator Course.

1047. Where may information concerning the Air Weather Service be obtained? For details and additional information, inquiries should be addressed to the Commander, Air Weather Service, An-drews Air Force Base, Washington 25, D.C.

1048. What is the World Meteorological Organization? The Earth's atmosphere, weather and climate do not respect political frontiers and they create many similar problems for everybody all over the world. Realizing that large-scale international co-operation was necessary to solve these problems, the nations of the world have made a common effort to co-ordinate, standardize and improve the services rendered by meteorology throughout the world to various human activities. The World Meteorological Organization (WMO) was formed as a specialized agency of the United Nations to help solve the common national problems of weather and climate.

1049. When did the World Meteorological Organization originate? As far back as 1853, efforts were made to establish a program of weather observations over the oceans, based on the collaboration of shipping belonging to most of the maritime countries. In 1878, the International Meteorological Organization was created, composed of the directors of national meteorological services. The tremendous developments in technology during the twentieth century and the increased requirements of modern economic activity emphasized the importance of meteorology. A more effective and vigorous interna-tional approach to meteorology seemed necessary and in 1947, fol-lowing a conference of directors of the national weather services, the World Meteorological Convention was adopted establishing a new organization. This convention was ratified and the new WMO com-menced activity on April 4, 1951, the former organization having been dissolved.

1050. What are the main functions of the WMO? The purpose of WMO is to:

(1) facilitate international co-operation in the establishment of networks of stations and centers to provide meteorological services and observations.

(2) promote the establishment and maintenance of systems for the rapid exchange of weather information.

(3) promote standardization of meteorological observations and insure the uniform publication of observations and statistics.

(4) further the application of meteorology to aviation, shipping, agriculture and other human activities.

(5) encourage research and training in meteorology.

1051. How many members does WMO have? At the end of 1955, WMO had 92 members. Any state or territory which administers a meteorological service of its own can become a member of WMO.

1052. How can further information be obtained about WMO? Inquiries concerning WMO may be sent to the Office of the Secretariat, World Meteorological Organization, Campagne Rigot, Ave. de la Paix, Geneva, Switzerland. Descriptive literature will be sent on request. A complete description of the functions, organization and activities of WMO is included in "Everyman's United Nations," published by the UN and available on sale at the United Nation's Book Shop, United Nations Building, New York, New York. (Mail orders permitted.)

XI. HISTORY OF WEATHER STUDIES

Introduction. The beginnings of meteorology are lost in the haze of antiquity. From the studies of anthropologists, we might guess that man's attitude about weather thousands of years ago evolved through two phases. In the first phase, his naïve self-importance prompted the belief that he could control the forces of nature by magical practices. When the passage of time and the moods of nature obviously did not respond to his boastful efforts, he reluctantly abandoned his sovereignty over nature. Then followed a period of several thousands of years, almost to the present, in which the caprices of weather were attributed to the gods. He assumed that his actions either provoked the anger of the weather gods or earned their blessings.

A few enlightened men of ancient Greece provided the beginning of a scientific attitude toward weather. But the thought-inhibiting era of the Dark Ages which followed was a clamped lid on scientific theory.

The era of instrumentation in the sixteenth and seventeenth centuries opened new vistas of advancement to weather science. Scientists like Galileo began important physical measurements of the atmosphere. The development of instrumentation was matched by theoretical considerations of physicists, chemists, mathematicians and astronomers of the seventeenth and eighteenth centuries, working under nearly infinite difficulties rarely recognized by the modern student in this scientific age.

Today, devices are probing the atmosphere to its outer reaches. Man stands on the threshold of an era in which he may completely escape his aerial envelope and travel through space. He hopes, by studying the atmosphere and its phenomena, someday to control much of the weather itself, thus making the dream of the earlier weather prophets a reality.

1053. Who were the earliest weathermen? The public magicians of early savage communities might be considered as the earliest weathermen. Of the many duties which the tribal wizard undertook for the good of the tribe, one of the most important was to control the weather and especially to insure an adequate fall of rain. This rainmaker was a very important personage, often equal in power to the tribal leader himself. Thrust upon him by his primitive society was not only the burden of regulation of weather but the healing of diseases, the settlement of quarrels, the forecasting of the future or any other activity of general utility. It was thus their duty and interest to know more than their fellows, to acquaint themselves with everything that could aid man in his arduous struggle with nature. To quote Sir James Frazer: ". . . in many parts of the world the king is the lineal successor of the old magician or medicine man." And again: ". . . it is probable that in Africa the king has often been developed out of the public magician, and especially out of the rainmaker."

1054. What methods were employed by primitive weather wizards? The methods by which they attempted to discharge their duties were usually based on the principle of *imitative magic*. If the magician wished to produce rain, for example, he simulated it by

sprinkling water, mimicking clouds or imitating the noises of a storm, hoping that nature would respond to these imitative suggestions. If the object was to stop excessive rain, he avoided water and resorted to a variety of rites involving warmth and fire for the sake of drying up the too abundant moisture.

1055. How was sympathetic magic employed to produce rain?
If the tricks of imitative magic described in the preceding question
were insufficiently effective, the rain magician sometimes would try
to bring on rain by the use of *sympathetic magic*. This might be accomplished by such activities as digging a pit in a clearing and placing one of the pet dogs of the tribe into it. He would bury the fright-
ened animal up to its head and pour heated oils or water into its
ear. The howls of anguish let loose by the poor creature would, he
hoped, invoke the pity of the rain god who would respond by dropping rain from the skies. (An ancient Indian rite.)

1056. What magical rites were used to control the wind?   The
control of wind was apparently practiced in scores of different magical
ways by earliest man all over the world. Many rites to cause wind
involved the strong blowing of air with the mouth, or striking or
whirling a "wind-stone." To stop wind, some wizards threw shells
or stones against the wind. Others fought strong winds off by bran-
dishing whips or firebrands so as to frighten the wind away. From
many anthropological records, it can be deduced that sometimes
whole tribes conducted physical warfare against windstorms which
were believed to contain whirling demons.

1057. Who were the following ancient weather gods? Zeus   In
Greek mythology, Zeus was the great sky god who wielded lightning,
thunder and rain. He was the chief of the Olympian gods; the god
of the elements, moral law and order, and punisher of guilt. His sign
was the thunderbolt and Iris, the rainbow, his messenger. Zeus was
said to dwell on the summit of Mt. Olympus and the chief shrine for
worship was at Dodona where frequent thunderstorms raged. In time
of drought, Athenians on the Acropolis would pray, "Rain, rain, O
dear Zeus, on the cornland of the Athenians and on the plains."
Spots which were struck by lightning were frequently fenced in by
the Greeks and consecrated to Zeus the Descender, that is, to the
god who came down in the flash from heaven.

1058. Jupiter   Jupiter was the Roman counterpart of Zeus (preceding question). He was the chief god of the Romans, ruler of the
heavens, judge of men and special guardian of Rome. Worship of
Jupiter as the god of rain and thunder was regularly conducted on

the Capitol of ancient Rome. One Roman writer contrasted the piety of the ancient Romans with the skepticism of its declining age. He pointed out that in former days noble matrons went up the long Capitoline slope with bare feet, praying to Jupiter for rain. Straightway, he wrote, it rained bucketfuls and the matrons returned dripping wet. "But nowadays," says he, "we are no longer religious, so the fields lie baking."

1059. Thor In Norse mythology, Thor was the chief of the deities, the god of thunder and the thunderbolt (the hammer of Thor) and of agriculture and the seasons. In Norse writings there appears: "Thor presides in the air; it is he who rules thunder and lightning, wind and rains, fine weather and crops." In these respects, therefore, the Norse god resembled his southern counterparts, Zeus and Jupiter.

1060. Donar Donar, or Thunar, was the god of thunder in the religion of the ancient Germans. He was the equivalent of the Norse god, Thor, and was similarly identified with the Italian thunder god, Jupiter. That the teutonic thunder god Donar, Thunar, Thor was identified with Jupiter appears from the word Thursday, Thunar's Day, which is merely a rendering of the Latin *dies joris*.

1061. Perun Amongst the ancient Slavs, Perun was the thunder god, the counterpart of Zeus and Jupiter. Like these southern European gods, Perun appears to have been a chief god to his people. It is said that at Norgorod there stood an image of Perun in the likeness of a man with a thunder stone in his hand. An historian writes that the Slavs "believed that one god, the maker of lightning, is alone lord of all things, and they sacrifice to him oxen and every victim."

1062. Perkunas The chief deity of the ancient Lithuanians was Perkunas or Perkuns, the god of thunder and lightning whose resemblance to Zeus and Jupiter has often been noticed. In time of drought, when rain was wanted, a black heifer, goat or cock was sacrificed to Perkunas who lived in the depths of the woods.

1063. Indra Indra was the great wind and thunder god of ancient India. His might was expressed by the thunderbolt. Indra and Parjanya, another Indian rain god, were believed to respond favorably to the supplications and special rites of a few selected men. Certain Brahmans would train for many years and practice elaborate rules which would bring them into union with water to make them, as it were, allies of the water powers and to guard them against the hostility of the rain gods.

1064. Lung-wong Lung-wong was the god of rain and wind in China. It was a common practice in China, when rain was wanted, to make a huge dragon of paper or wood to represent the rain god, and carry it about in procession. If no rain developed, the mock dragon would be torn to pieces in anger. If the rain fell, on the other hand, the god might be promoted to a higher rank by imperial decree. In the spring of 1888 the mandarins of Canton prayed to Lung-wong to stop an incessant downpour of rain. When he did not respond, they locked him up for five days and restored him to liberty when the sky cleared! Similarly, when the ancient Siamese needed rain, their idols would sometimes be placed outside in the burning sun to feel for themselves the terrible effects of dry hot weather.

1065. Fujin Fujin was a god of ancient Japan who controlled the winds. Like many other groups of people the Japanese had a variety of local gods who were concerned with weather conditions. When patience was worn thin and prayers for rain went unheeded, the people of some Japanese communities would take their local rain god image and throw it into the dry fields saying, ". . . see how you like it!".

1066. Aeolus According to Greek mythology, the north, south, east and west winds (*Boreas, Notus, Eurus* and *Zephyrus*) were under the rule of Aeolus, the king of winds, who kept them confined in a cave on Mount Haemus in Thrace. Winds of a more destructive nature were the brood of Typhoeus. It was Aeolus who tied all the winds up in a bag and presented them to Ulysses so that they might be used when necessary. When Ulysses' crew opened the bag in the belief that it contained a treasure, they unwittingly gave the winds their irrevocable freedom.

1067. What is the Tower of Winds? A tower which immortalized the winds was erected at the Acropolis in ancient Athens. It is octagonal in plan and at the top of each face is a sculptured figure of one of the winds with its name. All the figures are winged, have suggestive clothing and carry symbolic gifts. Notus, the south wind, is shown emptying a jar of water pursuing Eurus, the east wind, followed in turn by Zephyrus, the west wind, carrying fruits in his mantle. The Tower of Winds represented one of the first methods of orientation. It housed a water clock and showed by sundials the time of year as well as the time of day.

1068. What is the rain stone of Rome? In virtually every ancient land, some special object existed—a stone, edifice or image—which was supposed to contain magical weather control power. Such an object was the *lapis manalis,* a stone which was kept outside the walls of Rome near a temple of Mars. In time of drought, the stone was dragged into Rome and this was supposed to bring down immediate rain.

1069. Is the word weather used in the Bible? It is interesting to note that in the Authorized Version of the English Bible, the word *weather* occurs only four times. The word *season* occurs sixty times and the word *time,* apparently meaning season, about sixty times. This reflects the remarkable fact that the Mediterranean countries have no general word for weather as inclusive of any atmospheric phenomena. There are many references to separate phenomena of weather. Whirlwind appears frequently in the Old Testament—"Out of the south cometh the whirlwind" as Job is told by Elihu. Thunder, lightning, tempest, storm, rain, showers, clouds, rainbows, snow and hail are all referred to and indicate a common experience.

1070. When and where did astrology originate? Astrology can be traced back to ancient Babylon, about 3000 B. C., when that great city was the culture center of the world. The Babylonians, and somewhat later, the Chaldeans, were distinguished as astronomers and mathematicians. They held to the theory that man's destiny, as well as conditions on the Earth, were influenced by the phenomena of the heavens, especially the relative positions of the planets or *wandering stars*. The close relationship of seedtime and harvest to the positions in the heavens of certain planets, stars or constellations may have supported this theory. But whatever its origin, astrology played a profound role in human thought and action.

1071. Where did astrology spread?   Astrology passed from the east to the west, became established in Greece about 400 B. c. and reached Rome before the beginning of the Christian Era (although it was strongly opposed by the authorities of ancient Rome). It was actively cultivated in the region of the Nile during the Hellenistic and Roman periods, and by the Arabs between the seventh and thirteenth centuries. Astrologers held high positions and dominated the courts of Europe during the Middle Ages. The scope of astrology, actually, was very wide, borrowing from all the then known sciences such as botany, chemistry, anatomy, medicine and mathematics.

With the growth and development of the natural sciences after the Middle Ages, astrology lost its status. Today it seems to be almost wholly in the hands of people who exploit it commercially. It has become an occult pastime with a number of people who pay it serious credence or who use it as an amusing diversion. Although it was once significant in man's attempt to understand himself and the universe, the irrational phases of current astrology do not stand up under scientific scrutiny.

1072. How was astrology related to weather?   Study of the Sun, Moon, planets and stars gave rise to astrological forecasts. With the development of printing, these predictions were distributed in regular issues of almanacs and contained the salient aspects of weather throughout the year as auxiliary information for the calendar with its astronomical features and religious festivals. This blending of meteorology and astrology (astrometeorology) still has its present-day advocates and regular almanacs and journals, the distribution of which is amazingly large despite its lack of scientific evidence.

1073. What weather element was the first to be measured?   The amount of rainfall was the first weather element to be accurately measured. Nothing more was required than a bucket and ruler. Exactly when, where and by whom the first rain gauge was set up is not known. The Greeks kept some rainfall records as far back as the fifth century B. c. Rainfall was measured in India in the fourth century B. C. Rainfall measurements were made in Palestine in the first century A. D. and included in some Jewish religious writings. Heavy bronze cylindrical vessels set up on stone blocks were used by the Koreans in 1442 A. D. to measure rain.

1074. When was wind direction first measured? Like the measurement of rain, the exact beginning of wind direction measurement is not known. Wind vanes were used in Athens as far back as the second century B. C. During the early Christian Era, many Roman villas were ornamented with wind vanes, some of which were connected by a turning vertical rod to a hand which indicated the wind direction indoors on a dial on the ceiling. Rainfall and direction of the wind were the only weather elements to be *accurately* measured until the development of instruments in the sixteenth century.

1075. What did the ancient Babylonians contribute to the study of weather? The Babylonians enjoyed a mild climate, although not as equable as that of the Egyptians. They developed a system of a wind rose of eight rhumbs at a very early date. From about 700 to 900 B. C., Babylonian knowledge of weather phenomena was molded into somewhat of a profession by the priests and, correlated with their knowledge of astronomy, resulted in a system of astrometeorology (See Questions 1070–1072).

1076. What did the ancient Egyptians contribute to the study of weather? The Egyptians did very little to develop a science of the weather. Their indifference to the study of weather was due to their equable climate and to the presence of the Nile for irrigation purposes. However, the periodic overflowing of the Nile, a most important event, stirred them to investigations of the causes for this overflowing and, in turn, for the periodic renewal of the seasons. They attached great importance to the fact that the bright star, Sirius (the Dog Star of the constellation Canis Major), was visible in the sky in the early morning coinciding with the swelling waters of the Nile. The beginning of the Egyptian year, in fact, was determined by the appearance of Sirius and the length of the year was the interval between two successive appearances. The term *dog days,* which is used today to describe a hot spell of weather in August, is directly related to the appearance of Sirius, the Dog Star, which was believed to add its presence and heat to the Sun, causing exceptionally hot weather.

1077. What did the ancient Greeks contribute to the study of weather as a science?   The beginning of a scientific attitude toward weather study is traced to a few outstanding philosophers of ancient Greece who where not content with the accepted beliefs that weather was dispensed at the whims of deities. Hesiod, for example, as far back as 750 years B. C., offered some sound advice to farmers and sailors based on realistic weather observations. The Greek philosopher Anaximander of Ionia, around 600 B. C., defined wind as a *flowing of air*—a definition which has hardly been improved upon. The Greeks were making regular weather observations as far back as the fifth century B. C. Hippocrates, the Father of Medicine, made some extremely important observations regarding the effects of climate on human health and comfort in the fifth century B. C.

1078. What was Aristotle's great contribution to meteorology?   A monumental study of the weather was completed by Aristotle in about the year 350 B. C. This great philosopher of ancient Greece published four books under the general title *Meteorologica* in which were collected all the previous writings on meteorology as well as on

astronomy. For nearly 2,000 years *Meteorologica* remained the standard textbook on weather. The books contained a much wider range of subjects than is now believed to be within the scope of weather science. Everything of a physical nature pertaining to the Earth, sea or air was included in this giant geophysical compendium. Among many others, quotations and criticisms of Pythagoras, Anaxagoras, Socrates, Hippocrates and Democritus, his predecessors, were included.

1079. Who was Theophrastus? Despite the existence of *Meteorologica* described in the preceding question, the public then, as now, was far more interested in knowing what the weather was going to be than in any understanding of the how and why of it. In response to this demand, Theophrastus (about 375–285 b. c.), presented a treatise entitled the *Book of Signs* in the form of numerous rules for foretelling the coming weather. Theophrastus (his real name was Tyrantus) was one of Aristotle's favorite disciples in the famous peripatetic, or walking-about, school of philosophy at Athens. Like many of the early Greek philosophers, his interests covered a broad field of subjects: botany, psychology and physics. His *Book of Signs* contained at least 80 signs of rain, 45 of wind, 50 of storm, 24 of fair weather and seven of the weather for a year or more. His quotations were a mixture of science and folklore.

1080. What did the ancient Romans contribute to the study of weather? As with most of the sciences and arts, there was a wane and decline in weather science with the breakup of the philosophical ancient Greek civilization. Although the Romans were unexcelled in such practical demonstrations as law, politics, industry and engineering, they showed a surprising lack of scientific and artistic development. They added little to meteorology and even their weather folklore was borrowed from Theophrastus of the Greek school before them. The most notable Latin treatise on weather was written by Virgil in his first book of the *Georgics*. It was mostly a slight revision of folklore as handed down by the Greeks.

1081. What did the Arabians discover concerning the atmosphere? During the eleventh century, Arabian astronomers made the first rough estimation of the height of the upper atmosphere by watching the sunrise and sunset in the clear desert air. From the duration of twilight, they deduced the fact that the atmosphere extended in *appreciable* density to heights around 50 miles. Numerous later experiments confirmed this. The greatest height at which the density of air still is sufficient to scatter a perceptible amount of incident sunshine is found to be about 44 miles (twilight limit).

1082. What was the general attitude toward weather study in the Medieval period?   In Medieval Europe the tendency was to maintain the authenticity of the old proverbs and superstitions about the weather. Under the blight of astrology and the general belief that weather phenomena were the immediate acts of a supreme being, inquiries about weather understanding were futile, considered presumptious and even wicked. However, near the end of the twelfth century, a Latin version of Aristotle's great compendium, *Meteorologica,* appeared in Europe, taken from an Arabic translation of the original Greek. This led to some limited teaching of meteorology in certain universities. But teaching was by authority and Aristotle was complete and infallible gospel and, therefore, a block to new investigations about the weather until a few bold thinkers, long afterward, dared to study nature for answers to questions rather than go to Aristotle's books.

1083. What were the Shepherd of Banbury's Rules?   They were a rather unique collection of astrometeorology, folklore and cumulative observations about weather. Although first published in 1744 in England by a Mr. Claridge, their authorship is believed to be much earlier. The origin is obscure. They are considered to exemplify the attitude toward weather in Medieval Europe. In a large sense the *Shepherd of Banbury's Rules* might be considered as an elaborate compilation of Theophrastus' *Book of Signs* and the works of Aratus and Virgil dating back to ancient Greece and Rome. Some of the Rules were based on realistic observation and some were pure folklore.

1084. Where did systematic climatology begin?   Although weather records of a sort were kept by the Greeks as far back as the fifth century B. C., a collection of day-by-day records and the *deduction* therefrom of the run of the weather was not attempted until the year

1337 in England. These records, which mark a beginning of systematic climatology, were made by William Merle, Fellow of Merton College, Oxford, from 1337 to 1344. They were finally published in 1891.

1085. What was the chief defect of the climatology records of William Merle? The greatest defect was the unavoidable absence of temperature values because the records were made about 300 years before the invention of any kind of a thermometer and about 375 years before the invention of a suitable type of thermometer for this purpose. It is interesting to note that Merle's observations predated by about 325 years the first records of that sort kept on the American continent. So far as is known, the first climate records in America were kept by the Reverend John Campanius at Swedes Fort, near Wilmington, Delaware, in 1664. Like Merle's records, the records of Campanius were necessarily of a descriptive, noninstrumental nature.

1086. Who formed the first theories of weather cycles? The persistent belief that every so many years the course of the weather substantially repeats itself seems to have developed in Holland where a 35-year cycle was suggested. Francis Bacon (1561–1626), an English statesman and man of letters, in his Essay LVIII, 1625, commented on the 35-year cycle by writing, "Come to think of it, I have noticed much the same thing myself." Bacon's *Novum Organum,* published in 1620, was a storehouse of scientific information and contributed somewhat to the adoption of the scientific method in meteorological circles. His reflections about weather cycles are considered to mark the beginnings of a long train of substantially fruitless studies of weather cycles, still going on and still encountering frustrating obstacles.

1087. Upon what two instruments is modern meteorology founded? The invention of the thermometer and its sister instrument, the barometer, marked the beginning of a long era of instrumental meteorology. The importance of the invention of these two instruments to the status of meteorology as an independent and dignified science can hardly be overemphasized. These two instruments were the foundation for man's understanding of aerial forces which he could only imperfectly see or sense without sensitive instrumentation. As a doctor rarely can begin a diagnosis of his patient's illness without knowing the body temperature and blood pressure, so, too, does the weather observer find it difficult to chart weather symptoms without knowing the temperature of the air and its barometric pressure (weight or push of the air above him).

1088. Who invented the first air thermometer? Galileo Galilei, the great experimental scientist of Italy who lived from 1564 to 1642, was very much interested in the physics of the air around him. His interest in thermometry stemmed from the writings of Hero of Alexandria who, back in the third century, had devised a primitive thermoscope of very little practical value. Working on the theory that air expands when heated, Galileo constructed the first practical air thermometer at the University of Padua in 1592. Galileo and his associates were to make northern Italy the cradle of instrumental meteorology and instrumental physics.

1089. What did Galileo's first air thermometer resemble? His thermometer only slightly resembled the modern instrument. It consisted of a large bulb filled with air which, by its expansion when heated, forced downward the water level in a glass tube. In about 1612, Galileo devised, in essence, the modern thermometer by enclosing a liquid (alcohol, an early favorite and still used) in a round or cylindrical bulb free to expand into a sealed capillary tube. The use of mercury as a thermometer liquid was originated around 1709 by Gabriel Daniel Fahrenheit, a German physicist who settled in Holland.

1090. What major problem faced early experimenters in thermometry? One of the most difficult concepts to master concerning the early thermometers was that of obtaining reliable reference points so that a practical scale of measurement could be established. Numerous experiments were made to obtain high and low points in the sealed-tube thermometer and to establish in-between divisions. One scientist, Dalence, in 1688, used the temperature of air during freezing for his low point and the temperature of melted butter for the high point. Others used the heat of summer and the cold of winter or ice at its severest frost. Temperatures of the blood of different animals were used at various times. The obvious fault with these reference points lies in the fact that they vary from one time to another and have no constant value.

1091. When was the Fahrenheit scale determined? In 1714, Gabriel Daniel Fahrenheit invented the thermometer scale which bears his name. He set the reference of zero at a point where the mercury sank in his thermometer on a particularly cold day in Danzig. This point was later specified as the temperature of a given mixture of ammonium chloride, common salt and snow. The opposite reference point of the scale was taken at what Fahrenheit erroneously believed to represent the normal temperature of the human body. The interval between these two reference points was ultimately divided into 96 divisions or degrees (See Questions 74, 79–82).

1092. Who designed the centigrade temperature scale? The centigrade scale of temperature measurement was proposed by Anders Celsius, a Swedish astronomer, in 1742. His scale was based on the reference of the zero point as the temperature of a mixture of ice and water and the reference point of 100 degrees as the temperature of boiling water (See Questions 75, 79–82).

1093. Who devised the absolute temperature scale? The absolute or Kelvin scale of temperature was developed by the English scientist, William Thomson (Lord Kelvin), about 1850. It represents the most exacting scale for measurement of air temperature. The zero point on this scale represents a theoretical temperature at which molecular motion, hence heat, ceases. (See Question 76, 81–82)

1094. What event stimulated thoughts about the measurement of atmospheric pressure? The natural philosophers in the Middle Ages accounted for the rise of water in a pump by saying "nature abhors a vacuum." The Grand Duke Ferdinand II of Tuscany, when ordering that a deep well be dug near Florence in 1640, was surprised by the fact that he could not pump water from the well pipe, even when the air was drawn out of it, to a height greater than about 32 feet. The well was about 50 feet deep. Galileo, investigating the matter for the grand duke, was said to have remarked rather wryly that "nature does not seem to abhor a vacuum above 32 feet." Before Galileo died, he suggested that liquids of different densities

should rise to different heights in an evacuated tube. He passed
the suggestion that air had weight to his friend and disciple, Evangelista Torricelli, for further study.

1095. Who invented the first liquid barometer? Evangelista Torricelli, professor of mathematics in the Florentine Academy in 1643,
invented the first liquid barometer after studying the problem which
was given to him by Galileo (See preceding question). Instead of
working with water, Torricelli put the experiment on a more convenient scale by adopting mercury, the heaviest of liquids. He noticed that mercury rose only about 30 inches in an evacuated tube
and concluded that the rise of liquids in exhausted tubes is due to
an outside pressure exerted by the atmosphere on the surface of the
liquid, and not to any mysterious sucking power created by the
vacuum.

1096. What did Torricelli's first barometer consist of? It essentially
consisted of a glass tube about 36 inches long, sealed at one end
and filled with mercury. He closed the open end with his thumb,
slowly inverted the filled tube and put the lower end below the surface
of some mercury in a dish. When he removed his thumb, the mercury in the tube sank until it stood about 30 inches above the level
of the mercury in the dish. He proved that the weight of air determined the height of the mercury column by placing his barometer
in a bell jar and pumping out the air, whereupon the mercury column
sank in the barometer tube. His barometer, unchanged in general
principle, though considerably refined in design, is used for all accurate atmospheric pressure measurements today.

**1097. Who helped make Italy the cradle of instrumental mete-
orology?** The Grand Duke Ferdinand II of Tuscany, though not
a scientist, showed great understanding and sympathy for the work
of scientists like Galileo and Torricelli at the University of Padua
and the Florentine Academy. He was extremely weather-conscious
and devoted much time and money to furthering meteorological
knowledge. In 1653, he established several weather observing stations throughout northern Italy—the first weather network ever
planned and the embryo concept for modern national weather networks. It was unsuccessful because nobody apparently knew how to

tie together the information gathered from the different weather stations.

1098. What experiment proved that air pressure lowers with altitude? In 1648, a young French mathematician, Blaise Pascal, had an idea that air pressure, like pressure in other fluids, diminishes as one goes higher into the atmosphere. He was anxious to carry a barometer up a mountain in order to test his theory. Lacking high enough terrain, he wrote to his brother-in-law, Perrier, in the south

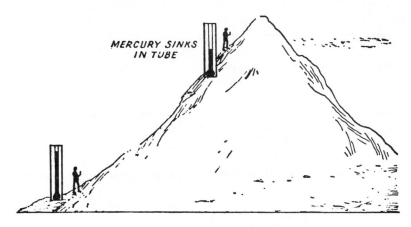

MERCURY SINKS IN TUBE

of France. Perrier carried a mercury barometer to the top of a sizeable mountain (Puy-de-Dome) and was astonished to observe the mercury column sinking in the tube. From this experiment, Pascal correctly inferred that the reading of the barometer is in proportion to the amount of air above it—a significant contribution to the general understanding of the atmosphere.

1099. When was the aneroid barometer invented? It was not until 1843, about 200 years after the invention of the mercury barometer, that the aneroid barometer was developed. Its inventor was Lucien Vidie, a French scientist, who reasoned that changes of atmospheric pressure could be shown by small, easily portable instruments which contained tiny metallic cells from which most of the air had been removed. Changes in atmospheric pressure caused the accordionlike cell to change in shape. These changes could be linked to a properly scaled dial to allow a reading of the pressure. This principle is the basis for the modern altimeter used in aircraft to indicate altitude by changes in atmospheric pressure.

1100. What is Boyle's law and its significance? Robert Boyle, born in Ireland, was educated in England and was extremely interested in experiments in the physics of gases. At Oxford, in 1662, he discovered and gave name to the law expressing the relation between the volume of a parcel of air and its pressure. Boyle's law, the first of all contributions to the dynamics of the atmosphere, states that, at constant temperature, the volume of a given quantity of air varies inversely as the pressure on it.

1101. What was the first type of anemometer? In 1667, Robert Hooke, an English physicist, devised the earliest type of anemometer which, for the first time, provided the means of obtaining a numerical measure of wind velocity. It was a simple device consisting of a rectangular plate hung by its top so that increasing wind pushed it out

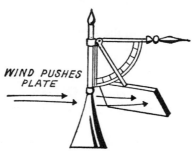

Early anemometer

to an increasing angle from the vertical. A suitable sight-reading scale was etched outwards from the perpendicular stand to show how far the plate was moved by the wind. Rotating types of anemometers came into use by about 1724 and were refined to further degrees in 1774 and 1790 by Swiss and German scientists.

1102. Who was the first to explain the mechanism of the trade winds? The trade winds had been known and used by maritime explorers and navigators of many nations for centuries before any-

HISTORY OF WEATHER STUDIES

one attempted a rational explanation of their behavior. In 1686, Edmund Halley, the British Astronomer-Royal, published an epochal paper in which the first attempt was made to attribute the general circulation of the atmosphere to differences of terrestrial heating around the world. He rightly concluded that various wind patterns around the globe were caused by the flow of cold, or dense, air toward regions of warmer, hence lighter, air in the process of coming into equilibrium. Halley's investigations of large-scale wind movements earned him the title of "father of dynamical meteorology."

1103. What lawyer contributed to the further understanding of wind flows? In 1735, George Hadley, a lawyer by profession, added an important factor of understanding concerning the winds of the world, overlooked by Edmund Halley (See previous question). He presented to the Royal Society of England the highly important theory that the Earth's rotation produced a deflective force on large-scale wind flows (coriolis force). This caused, he correctly surmised, the southward moving trades in the Northern Hemisphere (or any other large system of winds from the north) to turn westward. The same deflective effect on winds by the Earth's rotation, he added, caused northward air drifts in the Northern Hemisphere to turn eastward.

1104. How were kites first used to sound the free air? On a summer day in 1749, Alexander Wilson, of Glasgow, Scotland, sent aloft thermometers attached to several kites tied in tandem to a single long string. The highest kites disappeared in cumulus clouds about 2,500 feet above the Earth's surface. At intervals, a burning fuse along the upper kite string dropped the thermometers to the surface. Since the thermometers were tied to tassels of paper, their rapid descent was checked somewhat so that when recovered quickly, they provided approximate indications of existing temperatures aloft.

1105. What did Benjamin Franklin's famous kite experiment prove? In 1752, Benjamin Franklin performed perhaps the best-known experiment in electrical history by flying a kite in the teeth of a thunderstorm and disproving the superstitions which had been built up around lightning. Two important contributions to the study of lightning evolved from the kite experiment. He discovered that lightning is electricity and he developed a system of lightning rods for the protection of buildings. Franklin described the kite experiment to the chemist, Joseph Priestley, who wrote: "Struck with this promising appearance, he immediately presented his knuckle to the key (hung on the kite string) and . . . perceived a very evident electric spark. . . ." Franklin was very lucky on that auspicious day. If the

side flash that traveled down the wet kite string and sparked across his knuckle had been a full-sized bolt, his career would have been violently ended.

1106. What did Benjamin Franklin observe about Temperate Zone weather?   In 1743, a few years before his electrical experiment with kite and key, Franklin became the first American to recognize the general west-to-east movement of the weather in temperate regions of North America. Through simultaneous observations by himself in Philadelphia and a brother in Boston, Franklin found that a general storm may move in the opposite direction to that of rain-bearing winds. He found, by careful back-tracing of the correspondence, that despite a surface wind and rain from the east, the storm actually moved eastward over Philadelphia on to Boston and out to sea.

1107. Who proved that air was made up of several components? In 1756, Joseph Black, while a student of medicine at the University of Edinburgh, discovered the existence of carbon dioxide and proved that it is a permanent, though small, constituent of the atmosphere. This was a startling discovery because for some 2,000 years it was thought that the atmosphere was an all-alike substance consisting of just "air" and water vapor. Black's discovery was the first conclusive evidence that air consisted of more things than the simple element it was believed to be. Black's work incited further studies of the atmosphere by other investigators who soon found the various permanent gases contained in the atmosphere.

1108. When were nitrogen and oxygen discovered? Nitrogen, which accounts for about $\frac{4}{5}$ of the atmospheric makeup, was discovered in 1772 by Daniel Rutherford, an English chemist. Oxygen, the *gas of life,* which makes up most of the remaining $\frac{1}{5}$ of the atmosphere, was discovered by Joseph Priestly, one of the greatest of English chemists, in 1774. Discovery of the so-called *noble gases* (argon, neon, helium, krypton and xenon) was forecast as a chemical group by Cavendish in England in 1785, although their actual discoveries did not come about until various times in the mid to late nineteenth century. These gases are called *noble* because they are so inert chemically that they stand completely alone and are never found in any chemical compounds.

1109. Who invented the hair hygrometer? In 1783, the Swiss geologist, Horace Benedict de Saussure, invented the first hair hygrometer which indicates the amount of moisture in the atmosphere by the use of strands of human hair which expand when the air is moist and contracts when it is dry, irrespective of the air temperature. Saussure published an important paper in 1783 which explained the fact that at the same temperature and pressure moist air is lighter than dry air.

1110. Who was the first to show that rain can be produced by cooling saturated air? In 1784, the Scottish geologist, James Hutton, presented the fact through the Royal Society of Edinburgh that rain could be produced by cooling the water vapor in a saturated volume of air. This is a fundamental fact in meteorology, although Hutton did not mention that the cooling of ascending air, incident to expansion, is the cause of the heaviest percentage of all rain or snow.

1111. Who described the effects of compression on air? In 1800, John Dalton, a schoolmaster in England, delivered an account of the heating and cooling of gases by compression and expansion, effects which are linked to many weather phenomena. Cooling from expansion of humid air induces nearly all rainfall. Heating by compressed air causes many warm dry winds such as the Chinook or Santa Ana which in a few hours can convert a snow field to an area of dust.

1112. Who devised the first practical classification of clouds? In 1803, an English manufacturing chemist named Luke Howard made an important contribution to meteorology by devising the first practical and generally accepted classification of clouds with appropriate names (cirrus, stratus, cumulus, etc.). This made it possible to talk and write understandably about clouds and therefore paved the way for a long series of profitable cloud studies. Gradual revisions of Howard's original classifications have been made at various times by international cloud commissions. In essence, however, the cloud nomenclature he proposed is still the basis for present-day classification.

1113. When did the Beaufort wind scale originate? Sir Francis Beaufort, a rear admiral and hydrographer of the British Royal Navy, provided a most useful scale for winds in 1806. He devised a classification of winds, the relative velocities of which could be indicated at sea by waves, chiefly, and on land by a variety of familiar phenomena. Beaufort's scale of wind was originally based on the descriptive scales long used by sailors who specified wind force in such terms as *calm, air, breeze, gale, storm,* etc. Rather than depend on these descriptive terms, Beaufort wanted to define his scale in terms of some standard object, similar to the way that a standard measure might be used to check the length of another object. He therefore chose the typical British man-of-war of his time and assigned a scale of each of 13 terms of wind force describing their different effects upon his man-of-war (See Questions 229–234).

1114. Who first correctly described the formation of dew?   In 1814, an English physician, Charles Wells, put an end to the long-held idea that dew fell from the sky. He conclusively showed that when still air in contact with a cold surface cools to a temperature below the saturation point for the amount of water vapor present, condensation in the form of water (dew) or ice (frost) occurred on the surface object, depending upon the temperature of the surface.

1115. Who invented the psychrometer?   The invention of the psychrometer, a device for obtaining the dew point, and, in fact, practically all expressions for humidity, is credited to Richard Assmann of Germany, in 1825. The modern psychrometer is the simplest and most accurate of all practical humidity-measuring instruments. It consists merely of two thermometers mounted side by side. One thermometer bulb is moistened when a reading is desired. The resultant cooling by evaporation from the moistened or *wet bulb* causes a lower reading in that thermometer. The difference in temperature readings between the two thermometers provides an indication of the amount of moisture in the air.

1116. Where was the first system of weather stations established?   Nearly 200 years after the Grand Duke of Tuscany's aborted attempt to develop a network of weather stations in northern Italy, the idea of a weather-reporting network was again seriously tried in Europe. In the early part of the nineteenth century, the Chevalier de Lamark, working in France with P. S. La Place, A. L. Lavoisier and others, established a system of weather stations and published a series of yearly weather summaries. The biggest drawback to the project was the lack of rapid communication (the telegraph was still in an experimental stage). But the concept provided an extremely important foundation for later development of networks in France and other countries.

1117. When were the earliest weather charts used? In 1820, a German, Heinrich Brandes, conceived a profoundly important idea —that widespread weather phenomena could best be studied by plotting on a map various weather observations taken at the same time, covering as large an area as possible. The concept was a grand one and William Redfield, an American, joined in the idea of plotting synoptic (broad-view) maps. The trouble, as Brandes and Redfield realized in each of their respective countries, was that by the time slow mails came from distant places, the weather being studied had come and gone. Brandes's maps contained weather data which was 37 year old! The important thing, however, was that the beginnings of weather-map plotting techniques were made and were in existence by the time telegraphic communication developed.

1118. When was the telegraph first used for transmission of weather data? In 1844, the first commercial telegraph clicked the message, "What hath God wrought?" from Baltimore to Washington by Samuel F. B. Morse, 25 years after Oersted's discovery of the electromagnetic application to an electric telegraph. At the same time, many men in different nations were proposing that this device be used to send weather reports from outlying stations into central weather offices. The first actual telegraphic reports were sent in 1849 by the American electrophysicist and secretary of the Smithsonian Institution, Joseph Henry. He and an American weather theorist, James Espy, started the first regular weather service in the United States which was placed under the direction of the War Department. The first daily weather map, that is, one made and published the same day the information was received by electric telegraph, was drawn in England for 9:00 A. M., August 8, 1851.

1119. When was the telegraph transmission of weather observations first used for forecasting purposes? The use of the electric telegraph for transmitting weather data for forecasting purposes was first proposed by Carl Kreil, at Prague, in 1842. A similar, but more elaborate proposal, was made to the British Association by John Ball of England. Perhaps the most important of practical applications of the telegraph for uses in forecasting severe storms and other weather conditions was set up in 1854 by the French astronomer, Urbaine Le Verrier. He was instructed by the French Emperor, Napoleon III, to develop a system of weather forecasting following the sinking of many French and British vessels because of a storm in 1853. The vessels had been transporting valuable supplies for the allied armies in the Crimea. In 1860, an important service of forecasts and storm warnings in Europe was accomplished by Buys Ballot, professor of physics at the University of Utrecht in Holland.

1120. What problems faced the early national forecasting services? The early work of Le Verrier and others gave rise to a wave of enthusiasm and in the middle of the nineteenth century, networks of meteorological stations developed in many countries. Forecasting, however, was attempted mainly on rules of thumb and based on imperfectly understood theoretical considerations. These methods were found inadequate and early forecasting progress was followed by a profound lull, although descriptive and statistical meteorology continued at a good pace. Typical of the problem was the situation in England. The Meteorological Office in London was established in 1854 as a department of the Board of Trade, with Admiral Fitzroy placed in charge. The Meteorological Committee of the Royal Society started to function in 1867. The work of the early committee met many reverses in England because of the tendency for scientific circles to expect astronomical accuracy from weather forecasting. It was not until 1879 that a unified weather service finally got under way in England.

1121. Where were some of the earliest instrumental weather records kept in the United States? The first systematic instrumental records in the United States were made by John Lining at Charleston, South Carolina, beginning in 1738. Some older available records were made at New Haven, Connecticut, beginning in 1780; Baltimore, Maryland, in 1817; and Philadelphia, Pennsylvania, in 1825. In the western United States, instrumental weather records at St. Paul, Minnesota, and Leavenworth, Kansas, were begun in 1836 and at St. Louis, Missouri, in 1837. On the Pacific Coast, some of the longer records were begun in California, in 1849 at Sacramento and San Francisco, and in 1850 at San Diego. Most of these records were made at military posts by the Medical Department of the Army.

1122. What institution in the United States made important contri-

butions to national weather observations? In 1847, Joseph Henry, secretary of the Smithsonian Institution, submitted a program of organization and work for that institution, including a system of meteorological observations entitled "Solving the Problem of American Storms." In 1849, the Institution entered the field of observational work, supplementing records previously made by the Army's Medical Department. Careful climatological records were made from Smithsonian observing stations varying from less than 100 in about 1850 to about 350 in 1869, most stations being located in the eastern half of the country.

1123. What was the first official U.S. government forecast? On November 9, 1870, the first official forecast was released from Chicago by Increase Lapham, the country's first official forecaster when the government established the national weather service under the Army's supervision. The forecast called for "High winds probable along the Lakes." Included in the report were remarks concerning high winds at Cheyenne and Omaha, a lowering barometer and strong winds at Chicago and Milwaukee and a rising barometer farther east.

1124. When did weather services originate in Canada? In 1839, a combined meteorological and magnetic observatory was established in Toronto. During the 1850s, grammar-school principals were compelled by law to take weather observations. The reports were generally poor and equipment sent from England was often useless—thermometers, for example, would not read low enough. In 1872, the present Meteorological Division was set up and a few years later it became part of the Department of Marine and Fisheries in the Dominion government. The first storm warnings were issued in 1876 and the next year the distribution of *probabilities* began (the word *forecast* was not used in earlier years of weather services in Canada, England and the United States).

1125. Under what governmental agency does the weather service operate in Canada? In 1936, it was transferred to the Air Services Branch of the Department of Transport. The Meteorological Division is the only organization in Canada for providing weather information to aviation, industry and the general public. There are about 1,500 fulltime employees in Canada's weather service and about as many more doing parttime work as weather observers. Besides general public forecasts, the government weather service provides forecasts for all flying in Canada and neighboring waters as well as for transatlantic flights from eastern Canada and Newfoundland. Special services include frost-warnings service for fruit growers, storm warnings for fishermen, weather information for the forestry service to help them control forest fires and weather advice to shippers to assist them in shipping perishable goods such as fruit and vegetables. Chief centers are at Vancouver, Victoria, Edmonton, Whitehorse, Calgary, Winnipeg, Toronto, Hamilton, Ottawa, Montreal, Moncton, Gander, Goose Bay and Halifax.

1126. When did international co-operation for weather studies begin? When the Meteorological Society was formed in London in 1823, John Ruskin, a celebrated English authority on art, expressed the hope that a world-wide organization be developed as a necessary structure to deal with the peculiar character of weather science. By international agreement in 1853, the maritime nations began an international co-operation by exchanging systematic information about weather at sea. From a conference in 1872 of the various directors of different national weather services, an International Meteorological Organization resulted for the purpose of developing an exchange of information about weather from every habitable part of the globe and also for the distribution of information by wire or wireless from stations on land and ships at sea. (See World Meteorological Organization, Section X.)

1127. When were the first world climate maps charted? The first maps showing world-wide distribution of atmospheric pressure were made by Alexander Buchan, secretary of the Scottish Meteorological Society, in 1869. An atlas of storms was prepared in 1870 by H. Mohn of Germany. In 1882, the first map to show world-wide distribution of precipitation was made by Elias Loomis, an American mathematician. The first general world meteorological atlas was prepared by Julius von Hann, director of the Austrian Meteorology Center in 1887.

1128. Who first described the general characteristics of a low-pressure area? In 1841, an American student of meteorology,

James Pollard Espy, published the *Philosophy of Storms* which became the basis for many scientific discussions and papers concerning the structure and action of storms. Espy was the first one to show that convection (large-scale rising of air) takes place in areas of low pressure or cyclones. He expressed the theory that the low-pressure center of the storm was caused by the removal of air from that region by thermal convection, with some of the heat set free by condensation of water vapor. He wrote that the low pressure induces a spiral inflow of air and that the entire storm is carried forward by the general circulation of the atmosphere. While Espy's ideas did not conform to the origin, shape and behavior of the Temperate Zone low (caused by the meeting of opposed air masses), they did, however, account reasonably for the origin, structure and progress of the tropical storm (hurricane).

1129. Who wrote the Law of Storms?   Around 1850, Heinrich Wilhelm Dove, director of the Royal Meteorological Institute at Berlin, published the celebrated book, *Law of Storms*. In this profoundly important work he indicated that storms originate when polar and tropical air are brought together. The treatise marked what was probably the earliest forerunner of modern methods. He also gave a fuller explanation of the great wind systems of the world, suggesting two main circulations in each hemisphere; one between the equator and the subtropics, the other between the subtropics and the poles.

1130. What was the first major collection of wind and ocean current statistics?   In 1855, Matthew Fontaine Maury, an American Navy officer and hydrographer, published his classical work, *The Physical Geography of the Sea,* based on thousands of wind and ocean current entries in ship's logs. Amazingly complete charts of the winds of the seven seas were prepared by Maury which were invaluable aids to ship's captains of every maritime nation who found that they could cut the time of long voyages by as much as 25%.

1131. Who emphasized the importance of the Earth's rotation on a system of winds?   At the same time that Maury of the American Navy published his monumental work on the observational side of wind and ocean currents, William Ferrel, another American, was interested in the theoretical side of wind patterns. In 1856, he published his "Essay on the Winds and Currents of the Ocean." In this essay and later papers, he showed by rigorous mathematics that winds are affected by rotation of the Earth so that instead of a wind blowing in the direction it starts, it tends to turn to the right of that direction. In 1735, George Hadley had superficially discussed this theory and, in 1835, G. G. Coriolis, a French mathematician, had generally anticipated the ideas expressed by Ferrel. In fact, this phenomenon of air movement has been termed the *Coriolis Force*. But it was Ferrel who applied the theory to the winds of the Earth and impressed scientists with the importance of the laws involved.

1132. Who made important discoveries concerning hurricanes? In the 1860s, William C. Redfield, an American meteorologist who had contributed earlier to an understanding of weather map plotting, studied West Indian hurricanes and Atlantic Coast storms. He found that "a cyclone is constituted by a considerable mass of air endowed with a rapid movement of rotation in the direction opposite to the hands of a watch."

1133. Who first recognized the existence of high-pressure areas? While the American Civil War was in progress, Sir Francis Galton of England contributed a significant paper in which he wrote: "I have deduced . . . the existence not only of cyclones, but of what I dare to call *anticyclones* . . ." Today, the anticyclone or high-pressure area is recognized as the controlling center of an air mass. The movements of high-pressure areas probably assume even more importance in modern weather forecasting than the formerly overemphasized cyclone or low.

1134. What was one important discovery about condensation in the atmosphere? In 1880, John Aitken, a Scottish marine engineer, discovered a most important and surprising fact. He found out that condensation in the form of fog, cloud and rain occurs chiefly in the presence of certain foreign microscopic or submicroscopic particles in the air, such as sea salt, pollen, dust, etc. He showed that the presence of such particles in the atmosphere (called hygroscopic nuclei) represents a necessary condition for the occurrence of rain, although not in themselves a sufficient condition. Many modern rainmaking experiments are based on the idea of introducing minute foreign substances to the air.

1135. When were the first ascension balloons made?   The theory of using hot air or a gas as a lifting agent seems to have sprouted with many persons at unknown periods of time. Leonardo da Vinci is said to have filled some very thin wax figures with hot air which were sent aloft at the time of the coronation of Pope Leo X. In 1670

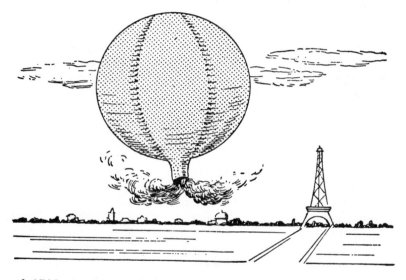

and 1709, two European Jesuits experimented with the concept of developing a type of airship on the hot-air principle, but their theories were unfulfilled. In 1783, the Montgolfier brothers, who were in the paper-making craft in France, created a sensation by building a balloon made of linen and paper, 105 feet in circumference. They filled the balloon with hot air by making a fire under it, cut the binding cords and watched the balloon rise impressively for a few thousand feet, landing ten minutes later nearly two miles away..

1136. When was hydrogen first used in a balloon?   The Montgolfier experiment described in the previous question electrified European scientists. The Paris Academy commissioned a physicist, J. A.

Charles, to develop further balloon studies. Charles designed a 13-foot diameter balloon employing hydrogen, then known as *inflammable air,* which had been a recent discovery of Cavendish in England. On August 27, 1783, the hydrogen-filled balloon was successfully released from the Champ de Mars before a crowd of "philosophers, officials, students and loiterers." One observer turned to the aged American ambassador, Benjamin Franklin, who was among the startled spectators, and asked, "Of what use is a balloon?" Replied Franklin, "Of what use, sir, is a newborn baby?" By the fall of that same year, human beings were successfully carried aloft in a large balloon, following an experimental balloon ascension containing barnyard animals.

1137. When were balloons first used for weather-measuring purposes? In the late eighteenth and early nineteenth centuries, the free balloon, or *aerostat,* was in full use following the Montgolfier and Charles demonstrations. In 1784, two Franco-American astronauts made the first accurate temperature measurements of the upper air. Soon after, the man-carrying balloons were in fairly common use for sounding the atmosphere with thermometer and barometer. In 1809, Thomas Foster of England began to study the direction and patterns of lower-level winds by observing small balloons which drifted freely upward and away. The major use of balloons today is not directed toward transportation as conceived at the time of its first uses, but rather toward a variety of ways of obtaining weather information from air levels up to 25 miles above the Earth's surface.

1138. When did balloons first carry self-recording weather instruments aloft? By 1892, the refinement of self-recording instruments which included changes of temperature, pressure and humidity (called *meteorographs*), made it unnecessary for men to go aloft after weather data. Kites, kite-balloons and captive balloons began to be widely used for carrying these instruments high into the air. Small sounding balloons, called *ballon-sondes,* were most successfully employed by Teisserenc de Bort, a talented French amateur meteorologist, who sent up his balloons from various points in Europe and from his yacht in the South Atlantic. A. Lawrence Rotch, an American pioneer in upper-air experiments, collaborated with de Bort in studies which showed the variation in height of the stratosphere base from pole to equator in the early 1900s. These unmanned balloons were designed so that after bursting, because of reaching thin air, the instrument came down with its gear acting as a parachute; at times two balloons would be harnessed, one bursting and the other acting as a parachute. The instrument-carrying balloons offered a reward to the finder if returned.

1139. When was the airplane first used for sounding the upper air? During the early 1900s, self-recording instruments used with balloons and kites underwent several refinements. In 1917–18, such meteorographs were carried by airplanes and used to explore the air levels above the fighting fronts of France in World War I. These *aerometeorographs* were further developed and used widely in the 1920s, 1930s and early 1940s, until replaced gradually by more practical and effective radio-transmitting devices.

1140. When were radio-transmitting weather instruments first used to explore the upper air? As far back as 1917, a method was suggested for transmitting signals from a meteorograph along a kite wire to a ground observer. The method never gained wide recognition. In 1927, Messrs. Bureau and Idrac of France first actually received weather signals from a small radio transmitter attached to a balloon that had reached some eight miles above the earth's surface. In 1930, P. A. Moltchanoff of Russia developed a practical miniature radio outfit which, when sent aloft with a small balloon, sent back reports of the state and condition of the atmosphere at various desired levels. This marked the beginning of the present-day *radiosonde,* the instrument-carrying balloon that transmits temperature, pressure and humidity data through levels up to 120,000 feet to a receiving station on the Earth's surface. Information regarding the wind force and direction of these upper-air levels is obtained by tracking the balloon's flight path with radar.

1141. How did World War I lead to an important development in weather understanding? During 1914–18, the Allied powers cut off public weather reports and forecasts in an attempt to block the enemy from keeping weather-posted. Neutral Norway, particularly, felt the pinch of the stoppage of weather reports from the west.

Turned inward by the force of circumstances, Norwegian meteorologists began to seek newer methods of weather understanding and forecasting. Out of this Norwegian school (Bergen school) of meteorology was born one of the most significant methods of weather study —air-mass and frontal analysis. Words that reflected the language of the war were used to describe weather activity—*fronts, advances* and *retreats*. This was natural because the Norwegian weathermen began to consider weather as the result of never-ending wars between *armies of air*.

1142. Who originated the concept of fronts and air masses? It was Jacob Bjerknes, son of the famous Norwegian physicist, Vilhelm Bjerknes, who headed the group which started an important departure in applied meteorology. Bjerknes studied and enlarged upon the theories of Heinrich Dove of Germany and Frank Bigelow of the U.S. Weather Bureau before him. He interpreted the Temperate Zone as the scene of constant strife between cold air masses pouring southward from the polar regions and warm air masses streaming northward from the tropics. He deduced correctly that these swirling and interacting air masses formed the cyclones and anticyclones which moved upon Norway in a relatively rhythmic procession across the Atlantic. He developed the modern concept of keeping close track of moving masses of air that differ from one another, and noting the positions and movements of their borders (fronts). This set the stage for a sharpening of short-range forecasting techniques.

1143. What characterized meteorological development between World Wars I and II? In general, the two decades between the wars marked a gradual transitional stage in meteorology—from the study of weather as a relatively two-dimensional problem to the treatment of weather as a complicated three-dimensional series of phenomena. In the 1920s, the ideas of air-mass analysis as developed by Norwegian meteorologists began to take root. The rapid development of aviation in the 1930s, with its needs for upper-air information, spurred meteorologists to drop empirical surface-weather forecasting techniques. As instrumental probing of the upper air developed, newer applications of thermodynamic and hydrodynamic principles were applied to forecasting. Studies of the upper air became the focus point for meteorologists. Important contributions during this period were made by Jacob Bjerknes, Halvor Solberg, Tor Bergeron, Sverre Petterssen, Carl-Gustave Rossby, Hurd C. Willett, George Stüve and Robert A. Millikan.

1144. What major problem faced meteorologists during World War II? The large-scale military operations on land, sea and in the air during World War II created a sudden need for the world's biggest corps of meteorologists. Long-range bomber and fighter craft were flying higher, faster and farther than ever before. Naval and land forces were deployed on a world-wide scale. The need for forecasting techniques and application of climatological studies to a wide range of military problems became desperately accute.

1145. How was the need for meteorologists recognized in the United States? The government's solution for solving the pressing need for meteorologists was to institute a series of high-pressure meteorological training programs. A few key topnotch meteorologists such as Dr. Robert A. Millikan and Professor Carl-Gustave Rossby were asked to create "crash" training methods at various university units across the country such as at Caltech, Massachusetts Institute of Technology, New York University and UCLA. Within months, hundreds of students received a crash-grounding in the sort of meteorology that would be most useful in war. The number swelled to thousands who were soon occupied with a host of weather problems created by global warfare. They learned practical methods, for example, for predicting whether the sky over an enemy target would be clear enough at a certain hour for high-altitude, visual bombings. Similar methods predicted days when dirty weather would protect ground troops from enemy air attack or when weather would be right for beachhead invasions.

Under the stimulus of military weather training, many meteorologists after the war continued to apply their talents to a variety of useful pursuits such as in the Weather Bureau, universities, government and foundation research and to many commercial and industrial activities.

1146. What marked meteorological development after World War II? Following World War II, development of special instrumentation for probing the atmosphere scaled new heights. Radarscope observations and the use of rockets and huge high-altitude balloons are some of the technical developments which are stripping away many atmospheric mysteries. The years after World War II have also seen the rapid expansion of the teletype weather networks and the simultaneous facsimile transmission by wire of upper air and surface weather maps and charts. International studies of world-wide weather have marked the era, with observers of many nations operating from the arctic and antarctic.

1147. How is radar applied to weather studies? The sleepless eye of radar is one of the most promising meteorological developments of modern times. Although radar is applied to a wide variety of weather problems, its main function is to search for, locate and define approaching storms and precipitation areas. Such radar storm-detection sets are best used as part of a general network to track meteorological phenomena over large areas and to aid in forecasting for land, sea and air activities. The application of radar for the detection of hurricanes, severe thunderstorms and tornadoes is especially being developed extensively by co-operative efforts of the U. S. Weather Bureau, Armed Forces, universities and industrial groups. Besides the detection and tracking of storms, radar impulses are sent out to measure cloud depths, to aid pilots in landing aircraft during marginal weather conditions, and to track free balloon paths in the upper air so as to obtain information about upper-wind conditions.

1148. What spurred the development of radar for weather use? Perhaps more than any single weather circumstance, it was the threat posed by hurricanes to the eastern coastal area of the United States that made the application of radar to weather studies a severe necessity. Today an expensive coastal chain of high-powered weather radar stations stretches from Brownsville on the Gulf of Mexico, to Portland, Maine. Weather Bureau, military, civil defense and industrial groups are closing the gaps to complete a radar net against hurricanes. Once a storm is within range (about 300 miles away), it becomes clipped to an electronic leash. A continuous horizontal cross-section picture of the storm is caught. By tracking and timing the storm's image across the scope grid, calculations about the storm's location, speed of movement and direction may be made.

1149. How does radar work?   The term radar is an abbreviation of *radio detecting and ranging.* Essentially, a radar set is composed of several systems. A transmitting system sends out radio impulses through a rotating antenna and wave guide. A receiving system includes circuits which accept the returning echo signals and convert them to amplified video information. This information is transferred to a console consisting of different scope grids which visually indicate the storm's distance, bearing and size. Radar sets for tracking storms and precipitation areas are not only limited to ground use, but are installed in aircraft and ships.

1150. What is a constant-level balloon?   During World War II, Japanese ingenuity was forced to develop a balloon which could be automatically controlled to stay aloft for periods of several days, drifting with the upper winds at one general level, pilotless, by the use of electromechanical control circuits. These constant-level balloons were used by the Japanese in the delivery of incendiary bombs from Japan to the North American continent during the war.

1151. How are constant-level balloons used in weather studies? The wartime achievements of the Japanese described in the preceding question stimulated interest concerning the application of these high-soaring balloons to weather studies. U.S. Navy and Air Force research groups devised huge polyethylene constant level balloons such as the *Skyhook* and *Moby Dick* which drifted across the continent many miles above the Earth's surface to report the pattern of air currents there. Huge balloons, called *transosonde* balloons, have been developed to carry hundreds of pounds of instrumentation for thousands of miles, floating along at controlled levels, and transmitting weather data to receiving stations on the surface.

1152. What does the transosonde balloon system consist of?   The essential components of the system are:

(1) A load-carrying balloon with control equipment for maintaining level flight.
(2) An airborne station consisting of meteorological sensing instruments, radio transmitter and power supply.

(3) A network of radio direction finder stations for periodic position checking of the balloon and for the collection of meteorological intelligence.

1153. How far have these giant balloons traveled? U.S. Navy constant-level balloons, carrying transmitting weather instruments, have traveled average distances of 5,000 miles at levels about 30,000 feet. One transosonde balloon was launched at Minneapolis on April 24, 1953, and landed just off the North African coast on April 27. An Air Force research balloon was launched from Vernalis, California, in February, 1955, and was tracked by radio direction finders for 14 days, during which time the balloon had traveled ¾ of the way around the Earth. The information gained from such transcontinental and oceanic balloon flights is useful in perfecting theories regarding the general circulation of the atmosphere. This, in turn, is important in the development of long-range forecasting techniques.

1154. What are other balloon uses in weather studies? Other current balloon developments include air-launched balloons to be released in circular storms such as hurricanes to serve as radar-reflecting beacons for tracing storm movements. Special balloons are also being developed to serve as missile targets.

1155. What is an automatic weather station? One of the most valuable contributions to meteorology in recent years has been the development of unmanned weather stations designed to be placed in areas where there are no human observers. Such automatic stations were used on land during World War II in the Pacific and in the Aleutians. Since then, two other types of stations have been developed —for use at sea and in the air.

1156. How is an automatic weather station used at sea? The boat or buoy station is anchored in a desirable water area and is designed to ride out any storm. It can transmit weather information by means of coded radio signal for a period of several months, with signals having a range of from 700 to 1,000 miles. Both air and seawater temperatures are measured, as well as atmospheric pressure, wind speed and wind direction, by instruments contained in an aluminum hull.

Automatic weather station

1157. How is an automatic weather station used in the air? A parachute automatic weather station (dropsonde) has been developed whereby a signal-transmitting instrument can be dropped from an airplane. As this device floats downward over normally inaccessible territory, it transmits information to a receiving station miles away concerning changes of temperature, pressure and humidity.

1158. What is the infrared hygrometer? In recent years, the Instrument Division of the U.S. Weather Bureau has developed a new and superior method for determining the amount of water vapor in the air. This new instrument, the infrared hygrometer, utlizes a beam of light containing two separate wave lengths which pass through the air being sampled. One of the wave lengths is absorbed by water vapor, while the other passes through the air undiminished. The ratio of energy transmitted in the two wave lengths can be used as a direct index of the quantity of water vapor present in the light path. This accurate instrument allows a sampling of air along any desired path length of atmosphere from a few inches to thousands of feet. It is also able to measure moisture at extremely low temperatures and low concentrations of water vapor where practically all other field systems for measuring moisture do not respond.

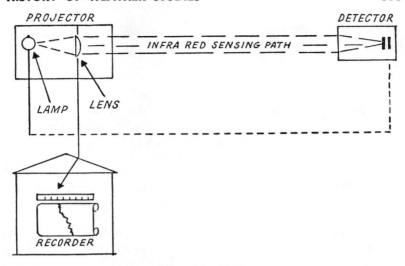

Infrared hygrometer

1159. What is meant by sferics?   Lightning strokes associated with thunderstorms and tornadoes radiate pulses of electromagnetic energy. This type of radio static can also be caused by snow and rain. The signals emitted as radio static by lightning or precipitation can be detected for long distances by a radio tuned to the same frequency of the weather-caused signals. Bearings of the storms can therefore be taken by several direction finding radio sets located as a network

Storm detection by sferics

several hundreds of miles apart. This method of tracking severe storms by radio is called *sferics* (a contraction of *atmospherics*). Current sferics studies may soon make possible the detection and tracking of thunderstorms as much as 2,000 miles away.

1160. How are the stars being used for upper-wind study? Typical of the host of modern meteorological studies is one based on the *twinkle* of stars as a measure of upper winds. The twinkle has been found to be caused by the movement of turbulent air parcels or cells having different densities and therefore different optical refraction. This turbulent movement is caused by high-altitude winds. Scientists believe that the frequency of the twinkle can be translated into accurate terms of wind speed and direction.

1161. What experiment involves the release of chemicals at high altitudes? U.S. Air Force geophysicists are making attempts toward controlling the release of energy received from the Sun and stored photochemically in the upper atmosphere. After laboratory

60 MI.

studies of the effect of radiation on gases at low pressure, clouds of nitric oxide have been released from rockets at about 60 miles above the Earth's surface. The nitric oxide augmented the natural release of energy and appeared as a glowing cloud in the night sky. It is conceivable that this effect may be used to produce reflectors for communication or to provide local illumination at night.

1162. What is meant by numerical prediction? It is the application of high-speed electronic computing machines to the problems of weather forecasting. Electronic weather forecasting is now being done with steadily increasing success by the Joint Numerical Weather Prediction Unit at Suitland, Maryland, where the U.S. Air Force, Navy and Weather Bureau have pooled their forces. Weather information flows into the machines from both ground stations and upper-air probes. Hundreds of punched cards cover both this information from North America and also other equally important information coming from the rest of the Northern Hemisphere including Russia and China. The electronic computers are programed so as to attempt to assort and put into coherent order these thousands of weather reports. By applying mathematical equations to current weather patterns, the machines can predict in some degree the future actions of the atmospheric patterns, especially those of the upper atmosphere.

1163. How successful is numerical forecasting? The machine's forecasts do not pinpoint ground-level weather for any locality. They concern the behavior of high-altitude patterns of air flow which control, in a broad sense, local surface weather. Authorities point out that computer forecasting should not be judged by its present performance but by its capacity to improve. Numerical forecasting is one of the really bright spots in the weather picture. Its full capabilities will gradually unfold as the machines are perfected, as men become more practiced in handling them, as new variables are brought into the equations, and as the flow of accurate surface and upper-air reports increase.

1164. What is meant by weather modification? Weather modification refers to the alteration of the atmospheric environment in some artificial way so as to gain a beneficial result. Weather modification includes such studies as the increase of rainfall by artificial means, the dissipation of fog or low clouds, and attempts to control, in a limited way, the energy of the Sun stored high in the atmosphere. While the term weather modication might logically be applied to various types of air conditioning, it is most often used as a reference to the stimulation of precipitation by artificial means. It should be mentioned that weather modification refers not only to an increase in rain formation, but to attempts to limit or break up storm and precipitation areas.

1165. What is the theory behind artificial rainmaking? Experiments for inducing an increase in rain by artificial means go back to the basic theory that for rainfall other than tropical, the presence of water droplets and ice crystals as a mixture is necessary in a cloud. In such deep clouds, at levels of critically low temperatures, it is be-

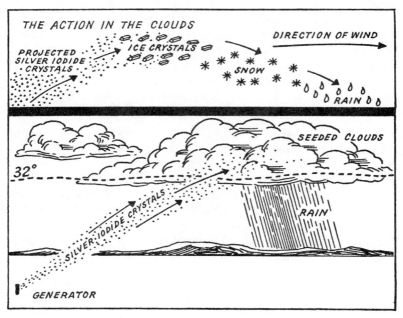

lieved that the minute ice crystals grow rapidly at the expense of the water droplets because of differences of vapor pressure between the water droplet and the ice crystal. The crystals grow large and heavy, start to fall and continue growth by collision. If the crystals fall through a warm air layer, they melt into rain (See Questions 158–165).

The object of the rainmaking experimenter is to introduce (seed) some substance into a cloud which will enhance and hasten this process of rain formation. Thus, under favorable conditions, the cloud is seeded with tiny particles of dry ice or chemicals to which water droplets may be attracted, such as silver iodide, salt particles, ammonium sulfate, solid carbon dioxide or methyl amine. This seeding may be accomplished by releasing the substance from a plane, or, in most experiments, by bombarding the base of a cloud with a chemical from generators on the ground.

1166. When did artificial rainmaking experiments begin? A scientific approach to weather modification was indicated as early as the 1890s, when patents were issued in the United States and Germany on devices for projecting dry ice into clouds. Limited knowledge of atmospheric conditions thwarted successful attempts. In 1931, a Dutchman, Veraart, dropped dry ice pellets into a cloud from an airplane and claimed to have caused measurable rain, although his appeal to the Dutch government for financial support to continue his experiments proved fruitless.

Current theories and experiments began in 1946 at the General Electric research laboratories under the direction of Irving Langmuir, Vincent Schaefer and Bernard Vonnegut. They introduced dry ice into a cold box filled with tiny supercooled water droplets, and this resulted in the formation of ice crystals.

1167. How successful are rainmaking experiments? Despite the fact that private meteorologists have taken on weather modification contracts from farmers and other industrial pursuits, reports which evaluate the effectiveness of the increase of rain through artificial means in the United States are marked with extreme caution and reserve. These investigations indicate that rainfall can be increased 9 to 17% under *special circumstances*. There are many *ifs* included in the reports, as indicated by the following statement of the Council of the American Meteorological Society: "The release of *substantial* amounts of precipitation by either natural or artificial means requires the pre-existence of an extensive moisture supply . . . for this reason, the meteorological conditions most favorable for the artificial release of precipitation are very much the same as those which usually lead to the natural release of precipitation. This makes the evaluation of seeding effects difficult and often inconclusive."

1168. What is the International Geophysical Year? The International Geophysical Year (IGY) is undoubtedly one of the most significant scientific undertakings in the history of man. The program, a co-operative venture involving scientists from some 56 nations, embraces observations and studies of the Earth's interior, its crust and oceans, the complex atmosphere reaching from the surface to heights of several hundred miles, and of the Sun, which virtually controls life and events on our planet. Although this gigantic international scientific investigation was originally conceived as being conducted over a period from July 1, 1957, to December 31, 1958, it seems possible that the studies will continue for at least a decade (International Geophysical Decade).

1169. How is IGY governed? It is governed by an international body resident in Brussels, Belgium, which is referred to as CSAGI, or the Special Committee for the International Geophysical Year. Each member country has a national committee which administers the scientific program of that country participating in the IGY. In the United States, the IGY program is administered by the U.S. National Committee for the IGY which is affiliated with the National Academy of Sciences. Its Chairman is Professor Joseph Kaplan, of the University of California at Los Angeles, and it maintains its headquarters in Washington, D.C.

1170. Is IGY the first study of its kind? International co-operation in the study of our physical environment is not new. The First International Polar Year occured in 1882–83 when meteorological, magnetic and auroral stations were established in arctic regions. Important studies at that time were made of the distribution of the auroras. A Second International Polar Year was held in 1932–33, fifty years later, which resulted in an increased knowledge of the upper air and greatly advanced the science of radio communications. The IGY program far surpasses in scope, intensity and geographical coverage these earlier programs which were limited largely to the north polar regions.

1171. How did IGY plans first develop? In 1950, a resolution was presented to the executive board of the International Council of Scientific Unions (ICSU), urging that a third international effort in geophysics be undertaken in 1957–58. The reason for the selection of this year was based in part upon the fact that this period corresponds with a period of maximum solar activity. Acting upon the resolution and similar recommendations from various international scientific unions, the executive board of ICSU in 1951 appointed a special committee, the Comite Special de l'Annee Geophysique Internationale (CSAGI) to co-ordinate the scientific planning of a world-wide cooperative program of geophysical observations. During succeeding years, various participating nations were asked to form national committees and to transmit their recommendations as to the content of the program to be included in the IGY. These recommendations and plans were gradually solidified by further meetings into a comprehensive international working plan.

1172. What are the studies of IGY? The principal fields of study during the IGY (or IGD if over a decade) include aurora and airglow, cosmic rays, geomagnetism, glaciology, gravity, ionospheric physics, longitude and latitude measurements, meteorology, oceanography, seismology, solar activity and upper-atmosphere studies using balloons, rockets and satellite vehicles. Stations for collection of data are spread strategically from pole to pole where scientists will explore almost every major land and sea area. They will study the Earth's core and crust. They will probe into its interior with explosion sound waves and send rockets and satellites to explore outer space. They will measure the deep ocean currents and the surging tides in the seas. They will observe and measure the many mysterious particles that continuously bombard the Earth from outer space.

1173. What questions may be answered by IGY studies? Answers to a thousand and more questions will be sought. Is the climate of the Earth changing? Are glaciers receding? Will melting ice sheets some day flood coastal lands? What is the exact nature of the atmospheric circulation around the Earth? Where do cosmic rays come from and what is their nature? What exactly causes the aurora? What is the relationship between sunspots and solar flares and long-range radio transmission? Can earthquakes be predicted? These and many other questions are the objectives of the IGY program. They are important to man's understanding of the Earth and the universe around him. The answers will provide him not only new basic knowledge, but applications in many fields of human society—from the raising of crops and transpolar air travel to better radio communications and navigation.

1174. How will IGY data be handled? The benefits mankind will realize will depend on the rapidity and extent to which scientists have access to IGY observations. Three World Data Centers to handle the mountain of information collected during the IGY program have been established. One center is located in the United States, one in the U.S.S.R., and the third divided between Western Europe and Japan. Each center is to receive original copies of IGY data from countries near the center. The receiving center immediately makes copies for other World Data Centers. Any institution or individual may request copies of data from a World Center, and these will be supplied at reproduction cost. Plans for a U. S. World Data Center include a series of primary archives located at appropriate institutions and agencies (probably in Washington, Boston, New York and Los Angeles) with a central co-ordination office for handling requests.

1175. What was the first operational phase of IGY? The first operational phase of the IGY was the exploratory expedition and establishment of stations in the antarctic during 1954–56. In 1954, U.S. Navy units surveyed ice conditions adjacent to the antarctic continent, and in 1955–56, a Naval Task Force was organized to install the Little America station and to transport equipment and supplies for construction of two forward stations at the South Pole and at Marie Byrd Land at 80° South, 120° West. Scientists from eleven nations will man stations so located as to encircle Antarctica. The antarctic program will permit daily weather maps of that huge continent for the first time. Sheathed in ice and snow, the six million square miles represent a unique region of cold weather, which is suspected to have a major influence on the world's weather.

1176. When were rockets first used for upper-air studies? As a form of automotion, the rocket antedates by centuries the airplane, the automobile and the steam engine. Essentially, through its history of development, the rocket represented, in various degrees, an instrument of war, from the Chinese gunpowder application to the present-day intercontinental ballistics missile. The use of a rocket vehicle for the scientific investigation of the upper air did not begin until just after World War II when the U. S. Army began assembling and testing captured German V-2 rockets at White Sands, New Mexico. The Army's primary purpose in the tests was to gain experience that would be helpful in the design and handling of future American guided missiles. But they had the vision to see that the rocket could be instrumented for upper-air research. With this in mind, the Army invited government agencies and universities to make high altitude measurements in the V-2s.

The first V-2, fully instrumented for upper-air research, was launched from White Sands on June 28, 1946, to an altitude of 67 miles. Included in the instrumentation was a Geiger-counter telescope to detect cosmic rays, a spectrograph, pressure and temperature gauges, and radio transmitters to probe the ionosphere.

1177. What studies are made by the use of research rockets?
High-altitude research rockets act as sensitive extensions of earthbound instruments. Upper-atmosphere rocket investigations gained and expected by present and future rocket probings include measurements of pressure, temperature, density and winds. Critical questions regarding the relative effects of solar ultraviolet light, solar x rays and incoming particle radiations on the upper atmosphere may be answered by rocket use. The vertical distribution of ozone and the presence of water vapor in the high atmosphere may be determined as well as the chemical composition. Relations between the aurora, ionospheric currents, high-altitude winds and fluctuations in the Earth's magnetic field are still to be clarified. These are just a few of the problems existing in the high atmosphere. They represent the very basis for current IGY rocket programs—the need for data which observations on the ground cannot provide.

1178. How is information sent from the rocket to the Earth?
Packed in the rocket are a variety of sensitive instruments which measure various elements such as pressure, temperature and cosmic rays. This information is relayed back to ground stations by a telemeter transmitted which flies in the rocket and consists of a closely packed maze of vacuum tubes, resistors, condensers and transformers. Telemeter signals are received and recorded at a ground block-

house station in the form of strip films, several inches wide and several feet long. Another method of obtaining the information from the rocket, especially photographs and motion-picture records made by automatic camera, is by parachute recovery of the instrument and camera sections.

1179. Where are rockets launched? Until the advent of the IGY, most upper-air research rockets, with very few exceptions, were flown in the Holloman-White Sands, New Mexico, area to various altitudes between 60 and 150 miles. In 1956, as part of the IGY, an important combined rocket program between Canada and the United States was started at Fort Churchill, Manitoba, Canada. Another important rocket site is located at Patrick Air Force Base, Cocoa, Florida.

1180. How many rockets will be flown during IGY? It is estimated that some 600 rockets will be launched, including Aerobees that reach an altitude of about 200 miles and rockoons (rockets launched from balloons) that reach about 60 miles in altitude. Most Aerobees will be fired from Fort Churchill in Canada and some from White Sands and Alamagordo, New Mexico. Rockoons will be launched from ships at sea off southern California and on the way to the Antarctic from Florida. The Thule, Greenland, area may witness small-rocket flights launched from aircraft (rockairs). Two-stage rocket combinations, like Nike-Deacons, will also supplement the basic program. British, Australian and French rockets will also roar skyward to add to the great international effort of unlocking the atmospheric mysteries.

1181. What is an artificial satellite? It is an unmanned vehicle instrument, designed to be launched from a multistaged rocket so as to circle some hundreds of miles above the Earth's surface in a continuous orbit, like an artificial Moon. The launching of such man-made Moons, one of the boldest and most imaginative steps taken by man, was planned as one of the many programs of the International Geophysical Year. It represents the first stage in man's acquisition of *direct* knowledge of the universe far beyond the Earth's surface and far beyond the scope of aircraft, balloons and even conventional research rockes.

1182. What advantage does an artificial satellite have over rockets for probing the upper air? While rockets provide crucial information about the upper atmosphere, they have two limitations. First, their total flight is extremely short, while the time spent in a particular altitude range is even shorter. Second, rocket coverage is limited to a small part of the Earth. Thus, in spite of the great value of rocket data, there is a need for a tool which can provide data over a long period of time, over considerable heights above the Earth and over large expanses of the atmosphere about the Earth. A satellite can achieve these objectives. The satellite, in effect, represents an extension of rockets. This led Professor Joseph Kaplan, Chairman of the U.S. National Committee for the IGY, to call the satellite program the *LPR*—long-playing rocket.

1183. How can an artificial satellite be launched? A three-stage rocket will be used to place it on orbit. The rocket will be about 72 feet long with a maximum width of 45 inches. It will be finless, using internal gyroscope controls for guidance. The first stage, providing

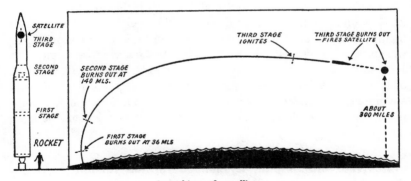

Launching of satellite

a thrust of 27,000 pounds, will start the system on its way and will operate for about 140 seconds. When its fuel is exhausted at some 40 miles altitude, the first stage will burn out and drop off, having attained a velocity of 3,000–4,000 miles per hour. The second stage rocket will then take over and will burn out at about 130 miles altitude, attaining a velocity of about 11,000 miles per hour and coasting onward. At about 300 miles altitude, the third motor will fire and accelerate the satellite vehicle to its orbital velocity, about 18,000 miles per hour, needed to overcome the Earth's gravitational pull.

1184. How long is the satellite planned to remain in its orbit? The intended orbit for the satellite is a nominal circle 300 miles above the Earth's surface, just within the outer edge of the atmosphere. If a satellite could be put into this desired orbit, it would stay up for a year. But the angle and velocity of firing cannot be controlled exactly so that the orbit will be elliptical, in which case it is intended that the nearest approach be not less than 200 miles and the farthest extension about 1,500 miles. While the atmosphere is extremely tenuous at these heights, drag is sufficient to take energy out of the orbit and cause the satellite to spiral to Earth. If the satellite had a circular orbit at 200 miles, its lifetime would be only 15 days, and were it 100 miles, then the lifetime would be less than one hour.

1185. What does a satellite consist of? After many considerations of size, sturdiness and weight factors, the instrumented satellite is planned to be a metallic sphere, at least 20 inches in diameter and 21½ pounds in weight. About half of this weight will be required for the instrument itself. The other half will be left for various lightweight measuring instruments and for telemetering equipment to relay information back to surface stations. The sphere will be made of highly machined magnesium, plated with gold, overlaid with aluminum and coated with a silicon compound. Much lighter *subsatellites,* made of aluminum-coated plastic, may supplement the main satellite.

1186. What orbital path is planned for the satellite? An inclined elliptical orbit of 40 degrees from the equator has been chosen. This means that the orbit of the satellite will shift within apparent latitudes of about 40 degrees on either side of the equator. Thus, as the satellite orbits around the rotating Earth once every hour and a half, many nations participating in the IGY will have an opportunity to make observations and measurements. In this orbit, the satellite will be observable from the United States, Central and South America, Africa, southern Europe and possibly some regions in mid-northern latitudes; the Balkans and Middle East; the Caspian Sea and part of the U.S.S.R.; Pakistan, China, Japan, India, Indonesia, Australia and New Zealand.

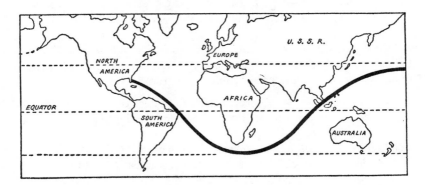

1187. How is a satellite tracked by radio? As the satellite will speed around the Earth, two methods for tracking it will be used; radio and optical. The radio method is known as Minitrack and involves the use of a miniature radio transmitter in the satellite which radiates a continuous signal to sensitive receiving equipment in a network of ground stations. Various participating IGY nations will obtain reasonably good position determinations by picking up the signals emitted from the satellite. The maximum range of these signals is expected to be between approximately 1,000 and 3,000 miles in all directions, but with an effective range of about 800 miles.

1188. How is a satellite tracked visually? Correlated with radio receiving stations tuned to the broadcasting devices inside the satellite will be a number of special high-speed cameras which will be operated by professional astronomers from several vantage points in different nations. Many amateur observers, using a variety of optical devices, are pooling their efforts in a plan to keep track of the satellite. If the radio transmitter fails, the work of these amateur volunteers will be extremely important. The satellite will be difficult to spot optically unless the observer has previous knowledge of the orbit's predicted position. At best, the satellite will be visible to optical devices a few hours only after sunset and before sunrise, weather conditions permitting. Observing the satellite with the naked eye may be possible under ideal conditions but extremely difficult. Its faint reflected image—equivalent to that of the dimmest star visible to the unaided eye—will move swiftly against a sky atwinkle with stars, not to mention the twilight glow.

1189. What studies are anticipated from the use of satellite vehicles? It is hoped that the following scientific experiments can be made as a result of the earth satellite program:

(1) *Air density.* By observing the drag effect on the satellite, calculations may be made concerning the density of the outer atmosphere.

(2) *Composition of the Earth's crust.* There are differences in the distribution of mass in the Earth's crust (the bulge at the equator, for example). Because the Earth's gravitational pull is an effect of the mass of the Earth, the geographic variations in the mass of the Earth will cause variations in the satellite orbit. Careful observations and calculations of these orbital changes will yield information on the mass distribution in the Earth. This, in turn, will tell something about the composition of the Earth's crust.

(3) *Geodetic determinations.* A better picture of the Earth's shape should be provided by measurements of the orbit related to (2). Also, synchronized observations may permit improvements in more exact determinations of longitude and latitude for mapping and navigational purposes.

(4) *Temperature.* Temperature within the satellite will be measured.

(5) *Pressure.* Pressure within the satellite will be measured.

(6) *Meteoritic impacts.* Instruments will measure the effects on the satellite of meteor penetration, erosion and impact.

(7) *Ultraviolet radiation.* Instruments will measure the extreme ultraviolet radiation from the Sun and, possibly, solar x-rays.

(8) *Cosmic rays.* Measurements of cosmic rays will be made, including the low-energy bombardments which are masked from the Earth by the atmosphere.

XII. HOW TO BE WEATHERWISE

Introduction. The chances are that no matter what newspaper you read, either a great metropolitan daily or a local gazette, a daily weather map is part of the reading fare. Weather is important news, affecting our daily activities and deserving of the widest dissemination. But the mere presence of the map guarantees neither its being studied nor its being understood by a majority of the reading public.

Perhaps the newspaper reader, who often doggedly pursues the most complicated cryptogram and crossword puzzle to their final solution, is not aware of the tale told by the weather map. For him, on a small, symbol-covered square, is placed a never-ending story of clashes of air thousands of miles around him and several miles deep. For him, thousands of observers from desert areas, from plains, mountains, oceans and aircraft, have collected and sent weather reports to be swiftly condensed and represented on a map.

The weather map speaks a language that can be learned and that can be most helpful in its application to work and play. Beyond its practical use, it can afford its interpreter the enjoyment that comes from participating in the solution of a puzzle—in this case a match of wits against the riddles posed by the restless atmosphere.

1190. Who prepares the daily newspaper weather map? The basic weather map in newspapers is prepared by various U.S. Weather Bureau offices. Two synoptic (broad-view) weather maps are given to newspapers daily, one for publication in evening editions and one for morning and afternoon editions. The weather map included in the morning newspaper usually represents a picture of the weather as it was at 1:30 P. M. of the previous day. The evening newspaper map represents the atmospheric picture as it was at 1:30 A. M. that same morning.

1191. Why do maps of different newspapers have different appearances? Although the weather map data prepared for newspapers is basically the same, the style or design of presentation is sometimes changed by a news agency for the sake of making it more simple and generally easier to understand visually. Some newspapers publish the map exactly as received from the Weather Bureau. Some strip it of its more technical symbols and present the information in a manner which is believed to register more quickly. Simplified weather maps are sometimes syndicated by a national wire service for inclusion in newspapers across the country. To help solve the time lag between the actual preparation of the map and its being seen by the reader, some newspapers publish simplified versions of the Weather Bureau's *prognostic* map which is a visual presentation of the national weather map as it will be *tomorrow*. Regardless of the method of presentation, the essential object is to provide the reader with a visual picture of the kind of weather occurring across the nation at a given time.

1192. What weather elements are usually shown on newspaper maps? The general map included in a newspaper for the public retains only a few of the heavy accumulation of data used on weather maps prepared for the guidance of a forecaster. The forecaster has before him charts which the public does not see; not only more elaborate surface weather charts, but also maps which show the pressure states at levels 2 and 3½ miles about the Earth's surface. Newspaper maps are surface maps only and are stripped, usually, of all but the starkest outlines of pressure centers, wind flow, temperature, precipitation, cloudiness and air-mass separations (fronts).

1193. What is meant by a synoptic weather map? The preparation of a weather map is a kind of atmospheric survey and is based upon the concept of observing the conditions near at hand at a large number of places, and then piecing these observations together on a map. This is called a *synoptic* method (from the Greek *synoptikos*), meaning the presentation of a general view or summary. A synoptic chart or weather map, therefore, is a chart showing meteorological conditions over a large region at a given time.

1194. How is weather information gathered for the weather map? At scheduled times every six hours, observers in all parts of the world simultaneously begin taking weather observations. The sky is observed and clouds are classified. Barometers are read and corrected to sea-level values. The direction and speed of the winds are noted. Rainfall or snowfall, if any, is measured. The current temperature and the extremes since previous observations are taken. The moisture of the air is calculated. The visibility and heights of clouds are determined, and all other phenomena such as thunderstorms, fog, smoke, haze, halos, etc., are carefully noted. Auxiliary information about winds, temperature, pressure and humidity of the upper air is also collected at various stations.

Each observer then condenses the information into a numerical message in the International Code. The message is composed of six or more five-figure groups which, if expanded into descriptive language, would comprise from 60 to 100 words. These messages are readily decoded in any country of the world. The job of condensing the weather information into code form takes about 20 minutes after which the coded messages are transmitted to collection centers and exchanged internationally.

1195. What happens to the transmitted coded weather information? At each receiving office the messages are immediately decoded and the weather conditions reported are transcribed by figures and symbols (the weatherman's international shorthand) upon an outline map covering the area from which reports are received. On the map, the location of each station is indicated by a small circle. Data in each station report are entered at the respective station circle in accordance with a definite pattern that is used throughout the world. When all the reported data have been entered on the map, the forecaster can see a broad picture of existing weather conditions over a very large geographical area.

1196. How is the surface weather map drawn? The forecaster studies the weather information which is transcribed in condensed symbol form on the map and begins to improve and develop the weather picture by drawing various lines or curves. He connects points on the map reporting equal barometric pressure readings with smooth solid lines called *isobars*. He marks boundaries or *fronts* between masses of air of different characteristics. Isobars are drawn at intervals of four millibars, that is, through all points reporting a barometric pressure of 1,020 millibars with successive isobars through points 1,024, 1,028 millibars, etc., toward regions of higher pressure and successive isobars through points reporting 1,016, 1,012 millibars, etc., toward regions of lower pressures. The centers of these regions are marked *HIGH* or *LOW* respectively. He makes the picture more graphic by shading or outlining areas over which precipitation is occurring and by drawing other curves which show temperature distributions, pressure tendencies, etc. (Lines connecting equal points of temperature are called *isotherms.*)

1197. How does the forecaster use the weather map? Having assembled a three-dimensional concept of the weather over a large geographical region, the forecaster begins the job of interpreting the gathered information before him. He compares the present maps with several preceding ones. He studies the movements and actions of pressure systems and notes the changes in temperature and moisture occurring within air masses as they move over different types of terrain. He studies the zones of meeting of these masses of air (fronts). He watches these movements and developments and attempts to anticipate the changes that will occur within the air masses and at the areas where they collide. He estimates the expanse of territory that will be covered and the time that given points will be reached by weather and temperature changes. He makes an estimate of weather conditions that may be reasonably expected in each locality during the next 12, 24, 36 or 48 hours. His results are sent out by radio, newspaper and TV in the form of a standard forecast for the public.

1198. What is meant by map types? Weather maps, like fingerprints, seem to come in an almost endless variety. It is doubtful if two maps have been found which may be classified as exact duplicates. However, like fingerprints, maps may be classified into certain types in which those belonging to each type will have general characteristics in common. A considerable amount of applied and experimental forecasting is being attempted by various meteorologists based on the collection and indexing of a very large number of maps. By selecting a series of maps which correspond in pattern to the current map, these meteorologists believe that a comparison of the weather as revealed by these corresponding map types provides a useful degree of probabilities which can assist in the solution of forecasting problems.

1199. Why is a knowledge of weather map symbols important? With a fundamental knowledge of the basic symbols which appear in most newspaper weather maps, the interested layman can, without

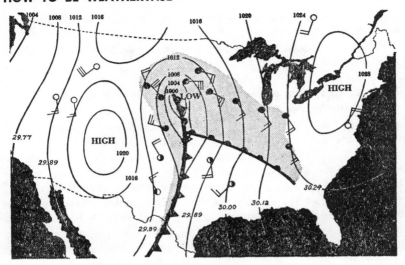

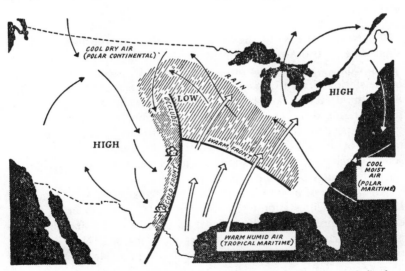

Lower map shows air masses, wind flows and precipitation that are symbolized on upper map

formal training, acquire a valuable tool in the understanding of weather processes. The symbols, in essence, provide a large picture story in abbreviated form of the manner in which weather is taking place.

1200. What symbol on the map shows the state of the sky? Various key cities across the country are represented on the map as small circles. The state of the sky (amount of cloudiness) is indicated by the way the circle is filled in. For example, if the sky is overcast, the circle will appear solid black. If clear, the circle will be empty of shading. If the sky is partly cloudy, only half of the circle will be black. If it is raining at the station at the time of the weather map, some maps show the letter R in the circle or S for snow. On other maps, rain or snow is shown over a general area by hatched lines or symbols indicative of raindrops or snowflakes. Many maps are embellished with a variety of artistic symbols, representing weather phenomena such as lightning, for example, as jagged lines on the map, or cloudiness as little individual cloud symbols.

1201. How is the wind shown by symbol on a weather map? A single line shaft drawn from the station circle represents the direction of the wind. To obtain the direction, it must be assumed that the top of the station circle (12-o'clock position) represents north. The shaft is drawn into the station *with* the wind. Thus, if the shaft with respect to the station lies due west, the wind direction is *from* the west. To tell the speed of the wind, featherlike barbs are drawn attached to the end of the direction shaft at right angles. Each long barb represents, roughly, a wind of about 10 miles per hour. A half barb is about five miles per hour.

An example of how to interpret state of the sky and wind symbols: If New York City station circle shows a shaft running south from the circle with three long barbs and a half barb attached and the circle is half black, this would indicate that the skies over New York are partly cloudy with the wind coming in from the south at about 35 miles per hour.

1202. What do arrow symbols on the weather map indicate? Some newspaper weather maps do not show the wind force and direction by the use of shafts and barbs as described in the previous

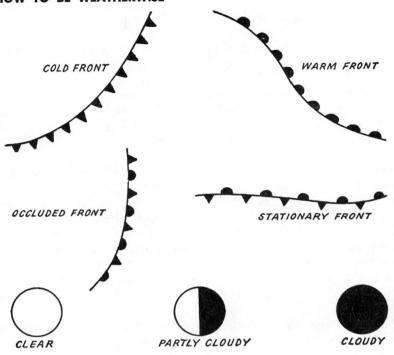

COLD FRONT

WARM FRONT

OCCLUDED FRONT

STATIONARY FRONT

CLEAR

PARTLY CLOUDY

CLOUDY

WIND DIRECTION AND FORCE

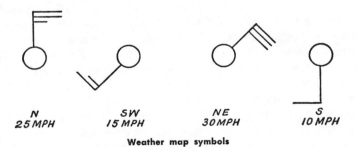

N
25 MPH

SW
15 MPH

NE
30 MPH

S
10 MPH

Weather map symbols

question. The wind direction, not including the wind speed, is simply shown by arrow symbols which indicate the general flow of wind over a large area. These arrows point *toward* where the wind is blowing.

1203. How is the temperature for each station indicated? Current temperatures (at the time the map was prepared) in degrees Fahrenheit are provided on most simplified newspaper weather maps as actual numbers placed near the station circle.

1204. What is an isobar? The solid curved black lines on a weather map are called isobars (from the Greek *iso,* meaning *equal,* and from *bar*ometer). These lines are drawn connecting stations reporting identical barometric pressures, after having been reduced to sealevel readings. Isobars are highly important since they show the contours or outlines of the pressure patterns which control the way the air flows. Isobar lines are labeled at one end in inches of mercury and at the other end in a corresponding value of millibars (isobars are drawn for every four millibars; 1,014, 1,018, 1,022, etc.).

1205. How do isobars indicate the wind flow on a map? When studying the wind shafts or arrows on a newspaper weather map, the reader will notice that winds blow clockwise and slightly outwards around a *HIGH* and counterclockwise and slightly inwards around a *LOW*. If there are a lot of isobars, this means that the pressure gradient is steep. As an analogy, isobars might be compared with contour maps of a hill; the steeper the hill, the more contour lines will be drawn to indicate the steepness. To continue the analogy somewhat, if water were to flow from a steep hill to a valley, it would flow rapidly because of the steepness. Air is motivated similarly. If the high-pressure dome is quite high with respect to the nearby low-pressure area (*valley*), the air will flow more strongly. Hence, the more isobars, the stronger will the wind be. If, on the other hand, there are just a few widely spaced isobars on the map over a region and they do not have a clear-cut sweeping pattern, the reader will notice that the winds are light or variable. This condition is known as *flat* pressure and, to return to the analogy, it is something like a plain lying between low hills.

1206. How should a high-pressure area be visualized? A high-pressure area might be considered as a clockwise-turning mass of air, heavier in weight than surrounding air, and heaviest (densest) in its central portion. The isobars surrounding the *HIGH* should be considered as contour lines showing the shape of the air dome or hill. Instead of the contour lines expressing the height of the dome of air in terms of feet or miles, they show the vertical gradations of weight or push of the air dome. The center marks the top of the dome or the area of greatest weight. The air in a *HIGH* descends or sinks (subsides) and flows outwards from the center much as sand tends to settle and flow out from the center of a sand pile. At the surface, the air flows slightly away from the center across the isobars and sweeps around the *HIGH* in a clockwise fashion (See Questions 310–318).

1207. How should a low-pressure area be visualized? A low-pressure area might be considered as a counterclockwise-turning system of winds, lighter in weight than surrounding air and lowest in weight (least dense) in its central portion (vortex). The isobars surrounding the *LOW* should be considered as contour lines analagous to showing the shape of a valley or depression. Instead of the contour lines expressing the depth of the depression or *trough* in terms of feet or miles, they show the gradations of vertical weight or push of the air in the *LOW,* with the center marking the area of lowest weight. Unlike air in a *HIGH* which is composed substantially of one large segment of air, the typical temperate zone *LOW* on a weather map is formed by the meeting together (convergence) of different air masses. The air in a *LOW* flows inward slightly across the isobars and sweeps around the center in a counterclockwise fashion (See Questions 297–301).

1208. How are cold fronts shown on a weather map? A cold front is the leading or forward edge of an advancing mass of cold air moving across country. As the colder air advances (usually southward and eastward), it burrows under and penetrates the warmer air like a wedge, forcing the warm air upwards and displacing it eastwards. On the map, the cold front is drawn as a solid black line with small triangles, the points of which face the direction of the front's movement. These triangles appear as icicle-like spikes attached to the solid line (See Questions 302–303).

1209. How are warm fronts shown on the weather map?   A warm front marks the leading edge of a moving mass of warm air, separating the warm air from a colder air mass. As the warm air advances (usually northward and eastward), it glides over the colder air and gradually displaces the colder air at the surface. Warm fronts on the map are drawn as solid lines with small, rounded, solid black semicircles, the base of each little hemisphere being on the front and the rounded portion in advance of the front. These little black semicircles have a bloblike or raindroplike shape.

1210. How is a stationary front shown on the weather map?   A stationary front marks a separation of two different air masses, neither of which is showing much movement. It is identified on the map by a drawn line which has alternate little black triangles and semicircles.

1211. What is meant by air-mass weather?   There are two principal types of weather, air mass and frontal. Air-mass weather develops within the heart of a single air mass. It is an expression of the character and personality of that air mass in its changing relations with the surface over which it is traveling. Air-mass weather is a spontaneous evolution of the air mass, untouched by other air masses. It depends on the temperature and moisture distribution within that air mass and it is likely to continue as a *spell* for several days at a time (See Questions 266–289).

1212. What is meant by frontal weather?   Frontal weather occurs along and near the discontinuity zones that separate unlike air masses. It depends not only on the characteristics of at least two air masses, but also on how they are brought together. Frontal weather is usually marked by rapid day-to-day or even hourly changes of weather. In the United States, summertime is mostly marked by air-mass weather, while wintertime is mostly featured by frontal weather.

1213. What is a fundamental principal of forecasting?   One fundamental principle on which all weather forecasting for middle latitudes is based is the fact that most weather in the Temperate Zones has a typical west-to-east drift, at an average speed of about 15 miles per hour in summer and 25 miles per hour in winter, carried along by the great lower- and upper-air streams known as the prevailing westerlies. Apart from the surface winds and isobars, there is an immense mass transport of the atmosphere from west to east across Canada and most of the United States. This is a part of the general or planetary circulation of the Earth's atmosphere. To a large extent, therefore, the pressure systems and fronts, which are merely disturbances in this general circulation, are carried along and show a progressive eastward movement from day to day.

1214. What should be done first when studying the weather map?
The reader should place himself mentally on the map according to his location. This will establish his position relative to any high- or low-pressure area which dominates his region. A simple wind test will provide an orientation for the observer with relation to any pressure center in the Northern Hemisphere. Starting with the basic concept that winds flow more or less with the isobars clockwise around a *High* and counterclockwise around a *Low,* the observer can check the direction of any high- or low-pressure center on the map. By mentally turning his back to the actual wind in his area, he will find that the center of the *Low* will be to his left. The direction of a high-pressure center will be to his right.

1215. How can the wind test provide a clue to the movement of a pressure center? For example, if the observer lies in the path of an east wind, by turning his back to the wind, he will find that the center of the *Low* is to his left, or south. If the wind should remain blowing from the east, he will know that the storm's center has remained relatively stationary. If, however, the wind should start to veer (change direction in a clockwise manner) and blow from the southeast, the back-to-the-wind test will show that the *Low* center has moved from a point due south of the observer to a point southwest. A continued veering of the wind means that the low-pressure center is moving northwards. The value of this test lies in the fact that it establishes a mental orientation of the observer with respect to pressure centers as they move across his area, attended by differences of weather.

1216. Why does an east wind precede a storm center to the south?

It will readily be seen that when a *Low* is approaching any locality, it will appear to the local observer to be coming from nearly the opposite direction, simply because the winds blow spirally around the center instead of with the direction of the *Low*'s movement. Thus, the *Lows* that are commonly called *northeasters* along the North Atlantic coast usually move up from the south or southwest.

1217. Why are good and bad weather associated with highs and lows respectively? An examination of areas marked *High* at the center will show that the winds are blowing in a direction opposite to those around the *Lows,* that is, spirally outward from the center with a clockwise rotation. Thus, considering the circulation of winds around both the *High* and *Low* centers, it is readily seen that there is a gradual transfer of air at the surface from areas of high pressure to those of low pressure.

This transfer of air from a high-pressure area to a low means that there is a slow sinking of air in *Highs* as air descends from upper levels to replace air at the surface as it flows outward. Conversely, there is a rising of air over *Lows* where it is pushed upward by converging winds. The descending or subsiding air in a *High* is warmed by the resultant increase in pressure. This compression of air has a cloud-dissipating effect. The rising air in a *Low* is cooled by expansion and this favors condensation of water vapor in the air. In a general way, therefore, there is a definite tendency for clear skies in a *High* and for cloudiness and precipitation to be found near a *Low*.

1218. What is meant by historical sequence of weather maps? To obtain a true measure of use from the weather map, a sequence of two, three or four *previous* weather maps should be kept. This will help to establish a picture of how the various pressure systems and fronts are moving. By the application of a few basic principles and individual correlations, these movements may be forecast to a reasonable degree. One practical method of forecasting by the layman involving an historical sequence of maps is called the *path method* and is briefly described in the following questions.

1219. What is the path method of forecasting? The path method may help to utilize the daily newspaper weather map as a tool for anticipating weather movement across the country. The method essentially relies upon the study of two or three consecutive weather maps kept on hand, including the current map. By selecting a given point on the map, such as the center of a low- or high-pressure area or a point along a cold or warm front, its speed and direction of movement is traced across the map through three consecutive days. The pattern of this movement is studied and can be projected into the future by an extension of the pattern.

1220. What are the first steps in using the path method? The first step is to study the weather map of 48 hours ago and to select any given point on the map which represents a particular weather situation. Since the map shows the location of weather-controlling *Highs* and *Lows* with cold and warm fronts associated with these systems, it is more useful to make the point the center of a *High* or *Low,* or to select a point on a cold or warm front near the observer. A pattern of movement and direction of this point is started by noting the point on the weather map of 48 hours ago. The point is marked on an onionskin overlay which is then placed on the weather map of 24 hours ago. The new position of the point is then marked on the overlay. Then, by placing the overlay on the current map, the movement of the point is marked accordingly. The observer now has three important points of reference. He can see at a glance the distance the point has moved in two consecutive 24-hour periods, its average speed of movement over that period and the pattern of direction of movement. He is ready now to make an estimation of where that point will be 24 hours from the current map. (See following question.)

1221. How is the path method applied to forecasting? Two examples of the path method application: (1) If point A has moved 500 miles due east in 24 hours and another 500 miles due east during the next 24 hours, it may be assumed that the point will continue in a similar direction and rate of speed. Therefore, point A can be displaced on tomorrow's weather map 500 miles to the east of its position on the current map. The reader can then place himself mentally on this projected weather map situation.

(2) If point A has moved 500 miles to the southeast the first day, then 250 miles to the east during the second day, the observer can note two things; first, its speed decelerated by one half; and second,

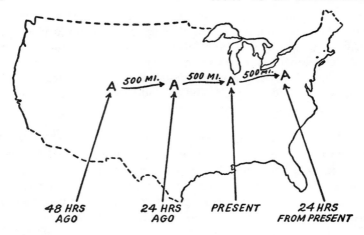

48 HRS
AGO

24 HRS
AGO

PRESENT

24 HRS
FROM PRESENT

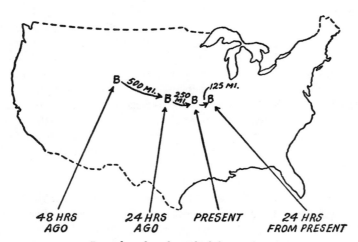

48 HRS
AGO

24 HRS
AGO

PRESENT

24 HRS
FROM PRESENT

Examples of path method forecasting

it curved, first going southeast and then east. Using the path and tendency method, it may be assumed that on the third day, tomorrow, it will continue to decelerate in the same manner and will continue to bear on the same curved path of direction. It can therefore be anticipated that tomorrow, or 24 hours from the current map, point A will have moved another *125* miles and be displaced that distance to the *northeast* from its position on the current map.

1222. How much stock should be placed on almanac weather forecasts? At best, an almanac does not provide a weather forecast. A few better almanacs include weather information for a specific area during the year based on average conditions experienced by that locality over a period of many years. To that extent, the weather information is climatological and has some value. If all almanacs would publish climatological data designed for a certain region, they could be usefully studied. The reader would then have a ready reference to the kind of weather that is *most frequently encountered* at any particular time of the year. It would not be right every time, particularly if a shorter period than a week is considered, but at least the odds would be in its favor. Unfortunately, most almanacs use a "buckshot" technique of forecasting, which by its very haphazard release and contradictory tone, is bound to appear as a correct forecast to *somebody, somewhere.* Time and again meteorologists have measured almanac weather statements against weather and have found them badly oriented.

1223. What are the difficulties in forecasting from local indications only? Weather changes are not usually heralded by local indications for periods longer than a few hours in advance. Indeed, many local storms give scarcely an hour's notice of their coming. Another difficulty is that very few local signs apply with equal force to all parts of the country. Indications along the Pacific Coast differ considerably from those along the Atlantic Coast, the Gulf Coast or the interior of the continent. In the interior there is a great difference between indications in the Rocky Mountain region, and the central valleys or the Great Lakes. Many signs which might be considered reliable in the Ohio Valley would be without much value in the dry regions of the southwest.

1224. How should the layman formulate rules of forecasting from local indications? In order that the layman may formulate such local rules as may be applicable to the locality in which he lives, he should proceed in a careful and systematic way to record and correlate his observations. A series of local observations, when compared and classified, will soon reveal certain well-defined relations that will enable the observer to begin compilation of a list of reliable indications.

1225. What indications offer the best clues to local weather changes? In studying local weather indications as signs for a change in the weather, certain basic elements should be studied in *connection* with one another, rather than separately. Pressure changes used in connection with the wind direction will provide the best key to the local situation. Used in concert, these two elements will give the layman an indication of where he is in relation to the moving high- and low-pressure areas above him. By studying a sequence of how the barometer acts in relation to wind force and direction, a pattern of reasonable forecasting accuracy may be accomplished.

1226. What wind-barometer indications are generally applicable to all parts of the United States? When the wind sets in from between south and southeast and the barometer falls steadily, a storm is approaching from the west or northwest, and its center will pass near or north of the observer within 12 to 24 hours, with wind shifting to northwest by way of south and southwest. When the wind sets in from points between east and northeast and the barometer falls steadily, a storm is approaching from the south or southwest, and its center will pass near or to the south of the observer within 12 to 24 hours, with wind shifting to northwest by way of north. The rapidity of the storm's approach and its intensity will be indicated by the rate and the amount of fall in the barometer. As a general rule, winds from the east quadrants and falling barometer indicate foul weather and winds shifting from the west quadrants indicate clearing and fair weather.

1227. How should the current aneroid barometer reading be determined? The barometer reading should be obtained by calling the local Weather Bureau office on the telephone or by listening to a radio or TV weather broadcast. Choose a day to set the reading on a home or office aneroid barometer which is rather settled with light winds, since on such days the pressure is most likely to be changing slowly. The readings will be given in inches and hundredths, for example; two nine point nine five (29.95), three naught point four eight (30.48), etc.

1228. How should the black hand of the barometer be set to the current reading? On the reverse side of the barometer there is a small opening about one quarter of an inch in diameter which permits access to a small set screw. Gently turning this screw will cause the *black* reading hand to move. The hand should be carefully adjusted so that the pointer rests directly over the reading obtained (See preceding question). No other adjustment is needed. A good barometer will continue indefinitely to read the current barometric pressure. This can be checked as frequently as desired.

1229. What is the purpose of the gold reference hand on the aneroid barometer? The only purpose of the gold reference hand is to provide a means for determining how much the barometric pressure has changed since the last reading. The small knob on the glass surface allows the gold hand to be moved independently from the black reading hand. Once or twice daily, it may be set to correspond with the underlying black reading hand. The black reading hand will move with the changing atmospheric pressure clockwise toward a higher reading when the pressure is rising, and counterclockwise toward a lower reading when the pressure is falling.

1230. Are clouds good indications of coming weather? Because clouds are tangible proof of a series of atmospheric changes, a study and understanding of them can be most helpful. It is important when studying clouds as clues to changing weather to observe their *historical sequence*. In themselves, clouds will not tell too much of a story unless the observer has an idea of how they have been building or developing, changing or dissipating over a period of time. Like a barometer, it's not the immediate reading, but the pattern of development and direction that counts.

1231. How should a study of clouds begin? To make useful application of clouds as a forecasting tool, the observer should get into the habit of identifying and classifying them, learning to estimate their heights above the Earth's surface, their direction of movement and whether they are thickening, building or gradually dissipating. A description of all major cloud groups is contained in Section II of this book. A handy cloud guide which can be used to identify clouds by matching cloud pictures with clouds observed is available at nominal cost from the Superintendent of Documents, Government Printing Office, Washington 25, D.C. The pamphlet is entitled *Circular S—A Cloud Manual.*

1232. What are some generalized cloud indications of approaching bad weather?

Chances for poor weather increase if . . .
Isolated high cloud patches thicken, increase and lower.
Fast-moving clouds thicken and lower.
Clouds are confused—move from different directions at different times.
A line of middle clouds darken the western horizon.
Isolated roll clouds fuse into sheetlike forms and lower.
Heavy piled-up clouds build to great vertical heights before noon on summer day.
Clouds develop dark pendulus bases.

1233. What are some generalized cloud indications for good weather?

Chances for better weather increase if . . .
Fog dissipates before noon.
Clouds decrease in number steadily.
Cloud bases become higher.
A fibrous, stratiform cloud wrinkles up and shows increasing breaks in the overcast.

1234. How can wind, barometer and cloud indications be summarized? For the use of the layman who is interested in studying changes in the barometer, winds and clouds as a guide for forecasting, several tables are presented following this Section. They are grouped as follows:

Table 1. Shows wind and barometer changes and the type of weather indicated from those changes.

Table 2. Shows the type of weather as indicated by the appearance of various clouds.

Table 3. Shows the various changes in weather elements which usually take place before, during and after cold- and warm-front passages.

1235. What are the limitations governing the use of these forecasting tables? It should be emphasized that the following forecasting guide tables represent a general summary of observations taken all over the United States. They are therefore broadly representative of average conditions and may not apply to a specific area without some alteration. Also, they represent conditions usually encountered in those temperate zones which lie in the normal tracks of moving high- and low-pressure areas. They are therefore more reliable as forecasting guides in some specific regions as against others. They may be generally applied most directly in the United States in those areas which are invaded alternately by cold air from the west and north and by warm air from the south. This includes most of the country, but less frequently in Florida, the Southwest States and the Pacific Coast.

These forecasting tables can be applied more effectively if the reader alters the sequence of indications according to careful correlations of weather in his own locality.

TABLE 1

WIND AND BAROMETER INDICATIONS FOR FORECASTING

Wind Direction	Sea-level barometric pressure in inches	Weather indicated

Note: A NW wind blows *from* the northwest

Wind Direction	Sea-level barometric pressure in inches	Weather indicated
SW to NW	30.10 to 30.20 and steady	Fair, with little temperature change for one to two days
SW to NW	30.10 to 30.20 and rising rapidly	Fair, followed within two days by warmer and rain
SW to NW	30.20 and above and stationary	Continued fair—no marked temperature change
SW to NW	30.20 and above and falling slowly	Fair with slowly rising temperatures for two days
S to SE	30.20 and falling rapidly	Wind increasing—rain within 12 to 24 hours
SE to NE	30.10 to 30.20 and falling slowly	Rain in 12 to 18 hours
SE to NE	30.10 to 30.20 and falling rapidly	Increasing wind and rain within 12 hours
E to NE	30.10 and above and falling slowly	In winter, rain within 24 hours. In summer, with light winds, rain may not occur for several days
E to NE	30.10 and above and falling rapidly	In winter, rain or snow with increasing winds. In summer, rain probable within 12 to 24 hours
SE to NE	30.00 or below and falling slowly	Rain will continue one to two days

Wind Direction	Sea-level barometric pressure in inches	Weather indicated

Note: A NW wind blows *from* the northwest

SE to NE	30.00 or below and falling rapidly	Rain with high winds followed in 36 hours by clearing, cooler
S to SW	30.00 or below and rising slowly	Clearing within a few hours and fair for several days
S to E	29.80 or below and falling rapidly	Severe storm soon followed within 24 hours by clearing and, in winter, colder
E to N	29.80 or below and falling rapidly	Severe northeast gale and heavy precipitation. In winter, heavy snow followed by cold wave
going to W	29.80 or below and rising rapidly	Clearing and colder

TABLE 2

FORECASTING FROM CLOUDS

A. HIGH CLOUDS. (Average lower level above 20,000 ft.)—*cirrus, cirrostratus, cirrocumulus*

Cloud appearance	*Weather indicated*
If cirrus, cirrostratus or cirrocumulus clouds remain thin, icy-white and in isolated patches with surface winds generally from a westerly quadrant . . .	Fair weather with little change
If they thicken or fuse together and seem to lose their fine whiteness—if they lower gradually and are followed by thicker clouds forming underneath and they move from the south or southwest while surface winds are veering from the east . . .	Precipitation probable within 12 to 24 hours with easterly to southeasterly winds and reduced visibility. Eventual change to warmer with winds freshening and shifting from the southwest

TABLE 2 (*Continued*)

B. MIDDLE CLOUDS. (Average lower level 6,500 ft.; average upper level 20,000 ft.)—*altocumulus, altostratus*

Cloud appearance	*Weather indicated*
If altocumulus patches seem to come together and become sheetlike or fused into more stratiform layers— if corona is present and shrinks with time—if their movement is from the south or southwest with surface winds veering and easterly . . .	Precipitation probable within 6 to 12 hours with southeast winds, reduced visibility and change to partly cloudy and warmer with southwest winds following
If altocumulus rolls or patches appear to be moving fast from the west or northwest—if surface winds are freshening from the south or southwest . . .	Watch for possible sudden wind shift from the northwest, hard brief precipitation, clearing and cooler. Squall possibilities
If altocumulus patches appear in midmorning on humid summer day and seem to have a crenellated appearance (castlelike tufts extending from a common base) . . .	Thunderstorms and strong local gusts probable during the day
If altocumulus masses remain isolated with open sky observed clearly through breaks or if patches appear in long whitish lens shapes—if they are generally moving in the same direction as surface winds . . .	Weather should remain fair with little immediate change expected
If altostratus layers become darker and lower from the south or southwest—if surface winds tend to veer from northeast, east and southeast— if the Sun's outline becomes faint and then obscure . . .	Poor weather approaching in few hours with steady type of precipitation and reduced visibility. East or southeast winds. Warm front will pass followed by warmer southwest winds and partly cloudy skies

C. LOW CLOUDS. (Average lower level close to the surface; average upper level, 6,500 ft.)—*stratocumulus, stratus, nimbostratus*

Cloud appearance	*Weather indicated*
If stratocumulus rolls retain rounded look and open sky breaks are apparent—if clouds drift with the general direction of the surface winds . . .	Generally unchanged weather; conditions should prevail for at least a day
If they fuse and become lead-gray, fibrous and form an overcast . . .	Unsettled and rainy weather will develop
If a bank of stratocumulus appears in the west or northwest as a long roll cloud and surface winds are fresh or strong from the south or southwest . . .	Watch for sudden squalls, shifting winds, brief hard precipitation, followed by clearing and cooler
If a thick stratocumulus overcast begins to wrinkle up with increasing breaks and the cloud bases become higher—if winds are steady and westerly . . .	Gradually clearing and cooler
If stratus exists in the morning and winds are light . . .	Fair weather will remain. Cloud will break up with Sun's heat leaving clear sky and light, variable winds
If dark nimbostratus overcast has developed from a lowering altostratus and the cloud direction is from the south or southwest with surface winds from the southeast . . .	Steady precipitation and reduced visibility imminent. Warmer weather, breaking skies and windshift from the south or southwest in a few hours

D. CLOUDS OF VERTICAL DEVELOPMENT. (Average lower level about 1,500 ft., average upper level to cirrus heights)—*cumulus, cumulonimbus*

If cumulus stand alone in the sky and do not develop until afternoon is underway . . .	Fair weather ahead with steady prevailing winds and good visibility. Clouds will usually dissi-

TABLE 2 (*Continued*)

D. CLOUDS OF VERTICAL DEVELOPMENT (Continued)

Cloud appearance	*Weather indicated*
	pate at day's end, leaving clear sky
If heavy cumulus form during late morning hours of humid day and build vertically throughout afternoon, increasing in amount . . .	Scattered showers probable by late afternoon. Freshening winds from the southwest. Good visibility
If cumulonimbus develop in isolated cells or masses and move separately from the south or southwest . . .	Scattered thundershowers will prevail throughout afternoon or evening. Winds will be generally fresh from the south or southwest. Possible lightning and hail and brief hard windshifts when cloud is in vicinity
If cumulonimbus or heavy cumulus appear in more or less solid banks on a line from the west, northwest or north and surface winds are fresh to strong from the south or southwest . . .	Squally weather change imminent. Hard thunderstorms in the offing with sharp windshift from the west or northwest followed by cooler, drier and clearing skies
If heavy cumulonimbus clouds develop blackish bulbous or pendulus bases which appear to build or boil downwards (mammatus) . . .	Severe thunderstorm and hail possibilities imminent. Watch for possible tornadic development

TABLE 3

FRONTAL CHARACTERISTICS

Element		Warm Front	Cold Front
Clouds	a. Before the front	Cirrus to cirrostratus to altostratus	Cirrostratus, altostratus or altocumulus, rapidly changing to cumulonimbus
	b. At the front	Altostratus to nimbostratus	

FRONTAL CHARACTERISTICS

Element		Warm Front	Cold Front
	c. Behind the front	Fractostratus to cumulus, or clear	Fractocumulus to altocumulus to clearing
Precipitation	a. Before the front	None, then steady rain or snow	None, then heavy, brief showers
	b. At the front	Rain	Squalls or thunder-storms
	c. Behind the front	Clear or scattered showers	Rapid clearing
Pressure	a. Before the front	Falling	Falling
	b. At the front	Steady or falling less rapidly	Slight rise
	c. Behind the front	Steady	Abrupt rise
Relative Humidity	a. Before the front	Constant, then increase	Fairly constant
	b. At the front	Near saturation	Increase near saturation
	c. Behind the front	Decrease from saturation to constant high value	Rapid drop
Temperature	a. Before the front	Steady, then rising gradually	Rising, then falling slightly
	b. At the front	Rising	Sudden change
	c. Behind the front	Steady—higher than advance of front	Abrupt drop
Visibility	a. Before the front	Fair	Below normal
	b. At the front	Poor	Poor
	c. Behind the front	Below normal	Very good
Winds	a. Before the front	S or SE	S, SW or W
	b. At the front	S, SE or SW	S or SW
	c. Behind the front	S or SW	SW, W or NW

A SELECTED BIBLIOGRAPHY
IN METEOROLOGY

Popular

Fisher, Robert M., HOW ABOUT THE WEATHER (Harper & Bros., N.Y., 1951). A good first reader on the cause and behavior of weather phenomena. Includes material on forecasting from newspaper maps and local signs.

Grant, H. D., CLOUD AND WEATHER ATLAS (Coward-McCann, N.Y., 1944). Heavily pictorial—detailed cloud descriptions and unusual phenomena illustrated.

Hare, F. K., THE RESTLESS ATMOSPHERE (Hutchinson's University Library. Longmans, Green and Co., N.Y., 1953). Contains brief review of physical laws governing weather processes, then treats the principal regions of the Earth in terms of dynamic meteorology.

Haynes, B. A., TECHNIQUES OF OBSERVING THE WEATHER (John Wiley & Sons, N.Y., 1947). Descriptions and illustrations of principal Weather Bureau instruments, with instructions for making and recording observations.

Humphreys, W. J., WAYS OF THE WEATHER (Jacques Cattell Press, Lancaster, Pa., 1942). Written in a simple popular style, the book is intended as a complete handbook about the weather for the layman and student.

Inwards, R., WEATHER LORE, 4th. ed. (Edited by E. L. Hawke, Rider and Co., London, 1950). The classical collection of sayings and proverbs on weather lore. Republished for the Royal Meteorological Society.

Krick and Fleming, SUN, SEA AND SKY (J. B. Lippincott Company, Philadelphia, 1954). The authors describe the basic causes of weather and stress the mechanics of clouds and artificial precipitation theories.

Longstretch, T. M., UNDERSTANDING THE WEATHER (A revision of KNOWING THE WEATHER, 1943. The Macmillan Company, N.Y., 1953). A very lively and readable account of the behavior of the weather elements. Recommended as a first book on the weather.

Ludlum, D., ed., AMATEUR WEATHERMAN'S ALMANAC (Weatherwise, Franklin Institute, Philadelphia, Pa., 1952). An annual first published in 1952, that summarizes the interesting weather events of the past year. Fully illustrated. Handbook section for amateur observers. Contains monthly blank recording forms.

Miller, D., WIND, STORM AND RAIN (Coward-McCann, N.Y., 1952). A series of well-written essays on what makes up our daily weather.

Murchie, G., SONG OF THE SKY (Houghton-Mifflin, Boston, 1954). An award-winning treasure house of historical meteorology, offering a sensitive and near-poetical insight into the panorama of weather phenomena.

Schneider, H., EVERYDAY WEATHER AND HOW IT WORKS (McGraw-Hill Co., N.Y., 1951). An explanation of the daily weather processes with well-illustrated instructions for making simple weather instruments. Recommended for junior and senior high schools.

Sloane, Eric, ERIC SLOANE'S WEATHER BOOK (Little, Brown and Company, Boston, 1952). A co-ordinated collection of Sloane's well-known weather illustrations and cartoons with appropriate text. Will be appreciated by all ages.

Stewart, G. E., STORM (Random House, N.Y., 1941). A fictional account of the history of a storm as it crosses the country. Highly authentic and interesting.

Tannehill, I. R., WEATHER AROUND THE WORLD, 2nd. ed. (Princeton University Press, Princeton, N.J., 1951). A description of normal weather at sea and in port for the world traveler.

Wenstrom, W. H., WEATHER AND THE OCEAN OF AIR (Houghton-Mifflin, Boston, 1942). A comprehensive book for the layman and student. The author stresses the three-dimensional concept of weather in an interesting and highly readable manner.

Wilson, V. and Berke, J., WATCH OUT FOR THE WEATHER (Viking Press, N.Y., 1951). An account of how the weather and climate affect your health and daily life.

Yates, R. F.,WEATHER FOR A HOBBY (Dodd, Mead & Company, N.Y., 1946). Deals mainly with all types of amateur instruments and their manufacture, with full plans for a home observatory.

Basic Texts

Albright, J.G., PHYSICAL METEOROLOGY (Prentice-Hall, N.Y., 1942). An excellent basic text on meteorological principles.

Blair, T. A., WEATHER ELEMENTS, 3rd. ed. (Prentice-Hall, N.Y., 1948). School edition also available. A basic text of long standing.

Donn, W. L., METEOROLOGY WITH MARINE APPLICATIONS (McGraw-Hill Co., N.Y., 1946). Written from the point of view of a deck officer.

Kraght, P. E., METEOROLOGY FOR SHIP AND AIRCRAFT (Cornell-Maritime Press, Cambridge, Md., 1944). A good text for all-around purposes, although stresses application of weathei principles to sea and air traffic.

Neuberger, H., INTRODUCTION TO PHYSICAL METEOROL-OGY (Pennsylvania State College, State College, Pa., 1951). Provides a basic understanding of the physical principles where many texts are weak. Recommended for all basic courses.

Neuberger, H., and Stephens, F., WEATHER AND MAN (Prentice-Hall, N.Y., 1948). Introductory text is followed by a description of how weather affects man in his social and business life. Good for survey courses.

Petterssen, S., INTRODUCTION TO METEOROLOGY (McGraw-Hill Co., N.Y., 1942). An admirably concise presentation of the basic principles on which modern forecasting is based.

Taylor, G. F., AERONAUTICAL METEOROLOGY (Pitman Co., N.Y., 1941). A good basic text on descriptive meteorology with special emphasis on application of meteorology to aeronautics.

Taylor, G. F., ELEMENTARY METEOROLOGY (Prentice-Hall, N.Y., 1954). An excellent all-purpose text in descriptive meteorology.

Advanced Texts

Brunt, D., PHYSICAL AND DYNAMICAL METEOROLOGY (Cambridge University Press, Cambridge, Mass., 1952). The standard reference in this field. Highly mathematical.

Byers, H. R., GENERAL METEOROLOGY (McGraw-Hill Co., N.Y., 1944). A well-balanced college text with due emphasis on aeronautical aspects.

Hewson, E. and Longley, R., METEOROLOGY, THEORETICAL AND APPLIED (John Wiley and Sons, N.Y., 1944). A well-planned college text by two eminent Canadian scientists.

Humphreys, W. J., PHYSICS OF THE AIR (McGraw-Hill Co., N.Y., 1940). A comprehensive text for the serious student.

Johnson, J. C., PHYSICAL METEOROLOGY (Massachusetts Institute of Technology, Cambridge, Mass., 1954). A thorough mathematical investigation of physical meteorology.

Petterssen, S., WEATHER ANALYSIS AND FORECASTING McGraw-Hill Co., N.Y., 1940). A textbook on synoptic meteorology. Principles of the physical, dynamic and kinematic atmosphere are discussed with their application to forecasting.

Willett, H.C., DESCRIPTIVE METEOROLOGY (Academic Press, N.Y., 1944). Treats the theories of the general circulation of the atmosphere. Strong on air masses and North American weather types. Useful to the nonmathematically inclined as well as to the serious student.

Handbooks, Manuals and Workbooks

Berry, Bollay and Beers, HANDBOOK OF METEOROLOGY (McGraw-Hill Co., N.Y., 1945). An extensive survey of the entire field of meteorology.

Caudle, F. L., WORKBOOK IN ELEMENTARY METEOROLOGY (McGraw-Hill Co., N.Y., 1945). Problems, quizzes and map-drawing exercises.

Huschke, Ralph E., ed., GLOSSARY OF METEOROLOGY (American Meteorological Society, Boston, Mass., 1957). A complete glossary of weather words, extremely useful for all reference work.

Kraght, P., METEOROLOGY WORKBOOK WITH PROBLEMS (Cornell-Maritime Press, Cambridge, Md., 1943). Meteorology problems and map-drawing exercises.

Malone, T., ed., COMPENDIUM OF METEOROLOGY (American Meteorological Society, Boston, Mass., 1951). 1,334 pages containing 108 articles which survey the present status of the subject and make suggestions for future progress. Indispensable as a reference work for the student of meteorology.

Pulk, E. and Murphy, E., WORKBOOK FOR WEATHER FORECASTING (Prentice-Hall, N.Y., 1950). Contains kit of practice charts. Instructions for plotting and analyzing synoptic weather maps along with a discussion of forecasting methods.

Shaw, N., MANUAL OF METEOROLOGY, 4 vols. (Cambridge University Press, Cambridge, Mass., 1926–34). A survey of meteorology in the late 1920s. Contains much descriptive, historical and theoretical matter that is found nowhere else.

Stefansson, V., ARCTIC MANUAL (Macmillan Company, N.Y., 1944). A detailed manual on living and working conditions in the arctic, containing many worth-while weather observational remarks. Written for the U.S. Army.

Climatology

Brooks, C. E. P., CLIMATE IN EVERYDAY LIFE (Philosophical Library, N.Y., 1951). A discussion of climate factors and how they affect our daily life and economy.

Brooks, C. E. P., CLIMATE THROUGH THE AGES. A STUDY OF CLIMATIC FACTORS AND THEIR VARIATIONS (McGraw-Hill Co., N.Y., 1949). A survey of climate and weather conditions from geologic times to the present. Good material on current climate changes.

Conrad and Pollak, METHODS IN CLIMATOLOGY (Harvard University Press, Cambridge, Mass., 1950). An advanced text for the student of climatology.

Department of Agriculture, U.S., CLIMATE AND MAN (Government Printing Office, Washington, D.C., 1941). A Department of Agriculture yearbook and a comprehensive summary of the relationships between climate and a wide variety of man's activities.

Hadlow, L., CLIMATE, VEGETATION AND MAN (Philosophical Library, N.Y., 1953). A survey of round-the-world cultures and societies in different climatic settings.

Haurwitz, B. and Austin, J., CLIMATOLOGY (McGraw-Hill Co., N.Y., 1944). A survey of the elements that make up our climate and a regional account of the world's climate zones. The approach is from the dynamic climatology viewpoint with much on air masses and disturbances.

Huntington, E., CIVILIZATION AND CLIMATE (Yale University Press, New Haven, Conn., 1933). Correlations and speculations concerning the effects of climate and climatic changes upon various civilizations.

Kendrew, W. G., CLIMATOLOGY, 3rd. ed. (Oxford, Clarendon Press, London, 1949). A revision of the classical and standard work in the field.

Lee, D. H., CLIMATE; ECONOMIC DEVELOPMENT IN THE TROPICS (Harper & Bros., N.Y., 1957). Deals systematically with the effect of tropical climates on men, animals, plants, materials and industrial production. Published for the Council of Foreign Relations.

Markham, S., CLIMATE AND THE ENERGY OF NATIONS (Oxford University Press, London, 1942). An analysis of the relation of climatic changes and conditions to man's industrial and economic development.

Shapley, H., ed., CLIMATIC CHANGE (Harvard University Press,

Cambridge, Mass., 1953). A series of papers by distinguished scientists dealing with various theories on causative factors of past and present climatic conditions.

Visher, S., CLIMATIC ATLAS OF THE UNITED STATES (Harvard University Press, Cambridge, Mass., 1954). An excellent climatological analysis of the United States. Presented in map form —1,031 maps and diagrams in 34 chapters. Embraces all major elements of climate.

Special Subjects

Ellison, M., THE SUN AND ITS INFLUENCE (Macmillan Company, N.Y., 1955). A factual introduction to the study of solar-terrestrial relations.

Flora, S. D., HAILSTORMS OF THE UNITED STATES (University of Oklahoma Press, Norman, Oklahoma, 1956). A statistical and descriptive state-by-state summary of principal hailstorm occurrences and their effects on life and property.

Flora, S. D., TORNADOES OF THE UNITED STATES (University of Oklahoma Press, Norman, Oklahoma, 1953). A descriptive summary of all we know about tornadoes. Lists by states all principal tornado occurrences.

Geiger, R., THE CLIMATE NEAR THE GROUND (Harvard University Press, Cambridge, Mass., 1950). A translation of the famous work summarizing Dr. Geiger's work in Germany in micrometeorology.

Landsberg, H. E., ed., ADVANCES IN GEOPHYSICS, Volume III (Academic Press, New York, 1957). An outstanding series of reviews on selected geophysical subjects.

Linsley, Kohler and Paulhus, APPLIED HYDROLOGY (McGraw-Hill Co., N.Y., 1950). The story of rainfall and its effects on topography by three meteorologists.

Lovelace Foundation, PHYSICS AND MEDICINE OF THE UPPER AIR (University of New Mexico, Albuquerque, New Mexico, 1953). A comprehensive collection of articles dealing with various considerations of the make-up of the upper atmosphere as well as reports on aviation and space medicine.

Menzel, D. H., FLYING SAUCERS (Harvard University Press, Cambridge, Mass., 1953). A top-flight scientist offers scientific explanations for many natural phenomena described as "saucers."

Minnaert, M., THE NATURE OF LIGHT AND COLOR IN THE OPEN AIR (Dover Publications, N.Y., 1954—an unabridged republication—originally published by G. Bell & Sons, Ltd., London, in English from Professor Minnaert's work at the University of Utrecht). The how and why of shadows, reflections, rainbows, mirages and over 100 other phenomena of light and color.

Perrie, D. W., CLOUD PHYSICS (John Wiley and Sons, N.Y., 1950). An account of the physical characteristics of clouds with a limited amount of information on artificial modification of clouds.

Sutton, O. G., MICROMETEOROLOGY (McGraw-Hill Co., N.Y., 1953). A survey with emphasis on theory.

Tannehill, I. R., DROUGHT, ITS CAUSES AND EFFECTS (Princeton University Press, Princeton, N.J., 1947). An excellent summary of conditions which breed dangerous droughts and the resultant damages which accrue.

Tannehill, I. R., HURRICANES, 7th. ed. (Princeton University Press, Princeton, N.J., 1950). All the essential facts and theories regarding the tropical cyclone. A careful documentation of major hurricanes over a period of many years.

U.S. Weather Bureau, THE THUNDERSTORM (Government Printing Office, Washington, D.C., 1950). A summary of the findings of the Thunderstorm Project.

World Meteorological Organization, WMO-No. 55, IGY-1 (World Meteorological Organization, Geneva, Switzerland, 1957). A single volume collection containing all information about the IGY meteorological program. Describes historical development of the IGY and various aspects of the program.

U.S. Weather Bureau Publications

Note: Over fifty periodicals and publications are issued by the U.S. Weather Bureau. These deal with a wide range of subject matter such as climatological comparisons, local and national weather maps, agricultural weather reports, aviation weather, clouds, Weather Bureau operations and services, forecasting, thunderstorms, hurricanes, tornadoes, etc.

These items are available at nominal costs (many pamphlets are less than 25 cents) from the Superintendent of Documents, U.S. Government Printing Office, Washington 25, D.C. For information concerning a list of publications, prices and ordering, write to the superintendent.

Meteorological Periodicals

United States

BULLETIN OF THE AMERICAN METEOROLOGICAL SOCI-
ETY. Monthly (except July and August). American Meteorological Society, 3 Joy Street, Boston 8, Mass.

JOURNAL OF METEOROLOGY. Bimonthly. American Meteorological Society, 3 Joy Street, Boston 8, Mass.

METEOROLOGICAL ABSTRACTS AND BIBLIOGRAPHY.
Monthly. American Meteorological Society, 3 Joy Street, Boston
8, Mass.

MONTHLY WEATHER REVIEW. Monthly. Published by the U.S.
Weather Bureau. Subscription office: Superintendent of Documents, Government Printing Office, Washington 25, D.C.

TRANSACTIONS OF THE AMERICAN GEOPHYSICAL UNION.
Bimonthly. Published by the National Research Council. American
Geophysical Union, 1530 P Street, NW, Washington 5, D.C.

WEATHERWISE. Bimonthly. American Meteorological Society, 3
Joy Street, Boston 8, Mass.

Great Britain

THE MARINE OBSERVER, A QUARTERLY JOURNAL OF
MARITIME METEOROLOGY. Quarterly. Her Majesty's Sta-
tionery Office, York House, Kingsway, London, W.C.2, England

METEOROLOGICAL MAGAZINE. Monthly. Her Majesty's Sta-
tionery Office, York House, Kingsway, London, W.C.2, England

QUARTERLY JOURNAL OF THE ROYAL METEOROLOGI-
CAL SOCIETY. Quarterly. Royal Meteorological Society, 49
Cromwell Road, London S.W.7, England

WEATHER. Monthly. Weather, 49 Cromwell Road, London, S.W.7,
England

Canada

A variety of pamphlets and booklets is available upon request
from the Meteorological Division, Department of Transport-
Canada, Toronto, Canada. Information distributed includes: climatological data, weather map data information for schools, meteorological employment opportunities, interesting facts about
weather phenomena, Canadian national weather services, operation, etc.

Europe, Africa, Asia

GEOPHYSICAL MAGAZINE. Central Meteorological Observatory of Japan. Irregular, about four numbers per year. Central Meteorological Observatory of Japan, Tokyo, Japan.

INDIAN JOURNAL OF METEOROLOGY AND GEOPHYSICS. Meteorological Department of India. Quarterly. Manager of Publications, Government of India, Delhi, India.

NOTOS. Weather Bureau of South Africa. Irregular. South African Weather Bureau, P.O. Box 1135, Pretoria, South Africa.

TELLUS, A QUARTERLY JOURNAL OF GEOPHYSICS. Quarterly. World Meteorological Organization, Campagne-Rigot 1, Avenue de la Paix, Geneva, Switzerland.

INDEX

A CATALOG OF SELECTED
DOVER BOOKS
IN ALL FIELDS OF INTEREST

A CATALOG OF SELECTED DOVER
BOOKS IN ALL FIELDS OF INTEREST

CONCERNING THE SPIRITUAL IN ART, Wassily Kandinsky. Pioneering work by father of abstract art. Thoughts on color theory, nature of art. Analysis of earlier masters. 12 illustrations. 80pp. of text. 5⅜ × 8½.   23411-8 Pa. $3.95

ANIMALS: 1,419 Copyright-Free Illustrations of Mammals, Birds, Fish, Insects, etc., Jim Harter (ed.). Clear wood engravings present, in extremely lifelike poses, over 1,000 species of animals. One of the most extensive pictorial sourcebooks of its kind. Captions. Index. 284pp. 9 × 12.   23766-4 Pa. $12.95

CELTIC ART: The Methods of Construction, George Bain. Simple geometric techniques for making Celtic interlacements, spirals, Kells-type initials, animals, humans, etc. Over 500 illustrations. 160pp. 9 × 12. (USO)   22923-8 Pa. $9.95

AN ATLAS OF ANATOMY FOR ARTISTS, Fritz Schider. Most thorough reference work on art anatomy in the world. Hundreds of illustrations, including selections from works by Vesalius, Leonardo, Goya, Ingres, Michelangelo, others. 593 illustrations. 192pp. 7⅛ × 10¼.   20241-0 Pa. $9.95

CELTIC HAND STROKE-BY-STROKE (Irish Half-Uncial from "The Book of Kells"): An Arthur Baker Calligraphy Manual, Arthur Baker. Complete guide to creating each letter of the alphabet in distinctive Celtic manner. Covers hand position, strokes, pens, inks, paper, more. Illustrated. 48pp. 8¼ × 11.   24336-2 Pa. $3.95

EASY ORIGAMI, John Montroll. Charming collection of 32 projects (hat, cup, pelican, piano, swan, many more) specially designed for the novice origami hobbyist. Clearly illustrated easy-to-follow instructions insure that even beginning papercrafters will achieve successful results. 48pp. 8¼ × 11.   27298-2 Pa. $2.95

THE COMPLETE BOOK OF BIRDHOUSE CONSTRUCTION FOR WOOD-WORKERS, Scott D. Campbell. Detailed instructions, illustrations, tables. Also data on bird habitat and instinct patterns. Bibliography. 3 tables. 63 illustrations in 15 figures. 48pp. 5¼ × 8½.   24407-5 Pa. $1.95

BLOOMINGDALE'S ILLUSTRATED 1886 CATALOG: Fashions, Dry Goods and Housewares, Bloomingdale Brothers. Famed merchants' extremely rare catalog depicting about 1,700 products: clothing, housewares, firearms, dry goods, jewelry, more. Invaluable for dating, identifying vintage items. Also, copyright-free graphics for artists, designers. Co-published with Henry Ford Museum & Greenfield Village. 160pp. 8¼ × 11.   25780-0 Pa. $9.95

HISTORIC COSTUME IN PICTURES, Braun & Schneider. Over 1,450 costumed figures in clearly detailed engravings—from dawn of civilization to end of 19th century. Captions. Many folk costumes. 256pp. 8⅜ × 11¾.   23150-X Pa. $11.95

CATALOG OF DOVER BOOKS

STICKLEY CRAFTSMAN FURNITURE CATALOGS, Gustav Stickley and L. & J. G. Stickley. Beautiful, functional furniture in two authentic catalogs from 1910. 594 illustrations, including 277 photos, show settles, rockers, armchairs, reclining chairs, bookcases, desks, tables. 183pp. 6½ × 9¼. 23838-5 Pa. $9.95

AMERICAN LOCOMOTIVES IN HISTORIC PHOTOGRAPHS: 1858 to 1949, Ron Ziel (ed.). A rare collection of 126 meticulously detailed official photographs, called "builder portraits," of American locomotives that majestically chronicle the rise of steam locomotive power in America. Introduction. Detailed captions. xi + 129pp. 9 × 12. 27393-8 Pa. $12.95

AMERICA'S LIGHTHOUSES: An Illustrated History, Francis Ross Holland, Jr. Delightfully written, profusely illustrated fact-filled survey of over 200 American lighthouses since 1716. History, anecdotes, technological advances, more. 240pp. 8 × 10¾. 25576-X Pa. $11.95

TOWARDS A NEW ARCHITECTURE, Le Corbusier. Pioneering manifesto by founder of "International School." Technical and aesthetic theories, views of industry, economics, relation of form to function, "mass-production split" and much more. Profusely illustrated. 320pp. 6⅛ × 9¼. (USO) 25023-7 Pa. $9.95

HOW THE OTHER HALF LIVES, Jacob Riis. Famous journalistic record, exposing poverty and degradation of New York slums around 1900, by major social reformer. 100 striking and influential photographs. 233pp. 10 × 7⅞.
22012-5 Pa $10.95

FRUIT KEY AND TWIG KEY TO TREES AND SHRUBS, William M. Harlow. One of the handiest and most widely used identification aids. Fruit key covers 120 deciduous and evergreen species; twig key 160 deciduous species. Easily used. Over 300 photographs. 126pp. 5⅜ × 8½. 20511-8 Pa. $3.95

COMMON BIRD SONGS, Dr. Donald J. Borror. Songs of 60 most common U.S. birds: robins, sparrows, cardinals, bluejays, finches, more—arranged in order of increasing complexity. Up to 9 variations of songs of each species.
Cassette and manual 99911-4 $8.95

ORCHIDS AS HOUSE PLANTS, Rebecca Tyson Northen. Grow cattleyas and many other kinds of orchids—in a window, in a case, or under artificial light. 63 illustrations. 148pp. 5⅜ × 8½. 23261-1 Pa. $4.95

MONSTER MAZES, Dave Phillips. Masterful mazes at four levels of difficulty. Avoid deadly perils and evil creatures to find magical treasures. Solutions for all 32 exciting illustrated puzzles. 48pp. 8¼ × 11. 26005-4 Pa. $2.95

MOZART'S DON GIOVANNI (DOVER OPERA LIBRETTO SERIES), Wolfgang Amadeus Mozart. Introduced and translated by Ellen H. Bleiler. Standard Italian libretto, with complete English translation. Convenient and thoroughly portable—an ideal companion for reading along with a recording or the performance itself. Introduction. List of characters. Plot summary. 121pp. 5¼ × 8½.
24944-1 Pa. $2.95

TECHNICAL MANUAL AND DICTIONARY OF CLASSICAL BALLET, Gail Grant. Defines, explains, comments on steps, movements, poses and concepts. 15-page pictorial section. Basic book for student, viewer. 127pp. 5⅜ × 8½.
21843-0 Pa. $4.95

CATALOG OF DOVER BOOKS

BRASS INSTRUMENTS: Their History and Development, Anthony Baines. Authoritative, updated survey of the evolution of trumpets, trombones, bugles, cornets, French horns, tubas and other brass wind instruments. Over 140 illustrations and 48 music examples. Corrected and updated by author. New preface. Bibliography. 320pp. 5⅜ × 8½. 27574-4 Pa. $9.95

HOLLYWOOD GLAMOR PORTRAITS, John Kobal (ed.). 145 photos from 1926–49. Harlow, Gable, Bogart, Bacall; 94 stars in all. Full background on photographers, technical aspects. 160pp. 8⅜ × 11¼. 23352-9 Pa. $11.95

MAX AND MORITZ, Wilhelm Busch. Great humor classic in both German and English. Also 10 other works: "Cat and Mouse," "Plisch and Plumm," etc. 216pp. 5⅜ × 8½. 20181-3 Pa. $5.95

THE RAVEN AND OTHER FAVORITE POEMS, Edgar Allan Poe. Over 40 of the author's most memorable poems: "The Bells," "Ulalume," "Israfel," "To Helen," "The Conqueror Worm," "Eldorado," "Annabel Lee," many more. Alphabetic lists of titles and first lines. 64pp. 5³⁄₁₆ × 8¼. 26685-0 Pa. $1.00

SEVEN SCIENCE FICTION NOVELS, H. G. Wells. The standard collection of the great novels. Complete, unabridged. First Men in the Moon, Island of Dr. Moreau, War of the Worlds, Food of the Gods, Invisible Man, Time Machine, In the Days of the Comet. Total of 1,015pp. 5⅜ × 8½. (USO) 20264-X Clothbd. $29.95

AMULETS AND SUPERSTITIONS, E. A. Wallis Budge. Comprehensive discourse on origin, powers of amulets in many ancient cultures: Arab, Persian, Babylonian, Assyrian, Egyptian, Gnostic, Hebrew, Phoenician, Syriac, etc. Covers cross, swastika, crucifix, seals, rings, stones, etc. 584pp. 5⅜ × 8½. 23573-4 Pa. $12.95

RUSSIAN STORIES/PYCCKNE PACCKA3bl: A Dual-Language Book, edited by Gleb Struve. Twelve tales by such masters as Chekhov, Tolstoy, Dostoevsky, Pushkin, others. Excellent word-for-word English translations on facing pages, plus teaching and study aids, Russian/English vocabulary, biographical/critical introductions, more. 416pp. 5⅜ × 8½. 26244-8 Pa. $8.95

PHILADELPHIA THEN AND NOW: 60 Sites Photographed in the Past and Present, Kenneth Finkel and Susan Oyama. Rare photographs of City Hall, Logan Square, Independence Hall, Betsy Ross House, other landmarks juxtaposed with contemporary views. Captures changing face of historic city. Introduction. Captions. 128pp. 8¼ × 11. 25790-8 Pa. $9.95

AIA ARCHITECTURAL GUIDE TO NASSAU AND SUFFOLK COUNTIES, LONG ISLAND, The American Institute of Architects, Long Island Chapter, and the Society for the Preservation of Long Island Antiquities. Comprehensive, well-researched and generously illustrated volume brings to life over three centuries of Long Island's great architectural heritage. More than 240 photographs with authoritative, extensively detailed captions. 176pp. 8¼ × 11. 26946-9 Pa. $14.95

NORTH AMERICAN INDIAN LIFE: Customs and Traditions of 23 Tribes, Elsie Clews Parsons (ed.). 27 fictionalized essays by noted anthropologists examine religion, customs, government, additional facets of life among the Winnebago, Crow, Zuni, Eskimo, other tribes. 480pp. 6⅛ × 9¼. 27377-6 Pa. $10.95

CATALOG OF DOVER BOOKS

FRANK LLOYD WRIGHT'S HOLLYHOCK HOUSE, Donald Hoffmann. Lavishly illustrated, carefully documented study of one of Wright's most controversial residential designs. Over 120 photographs, floor plans, elevations, etc. Detailed perceptive text by noted Wright scholar. Index. 128pp. 9¼ × 10¾.
27133-1 Pa. $11.95

THE MALE AND FEMALE FIGURE IN MOTION: 60 Classic Photographic Sequences, Eadweard Muybridge. 60 true-action photographs of men and women walking, running, climbing, bending, turning, etc., reproduced from rare 19th-century masterpiece. vi + 121pp. 9 × 12.
24745-7 Pa. $10.95

1001 QUESTIONS ANSWERED ABOUT THE SEASHORE, N. J. Berrill and Jacquelyn Berrill. Queries answered about dolphins, sea snails, sponges, starfish, fishes, shore birds, many others. Covers appearance, breeding, growth, feeding, much more. 305pp. 5¼ × 8¼.
23366-9 Pa. $7.95

GUIDE TO OWL WATCHING IN NORTH AMERICA, Donald S. Heintzelman. Superb guide offers complete data and descriptions of 19 species: barn owl, screech owl, snowy owl, many more. Expert coverage of owl-watching equipment, conservation, migrations and invasions, etc. Guide to observing sites. 84 illustrations. xiii + 193pp. 5⅜ × 8½.
27344-X Pa. $8.95

MEDICINAL AND OTHER USES OF NORTH AMERICAN PLANTS: A Historical Survey with Special Reference to the Eastern Indian Tribes, Charlotte Erichsen-Brown. Chronological historical citations document 500 years of usage of plants, trees, shrubs native to eastern Canada, northeastern U.S. Also complete identifying information. 343 illustrations. 544pp. 6½ × 9¼.
25951-X Pa. $12.95

STORYBOOK MAZES, Dave Phillips. 23 stories and mazes on two-page spreads: Wizard of Oz, Treasure Island, Robin Hood, etc. Solutions. 64pp. 8¼ × 11.
23628-5 Pa. $2.95

NEGRO FOLK MUSIC, U.S.A., Harold Courlander. Noted folklorist's scholarly yet readable analysis of rich and varied musical tradition. Includes authentic versions of over 40 folk songs. Valuable bibliography and discography. xi + 324pp. 5⅜ × 8½.
27350-4 Pa. $7.95

MOVIE-STAR PORTRAITS OF THE FORTIES, John Kobal (ed.). 163 glamor, studio photos of 106 stars of the 1940s: Rita Hayworth, Ava Gardner, Marlon Brando, Clark Gable, many more. 176pp. 8⅜ × 11¼.
23546-7 Pa. $11.95

BENCHLEY LOST AND FOUND, Robert Benchley. Finest humor from early 30s, about pet peeves, child psychologists, post office and others. Mostly unavailable elsewhere. 73 illustrations by Peter Arno and others. 183pp. 5⅜ × 8½.
22410-4 Pa. $5.95

YEKL and THE IMPORTED BRIDEGROOM AND OTHER STORIES OF YIDDISH NEW YORK, Abraham Cahan. Film Hester Street based on Yekl (1896). Novel, other stories among first about Jewish immigrants on N.Y.'s East Side. 240pp. 5⅜ × 8½.
22427-9 Pa. $6.95

SELECTED POEMS, Walt Whitman. Generous sampling from Leaves of Grass. Twenty-four poems include "I Hear America Singing," "Song of the Open Road," "I Sing the Body Electric," "When Lilacs Last in the Dooryard Bloom'd," "O Captain! My Captain!"—all reprinted from an authoritative edition. Lists of titles and first lines. 128pp. 5³⁄₁₆ × 8¼.
26878-0 Pa. $1.00

THE BEST TALES OF HOFFMANN, E. T. A. Hoffmann. 10 of Hoffmann's most important stories: "Nutcracker and the King of Mice," "The Golden Flowerpot," etc. 458pp. 5⅜ × 8½. 21793-0 Pa. $8.95

FROM FETISH TO GOD IN ANCIENT EGYPT, E. A. Wallis Budge. Rich detailed survey of Egyptian conception of "God" and gods, magic, cult of animals, Osiris, more. Also, superb English translations of hymns and legends. 240 illustrations. 545pp. 5⅜ × 8½. 25803-3 Pa. $11.95

FRENCH STORIES/CONTES FRANÇAIS: A Dual-Language Book, Wallace Fowlie. Ten stories by French masters, Voltaire to Camus: "Micromegas" by Voltaire; "The Atheist's Mass" by Balzac; "Minuet" by de Maupassant; "The Guest" by Camus, six more. Excellent English translations on facing pages. Also French-English vocabulary list, exercises, more. 352pp. 5⅜ × 8½. 26443-2 Pa. $8.95

CHICAGO AT THE TURN OF THE CENTURY IN PHOTOGRAPHS: 122 Historic Views from the Collections of the Chicago Historical Society, Larry A. Viskochil. Rare large-format prints offer detailed views of City Hall, State Street, the Loop, Hull House, Union Station, many other landmarks, circa 1904–1913. Introduction. Captions. Maps. 144pp. 9⅜ × 12¼. 24656-6 Pa. $12.95

OLD BROOKLYN IN EARLY PHOTOGRAPHS, 1865–1929, William Lee Younger. Luna Park, Gravesend race track, construction of Grand Army Plaza, moving of Hotel Brighton, etc. 157 previously unpublished photographs. 165pp. 8⅞ × 11¾. 23587-4 Pa. $13.95

THE MYTHS OF THE NORTH AMERICAN INDIANS, Lewis Spence. Rich anthology of the myths and legends of the Algonquins, Iroquois, Pawnees and Sioux, prefaced by an extensive historical and ethnological commentary. 36 illustrations. 480pp. 5⅜ × 8½. 25967-6 Pa. $8.95

AN ENCYCLOPEDIA OF BATTLES: Accounts of Over 1,560 Battles from 1479 B.C. to the Present, David Eggenberger. Essential details of every major battle in recorded history from the first battle of Megiddo in 1479 B.C. to Grenada in 1984. List of Battle Maps. New Appendix covering the years 1967–1984. Index. 99 illustrations. 544pp. 6½ × 9¼. 24913-1 Pa. $14.95

SAILING ALONE AROUND THE WORLD, Captain Joshua Slocum. First man to sail around the world, alone, in small boat. One of great feats of seamanship told in delightful manner. 67 illustrations. 294pp. 5⅜ × 8½. 20326-3 Pa. $5.95

ANARCHISM AND OTHER ESSAYS, Emma Goldman. Powerful, penetrating, prophetic essays on direct action, role of minorities, prison reform, puritan hypocrisy, violence, etc. 271pp. 5⅜ × 8½. 22484-8 Pa. $5.95

MYTHS OF THE HINDUS AND BUDDHISTS, Ananda K. Coomaraswamy and Sister Nivedita. Great stories of the epics; deeds of Krishna, Shiva, taken from puranas, Vedas, folk tales; etc. 32 illustrations. 400pp. 5⅜ × 8½. 21759-0 Pa. $9.95

BEYOND PSYCHOLOGY, Otto Rank. Fear of death, desire of immortality, nature of sexuality, social organization, creativity, according to Rankian system. 291pp. 5⅜ × 8½. 20485-5 Pa. $8.95

A THEOLOGICO-POLITICAL TREATISE, Benedict Spinoza. Also contains unfinished Political Treatise. Great classic on religious liberty, theory of government on common consent. R. Elwes translation. Total of 421pp. 5⅜ × 8½. 20249-6 Pa. $8.95

CATALOG OF DOVER BOOKS

MY BONDAGE AND MY FREEDOM, Frederick Douglass. Born a slave, Douglass became outspoken force in antislavery movement. The best of Douglass' autobiographies. Graphic description of slave life. 464pp. 5⅜ × 8½.   22457-0 Pa. $8.95

FOLLOWING THE EQUATOR: A Journey Around the World, Mark Twain. Fascinating humorous account of 1897 voyage to Hawaii, Australia, India, New Zealand, etc. Ironic, bemused reports on peoples, customs, climate, flora and fauna, politics, much more. 197 illustrations. 720pp. 5⅜ × 8½.   26113-1 Pa. $15.95

THE PEOPLE CALLED SHAKERS, Edward D. Andrews. Definitive study of Shakers: origins, beliefs, practices, dances, social organization, furniture and crafts, etc. 33 illustrations. 351pp. 5⅜ × 8½.   21081-2 Pa. $8.95

THE MYTHS OF GREECE AND ROME, H. A. Guerber. A classic of mythology, generously illustrated, long prized for its simple, graphic, accurate retelling of the principal myths of Greece and Rome, and for its commentary on their origins and significance. With 64 illustrations by Michelangelo, Raphael, Titian, Rubens, Canova, Bernini and others. 480pp. 5⅜ × 8½.   27584-1 Pa. $9.95

PSYCHOLOGY OF MUSIC, Carl E. Seashore. Classic work discusses music as a medium from psychological viewpoint. Clear treatment of physical acoustics, auditory apparatus, sound perception, development of musical skills, nature of musical feeling, host of other topics. 88 figures. 408pp. 5⅜ × 8½. 21851-1 Pa. $9.95

THE PHILOSOPHY OF HISTORY, Georg W. Hegel. Great classic of Western thought develops concept that history is not chance but rational process, the evolution of freedom. 457pp. 5⅜ × 8½.   20112-0 Pa. $9.95

THE BOOK OF TEA, Kakuzo Okakura. Minor classic of the Orient: entertaining, charming explanation, interpretation of traditional Japanese culture in terms of tea ceremony. 94pp. 5⅜ × 8½.   20070-1 Pa. $3.95

LIFE IN ANCIENT EGYPT, Adolf Erman. Fullest, most thorough, detailed older account with much not in more recent books, domestic life, religion, magic, medicine, commerce, much more. Many illustrations reproduce tomb paintings, carvings, hieroglyphs, etc. 597pp. 5⅜ × 8½.   22632-8 Pa. $10.95

SUNDIALS, Their Theory and Construction, Albert Waugh. Far and away the best, most thorough coverage of ideas, mathematics concerned, types, construction, adjusting anywhere. Simple, nontechnical treatment allows even children to build several of these dials. Over 100 illustrations. 230pp. 5⅜ × 8½.   22947-5 Pa. $7.95

DYNAMICS OF FLUIDS IN POROUS MEDIA, Jacob Bear. For advanced students of ground water hydrology, soil mechanics and physics, drainage and irrigation engineering, and more. 335 illustrations. Exercises, with answers. 784pp. 6⅛ × 9¼.   65675-6 Pa. $19.95

SONGS OF EXPERIENCE: Facsimile Reproduction with 26 Plates in Full Color, William Blake. 26 full-color plates from a rare 1826 edition. Includes "The Tyger," "London," "Holy Thursday," and other poems. Printed text of poems. 48pp. 5¼ × 7.   24636-1 Pa. $4.95

OLD-TIME VIGNETTES IN FULL COLOR, Carol Belanger Grafton (ed.). Over 390 charming, often sentimental illustrations, selected from archives of Victorian graphics—pretty women posing, children playing, food, flowers, kittens and puppies, smiling cherubs, birds and butterflies, much more. All copyright-free. 48pp. 9¼ × 12¼.   27269-9 Pa. $5.95

PERSPECTIVE FOR ARTISTS, Rex Vicat Cole. Depth, perspective of sky and sea, shadows, much more, not usually covered. 391 diagrams, 81 reproductions of drawings and paintings. 279pp. 5⅜ × 8½. 22487-2 Pa. $6.95

DRAWING THE LIVING FIGURE, Joseph Sheppard. Innovative approach to artistic anatomy focuses on specifics of surface anatomy, rather than muscles and bones. Over 170 drawings of live models in front, back and side views, and in widely varying poses. Accompanying diagrams. 177 illustrations. Introduction. Index. 144pp. 8⅜ × 11¼. 26723-7 Pa. $8.95

GOTHIC AND OLD ENGLISH ALPHABETS: 100 Complete Fonts, Dan X. Solo. Add power, elegance to posters, signs, other graphics with 100 stunning copyright-free alphabets: Blackstone, Dolbey, Germania, 97 more—including many lower-case, numerals, punctuation marks. 104pp. 8⅛ × 11. 24695-7 Pa. $8.95

HOW TO DO BEADWORK, Mary White. Fundamental book on craft from simple projects to five-bead chains and woven works. 106 illustrations. 142pp. 5⅜ × 8. 20697-1 Pa. $4.95

THE BOOK OF WOOD CARVING, Charles Marshall Sayers. Finest book for beginners discusses fundamentals and offers 34 designs. "Absolutely first rate . . . well thought out and well executed."—E. J. Tangerman. 118pp. 7¾ × 10⅝. 23654-4 Pa. $5.95

ILLUSTRATED CATALOG OF CIVIL WAR MILITARY GOODS: Union Army Weapons, Insignia, Uniform Accessories, and Other Equipment, Schuyler, Hartley, and Graham. Rare, profusely illustrated 1846 catalog includes Union Army uniform and dress regulations, arms and ammunition, coats, insignia, flags, swords, rifles, etc. 226 illustrations. 160pp. 9 × 12. 24939-5 Pa. $10.95

WOMEN'S FASHIONS OF THE EARLY 1900s: An Unabridged Republication of "New York Fashions, 1909," National Cloak & Suit Co. Rare catalog of mail-order fashions documents women's and children's clothing styles shortly after the turn of the century. Captions offer full descriptions, prices. Invaluable resource for fashion, costume historians. Approximately 725 illustrations. 128pp. 8⅜ × 11¼. 27276-1 Pa. $11.95

THE 1912 AND 1915 GUSTAV STICKLEY FURNITURE CATALOGS, Gustav Stickley. With over 200 detailed illustrations and descriptions, these two catalogs are essential reading and reference materials and identification guides for Stickley furniture. Captions cite materials, dimensions and prices. 112pp. 6½ × 9¼. 26676-1 Pa. $9.95

EARLY AMERICAN LOCOMOTIVES, John H. White, Jr. Finest locomotive engravings from early 19th century: historical (1804–74), main-line (after 1870), special, foreign, etc. 147 plates. 142pp. 11⅜ × 8¼. 22772-3 Pa. $10.95

THE TALL SHIPS OF TODAY IN PHOTOGRAPHS, Frank O. Braynard. Lavishly illustrated tribute to nearly 100 majestic contemporary sailing vessels: Amerigo Vespucci, Clearwater, Constitution, Eagle, Mayflower, Sea Cloud, Victory, many more. Authoritative captions provide statistics, background on each ship. 190 black-and-white photographs and illustrations. Introduction. 128pp. 8⅞ × 11¾. 27163-3 Pa. $13.95

EARLY NINETEENTH-CENTURY CRAFTS AND TRADES, Peter Stockham (ed.). Extremely rare 1807 volume describes to youngsters the crafts and trades of the day: brickmaker, weaver, dressmaker, bookbinder, ropemaker, saddler, many more. Quaint prose, charming illustrations for each craft. 20 black-and-white line illustrations. 192pp. 4⅝ × 6. 27293-1 Pa. $4.95

VICTORIAN FASHIONS AND COSTUMES FROM HARPER'S BAZAR, 1867–1898, Stella Blum (ed.). Day costumes, evening wear, sports clothes, shoes, hats, other accessories in over 1,000 detailed engravings. 320pp. 9⅜ × 12¼. 22990-4 Pa. $13.95

GUSTAV STICKLEY, THE CRAFTSMAN, Mary Ann Smith. Superb study surveys broad scope of Stickley's achievement, especially in architecture. Design philosophy, rise and fall of the Craftsman empire, descriptions and floor plans for many Craftsman houses, more. 86 black-and-white halftones. 31 line illustrations. Introduction. 208pp. 6½ × 9¼. 27210-9 Pa. $9.95

THE LONG ISLAND RAIL ROAD IN EARLY PHOTOGRAPHS, Ron Ziel. Over 220 rare photos, informative text document origin (1844) and development of rail service on Long Island. Vintage views of early trains, locomotives, stations, passengers, crews, much more. Captions. 8⅞ × 11¾. 26301-0 Pa. $13.95

THE BOOK OF OLD SHIPS: From Egyptian Galleys to Clipper Ships, Henry B. Culver. Superb, authoritative history of sailing vessels, with 80 magnificent line illustrations. Galley, bark, caravel, longship, whaler, many more. Detailed, informative text on each vessel by noted naval historian. Introduction. 256pp. 5⅜ × 8½. 27332-6 Pa. $6.95

TEN BOOKS ON ARCHITECTURE, Vitruvius. The most important book ever written on architecture. Early Roman aesthetics, technology, classical orders, site selection, all other aspects. Morgan translation. 331pp. 5⅜ × 8½. 20645-9 Pa. $8.95

THE HUMAN FIGURE IN MOTION, Eadweard Muybridge. More than 4,500 stopped-action photos, in action series, showing undraped men, women, children jumping, lying down, throwing, sitting, wrestling, carrying, etc. 390pp. 7⅞ × 10⅝. 20204-6 Clothbd. $24.95

TREES OF THE EASTERN AND CENTRAL UNITED STATES AND CANADA, William M. Harlow. Best one-volume guide to 140 trees. Full descriptions, woodlore, range, etc. Over 600 illustrations. Handy size. 288pp. 4½ × 6⅜. 20395-6 Pa. $5.95

SONGS OF WESTERN BIRDS, Dr. Donald J. Borror. Complete song and call repertoire of 60 western species, including flycatchers, juncoes, cactus wrens, many more—includes fully illustrated booklet. Cassette and manual 99913-0 $8.95

GROWING AND USING HERBS AND SPICES, Milo Miloradovich. Versatile handbook provides all the information needed for cultivation and use of all the herbs and spices available in North America. 4 illustrations. Index. Glossary. 236pp. 5⅜ × 8½. 25058-X Pa. $6.95

BIG BOOK OF MAZES AND LABYRINTHS, Walter Shepherd. 50 mazes and labyrinths in all—classical, solid, ripple, and more—in one great volume. Perfect inexpensive puzzler for clever youngsters. Full solutions. 112pp. 8⅛ × 11. 22951-3 Pa. $4.95

PIANO TUNING, J. Cree Fischer. Clearest, best book for beginner, amateur. Simple repairs, raising dropped notes, tuning by easy method of flattened fifths. No previous skills needed. 4 illustrations. 201pp. 5⅜ × 8½.   23267-0 Pa. $5.95

A SOURCE BOOK IN THEATRICAL HISTORY, A. M. Nagler. Contemporary observers on acting, directing, make-up, costuming, stage props, machinery, scene design, from Ancient Greece to Chekhov. 611pp. 5⅜ × 8½.   20515-0 Pa. $11.95

THE COMPLETE NONSENSE OF EDWARD LEAR, Edward Lear. All nonsense limericks, zany alphabets, Owl and Pussycat, songs, nonsense botany, etc., illustrated by Lear. Total of 320pp. 5⅜ × 8½. (USO)   20167-8 Pa. $6.95

VICTORIAN PARLOUR POETRY: An Annotated Anthology, Michael R. Turner. 117 gems by Longfellow, Tennyson, Browning, many lesser-known poets. "The Village Blacksmith," "Curfew Must Not Ring Tonight," "Only a Baby Small," dozens more, often difficult to find elsewhere. Index of poets, titles, first lines. xxiii + 325pp. 5⅜ × 8¼.   27044-0 Pa. $8.95

DUBLINERS, James Joyce. Fifteen stories offer vivid, tightly focused observations of the lives of Dublin's poorer classes. At least one, "The Dead," is considered a masterpiece. Reprinted complete and unabridged from standard edition. 160pp. 5³⁄₁₆ × 8¼.   26870-5 Pa. $1.00

THE HAUNTED MONASTERY and THE CHINESE MAZE MURDERS, Robert van Gulik. Two full novels by van Gulik, set in 7th-century China, continue adventures of Judge Dee and his companions. An evil Taoist monastery, seemingly supernatural events; overgrown topiary maze hides strange crimes. 27 illustrations. 328pp. 5⅜ × 8½.   23502-5 Pa. $7.95

THE BOOK OF THE SACRED MAGIC OF ABRAMELIN THE MAGE, translated by S. MacGregor Mathers. Medieval manuscript of ceremonial magic. Basic document in Aleister Crowley, Golden Dawn groups. 268pp. 5⅜ × 8½.   23211-5 Pa. $8.95

NEW RUSSIAN-ENGLISH AND ENGLISH-RUSSIAN DICTIONARY, M. A. O'Brien. This is a remarkably handy Russian dictionary, containing a surprising amount of information, including over 70,000 entries. 366pp. 4½ × 6⅛.   20208-9 Pa. $9.95

HISTORIC HOMES OF THE AMERICAN PRESIDENTS, Second, Revised Edition, Irvin Haas. A traveler's guide to American Presidential homes, most open to the public, depicting and describing homes occupied by every American President from George Washington to George Bush. With visiting hours, admission charges, travel routes. 175 photographs. Index. 160pp. 8¼ × 11. 26751-2 Pa. $10.95

NEW YORK IN THE FORTIES, Andreas Feininger. 162 brilliant photographs by the well-known photographer, formerly with *Life* magazine. Commuters, shoppers, Times Square at night, much else from city at its peak. Captions by John von Hartz. 181pp. 9¼ × 10¾.   23585-8 Pa. $12.95

INDIAN SIGN LANGUAGE, William Tomkins. Over 525 signs developed by Sioux and other tribes. Written instructions and diagrams. Also 290 pictographs. 111pp. 6⅛ × 9¼.   22029-X Pa. $3.50

CATALOG OF DOVER BOOKS

ANATOMY: A Complete Guide for Artists, Joseph Sheppard. A master of figure drawing shows artists how to render human anatomy convincingly. Over 460 illustrations. 224pp. 8⅜ × 11¼. 27279-6 Pa. $10.95

MEDIEVAL CALLIGRAPHY: Its History and Technique, Marc Drogin. Spirited history, comprehensive instruction manual covers 13 styles (ca. 4th century thru 15th). Excellent photographs; directions for duplicating medieval techniques with modern tools. 224pp. 8⅜ × 11¼. 26142-5 Pa. $11.95

DRIED FLOWERS: How to Prepare Them, Sarah Whitlock and Martha Rankin. Complete instructions on how to use silica gel, meal and borax, perlite aggregate, sand and borax, glycerine and water to create attractive permanent flower arrangements. 12 illustrations. 32pp. 5⅜ × 8½. 21802-3 Pa. $1.00

EASY-TO-MAKE BIRD FEEDERS FOR WOODWORKERS, Scott D. Campbell. Detailed, simple-to-use guide for designing, constructing, caring for and using feeders. Text, illustrations for 12 classic and contemporary designs. 96pp. 5⅜ × 8½. 25847-5 Pa. $2.95

OLD-TIME CRAFTS AND TRADES, Peter Stockham. An 1807 book created to teach children about crafts and trades open to them as future careers. It describes in detailed, nontechnical terms 24 different occupations, among them coachmaker, gardener, hairdresser, lacemaker, shoemaker, wheelwright, copper-plate printer, milliner, trunkmaker, merchant and brewer. Finely detailed engravings illustrate each occupation. 192pp. 4⅝ × 6. 27398-9 Pa. $4.95

THE HISTORY OF UNDERCLOTHES, C. Willett Cunnington and Phyllis Cunnington. Fascinating, well-documented survey covering six centuries of English undergarments, enhanced with over 100 illustrations: 12th-century laced-up bodice, footed long drawers (1795), 19th-century bustles, 19th-century corsets for men, Victorian "bust improvers," much more. 272pp. 5⅜ × 8¼. 27124-2 Pa. $9.95

ARTS AND CRAFTS FURNITURE: The Complete Brooks Catalog of 1912, Brooks Manufacturing Co. Photos and detailed descriptions of more than 150 now very collectible furniture designs from the Arts and Crafts movement depict davenports, settees, buffets, desks, tables, chairs, bedsteads, dressers and more, all built of solid, quarter-sawed oak. Invaluable for students and enthusiasts of antiques, Americana and the decorative arts. 80pp. 6½ × 9¼. 27471-3 Pa. $7.95

HOW WE INVENTED THE AIRPLANE: An Illustrated History, Orville Wright. Fascinating firsthand account covers early experiments, construction of planes and motors, first flights, much more. Introduction and commentary by Fred C. Kelly. 76 photographs. 96pp. 8¼ × 11. 25662-6 Pa. $8.95

THE ARTS OF THE SAILOR: Knotting, Splicing and Ropework, Hervey Garrett Smith. Indispensable shipboard reference covers tools, basic knots and useful hitches; handsewing and canvas work, more. Over 100 illustrations. Delightful reading for sea lovers. 256pp. 5⅜ × 8½. 26440-8 Pa. $7.95

FRANK LLOYD WRIGHT'S FALLINGWATER: The House and Its History, Second, Revised Edition, Donald Hoffmann. A total revision—both in text and illustrations—of the standard document on Fallingwater, the boldest, most personal architectural statement of Wright's mature years, updated with valuable new material from the recently opened Frank Lloyd Wright Archives. "Fascinating"—The New York Times. 116 illustrations. 128pp. 9¼ × 10¾. 27430-6 Pa. $10.95

PHOTOGRAPHIC SKETCHBOOK OF THE CIVIL WAR, Alexander Gardner. 100 photos taken on field during the Civil War. Famous shots of Manassas, Harper's Ferry, Lincoln, Richmond, slave pens, etc. 244pp. 10⅝ × 8¼.
22731-6 Pa. $9.95

FIVE ACRES AND INDEPENDENCE, Maurice G. Kains. Great back-to-the-land classic explains basics of self-sufficient farming. The one book to get. 95 illustrations. 397pp. 5⅜ × 8½. 20974-1 Pa. $7.95

SONGS OF EASTERN BIRDS, Dr. Donald J. Borror. Songs and calls of 60 species most common to eastern U.S.: warblers, woodpeckers, flycatchers, thrushes, larks, many more in high-quality recording. Cassette and manual 99912-2 $8.95

A MODERN HERBAL, Margaret Grieve. Much the fullest, most exact, most useful compilation of herbal material. Gigantic alphabetical encyclopedia, from aconite to zedoary, gives botanical information, medical properties, folklore, economic uses, much else. Indispensable to serious reader. 161 illustrations. 888pp. 6½ × 9¼. 2-vol. set. (USO) Vol. I: 22798-7 Pa. $9.95
Vol. II: 22799-5 Pa. $9.95

HIDDEN TREASURE MAZE BOOK, Dave Phillips. Solve 34 challenging mazes accompanied by heroic tales of adventure. Evil dragons, people-eating plants, bloodthirsty giants, many more dangerous adversaries lurk at every twist and turn. 34 mazes, stories, solutions. 48pp. 8¼ × 11. 24566-7 Pa. $2.95

LETTERS OF W. A. MOZART, Wolfgang A. Mozart. Remarkable letters show bawdy wit, humor, imagination, musical insights, contemporary musical world; includes some letters from Leopold Mozart. 276pp. 5⅜ × 8½. 22859-2 Pa. $7.95

BASIC PRINCIPLES OF CLASSICAL BALLET, Agrippina Vaganova. Great Russian theoretician, teacher explains methods for teaching classical ballet. 118 illustrations. 175pp. 5⅜ × 8½. 22036-2 Pa. $4.95

THE JUMPING FROG, Mark Twain. Revenge edition. The original story of The Celebrated Jumping Frog of Calaveras County, a hapless French translation, and Twain's hilarious "retranslation" from the French. 12 illustrations. 66pp. 5⅜ × 8½.
22686-7 Pa. $3.95

BEST REMEMBERED POEMS, Martin Gardner (ed.). The 126 poems in this superb collection of 19th- and 20th-century British and American verse range from Shelley's "To a Skylark" to the impassioned "Renascence" of Edna St. Vincent Millay and to Edward Lear's whimsical "The Owl and the Pussycat." 224pp. 5⅜ × 8½.
27165-X Pa. $4.95

COMPLETE SONNETS, William Shakespeare. Over 150 exquisite poems deal with love, friendship, the tyranny of time, beauty's evanescence, death and other themes in language of remarkable power, precision and beauty. Glossary of archaic terms. 80pp. 5³⁄₁₆ × 8¼. 26686-9 Pa. $1.00

BODIES IN A BOOKSHOP, R. T. Campbell. Challenging mystery of blackmail and murder with ingenious plot and superbly drawn characters. In the best tradition of British suspense fiction. 192pp. 5⅜ × 8½. 24720-1 Pa. $5.95

THE WIT AND HUMOR OF OSCAR WILDE, Alvin Redman (ed.). More than 1,000 ripostes, paradoxes, wisecracks: Work is the curse of the drinking classes; I can resist everything except temptation; etc. 258pp. 5⅜ × 8½. 20602-5 Pa. $5.95

SHAKESPEARE LEXICON AND QUOTATION DICTIONARY, Alexander Schmidt. Full definitions, locations, shades of meaning in every word in plays and poems. More than 50,000 exact quotations. 1,485pp. 6½ × 9¼. 2-vol. set.
Vol. I: 22726-X Pa. $16.95
Vol. 2: 22727-8 Pa. $15.95

SELECTED POEMS, Emily Dickinson. Over 100 best-known, best-loved poems by one of America's foremost poets, reprinted from authoritative early editions. No comparable edition at this price. Index of first lines. 64pp. 5³⁄₁₆ × 8¼.
26466-1 Pa. $1.00

CELEBRATED CASES OF JUDGE DEE (DEE GOONG AN), translated by Robert van Gulik. Authentic 18th-century Chinese detective novel; Dee and associates solve three interlocked cases. Led to van Gulik's own stories with same characters. Extensive introduction. 9 illustrations. 237pp. 5⅜ × 8½.
23337-5 Pa. $6.95

THE MALLEUS MALEFICARUM OF KRAMER AND SPRENGER, translated by Montague Summers. Full text of most important witchhunter's "bible," used by both Catholics and Protestants. 278pp. 6⅝ × 10. 22802-9 Pa. $11.95

SPANISH STORIES/CUENTOS ESPAÑOLES: A Dual-Language Book, Angel Flores (ed.). Unique format offers 13 great stories in Spanish by Cervantes, Borges, others. Faithful English translations on facing pages. 352pp. 5⅜ × 8½.
25399-6 Pa. $8.95

THE CHICAGO WORLD'S FAIR OF 1893: A Photographic Record, Stanley Appelbaum (ed.). 128 rare photos show 200 buildings, Beaux-Arts architecture, Midway, original Ferris Wheel, Edison's kinetoscope, more. Architectural emphasis; full text. 116pp. 8¼ × 11. 23990-X Pa. $9.95

OLD QUEENS, N.Y., IN EARLY PHOTOGRAPHS, Vincent F. Seyfried and William Asadorian. Over 160 rare photographs of Maspeth, Jamaica, Jackson Heights, and other areas. Vintage views of DeWitt Clinton mansion, 1939 World's Fair and more. Captions. 192pp. 8⅞ × 11. 26358-4 Pa. $12.95

CAPTURED BY THE INDIANS: 15 Firsthand Accounts, 1750–1870, Frederick Drimmer. Astounding true historical accounts of grisly torture, bloody conflicts, relentless pursuits, miraculous escapes and more, by people who lived to tell the tale. 384pp. 5⅜ × 8½. 24901-8 Pa. $8.95

THE WORLD'S GREAT SPEECHES, Lewis Copeland and Lawrence W. Lamm (eds.). Vast collection of 278 speeches of Greeks to 1970. Powerful and effective models; unique look at history. 842pp. 5⅜ × 8½. 20468-5 Pa. $14.95

THE BOOK OF THE SWORD, Sir Richard F. Burton. Great Victorian scholar/adventurer's eloquent, erudite history of the "queen of weapons"—from prehistory to early Roman Empire. Evolution and development of early swords, variations (sabre, broadsword, cutlass, scimitar, etc.), much more. 336pp. 6⅛ × 9¼. 25434-8 Pa. $8.95

AUTOBIOGRAPHY: The Story of My Experiments with Truth, Mohandas K. Gandhi. Boyhood, legal studies, purification, the growth of the Satyagraha (nonviolent protest) movement. Critical, inspiring work of the man responsible for the freedom of India. 480pp. 5⅜ × 8½. (USO) 24593-4 Pa. $8.95

CELTIC MYTHS AND LEGENDS, T. W. Rolleston. Masterful retelling of Irish and Welsh stories and tales. Cuchulain, King Arthur, Deirdre, the Grail, many more. First paperback edition. 58 full-page illustrations. 512pp. 5⅜ × 8½. 26507-2 Pa. $9.95

THE PRINCIPLES OF PSYCHOLOGY, William James. Famous long course complete, unabridged. Stream of thought, time perception, memory, experimental methods; great work decades ahead of its time. 94 figures. 1,391pp. 5⅜ × 8½. 2-vol. set.
Vol. I: 20381-6 Pa. $12.95
Vol. II: 20382-4 Pa. $12.95

THE WORLD AS WILL AND REPRESENTATION, Arthur Schopenhauer. Definitive English translation of Schopenhauer's life work, correcting more than 1,000 errors, omissions in earlier translations. Translated by E. F. J. Payne. Total of 1,269pp. 5⅜ × 8½. 2-vol. set.
Vol. 1: 21761-2 Pa. $11.95
Vol. 2: 21762-0 Pa. $11.95

MAGIC AND MYSTERY IN TIBET, Madame Alexandra David-Neel. Experiences among lamas, magicians, sages, sorcerers, Bonpa wizards. A true psychic discovery. 32 illustrations. 321pp. 5⅜ × 8½. (USO) 22682-4 Pa. $8.95

THE EGYPTIAN BOOK OF THE DEAD, E. A. Wallis Budge. Complete reproduction of Ani's papyrus, finest ever found. Full hieroglyphic text, interlinear transliteration, word-for-word translation, smooth translation. 533pp. 6½ × 9¼. 21866-X Pa. $9.95

MATHEMATICS FOR THE NONMATHEMATICIAN, Morris Kline. Detailed, college-level treatment of mathematics in cultural and historical context, with numerous exercises. Recommended Reading Lists. Tables. Numerous figures. 641pp. 5⅜ × 8½. 24823-2 Pa. $11.95

THEORY OF WING SECTIONS: Including a Summary of Airfoil Data, Ira H. Abbott and A. E. von Doenhoff. Concise compilation of subsonic aerodynamic characteristics of NACA wing sections, plus description of theory. 350pp. of tables. 693pp. 5⅜ × 8½. 60586-8 Pa. $14.95

THE RIME OF THE ANCIENT MARINER, Gustave Doré, S. T. Coleridge. Doré's finest work; 34 plates capture moods, subtleties of poem. Flawless full-size reproductions printed on facing pages with authoritative text of poem. "Beautiful. Simply beautiful."—*Publisher's Weekly*. 77pp. 9¼ × 12. 22305-1 Pa. $6.95

NORTH AMERICAN INDIAN DESIGNS FOR ARTISTS AND CRAFTSPEOPLE, Eva Wilson. Over 360 authentic copyright-free designs adapted from Navajo blankets, Hopi pottery, Sioux buffalo hides, more. Geometrics, symbolic figures, plant and animal motifs, etc. 128pp. 8⅜ × 11. (EUK) 25341-4 Pa. $7.95

SCULPTURE: Principles and Practice, Louis Slobodkin. Step-by-step approach to clay, plaster, metals, stone; classical and modern. 253 drawings, photos. 255pp. 8⅛ × 11. 22960-2 Pa. $10.95

CATALOG OF DOVER BOOKS

THE INFLUENCE OF SEA POWER UPON HISTORY, 1660–1783, A. T. Mahan. Influential classic of naval history and tactics still used as text in war colleges. First paperback edition. 4 maps. 24 battle plans. 640pp. 5⅜ × 8½.
25509-3 Pa. $12.95

THE STORY OF THE TITANIC AS TOLD BY ITS SURVIVORS, Jack Winocour (ed.). What it was really like. Panic, despair, shocking inefficiency, and a little heroism. More thrilling than any fictional account. 26 illustrations. 320pp. 5⅜ × 8½.
20610-6 Pa. $8.95

FAIRY AND FOLK TALES OF THE IRISH PEASANTRY, William Butler Yeats (ed.). Treasury of 64 tales from the twilight world of Celtic myth and legend: "The Soul Cages," "The Kildare Pooka," "King O'Toole and his Goose," many more. Introduction and Notes by W. B. Yeats. 352pp. 5⅜ × 8½.
26941-8 Pa. $8.95

BUDDHIST MAHAYANA TEXTS, E. B. Cowell and Others (eds.). Superb, accurate translations of basic documents in Mahayana Buddhism, highly important in history of religions. The Buddha-karita of Asvaghosha, Larger Sukhavativyuha, more. 448pp. 5⅜ × 8½. ,
25552-2 Pa. $9.95

ONE TWO THREE . . . INFINITY: Facts and Speculations of Science, George Gamow. Great physicist's fascinating, readable overview of contemporary science: number theory, relativity, fourth dimension, entropy, genes, atomic structure, much more. 128 illustrations. Index. 352pp. 5⅜ × 8½.
25664-2 Pa. $8.95

ENGINEERING IN HISTORY, Richard Shelton Kirby, et al. Broad, nontechnical survey of history's major technological advances: birth of Greek science, industrial revolution, electricity and applied science, 20th-century automation, much more. 181 illustrations. ". . . excellent . . ."—Isis. Bibliography. vii + 530pp. 5⅜ × 8¼.
26412-2 Pa. $14.95

Prices subject to change without notice.
Available at your book dealer or write for free catalog to Dept. GI, Dover Publications, Inc., 31 East 2nd St., Mineola, N.Y. 11501. Dover publishes more than 500 books each year on science, elementary and advanced mathematics, biology, music, art, literary history, social sciences and other areas.